GSM, GPRS and EDGE Performance

GSM, GPRS and EDGE Performance

Evolution Towards 3G/UMTS

Edited by

Timo Halonen, Javier Romero and Juan Melero

All of Nokia Spain, S.A.

JOHN WILEY & SONS, LTD

Other Wiley Editorial Offices

John Wiley & Sons, Inc., 605 Third Avenue,
New York, NY 10158-0012, USA

WILEY-VCH Verlag GmbH,
Pappelallee 3, D-69469 Weinheim, Germany

John Wiley & Sons Australia Ltd, 33 Park Road, Milton,
Queensland 4064, Australia

John Wiley & Sons (Canada) Ltd, 22 Worcester Road,
Rexdale, Ontario, M9W 1L1, Canada

John Wiley & Sons (Asia) Pte Ltd, 2 Clementi Loop #02-01,
Jin Xing Distripark, Singapore 129809

British Library Cataloguing in Publication Data

A catalogue record for this book is available from the British Library

ISBN 0470 84457 4

Typeset in Sabon by Deerpark Publishing Services Ltd, Shannon, Ireland.
Printed and bound in Great Britain by T. J. International Ltd, Padstow, Cornwall.
This book is printed on acid-free paper responsibly manufactured from sustainable forestry, in which at least
two trees are planted for each one used for paper production.

Contents

Acknowledgements

We would like to thank all contributors for the "world-class" material produced to support this book. This book has been the result of a collaborative effort of a truly motivated and committed team.

We would also like to thank our listed colleagues for their valuable input and comments: Heikki Annala, Harri Jokinen, Mattias Wahlqvist, Timo Rantalainen, Juha Kasinen, Jussi Reunanen, Jari Ryynanen, Janne Vainio, Kari Niemelä, Simon Browne, Heikki Heliste, Petri Gromberg, Oscar Salonaho, Harri Holma, Fabricio Velez, Jyrki Mattila, Mikko Säily, Gilles Charbit, Jukka Kivijärvi, Martti Tuulos, Lauri Oksanen, Ashley Colley, Mika Kähkölä, Per Henrik Michaelson, Rauli Jarvela, Rauli Parkkali, Kiran Kuchi, Pekka Ranta, Brian Roan, James Harper, Raul Carral, Joe Barret, Paul Timothy and Javier Munoz. Special thanks to Jussi Sipola and Ashish Kohli for the material provided in chapter 7.

Thanks to Mark Keenan for the support, guidance and encouragement provided during the last years. We hope he feels as proud as we do about the result of this book.

Thanks to the authors of the forewords (Mike Bamburak and Chris Pearson) and the companies/institutions they represent (AT&T and 3G Americas) for the support provided.

Many thanks to the operators (CSL, Radiolinja, Sonofon, AT&T and many others) we have been closely working with since only through this tight collaboration the understanding of the technology capabilities contained in this book has been possible. Special thanks to Optus, for their collaboration, support and excellent technical team (Andrew Smith, Carolyn Coronno, Bradley Smith and the rest of the team).

Part of the studies presented in this book have been performed as the result of the cooperation agreement between Nokia and the University of Malaga. This agreement is partially supported by the "Programa de Fomento de la Investigación Técnica" (PROFIT) of the Spanish Ministry of Science and Technology. We want to thank Malaga University and the Ministry of Science and Technology for the provided support.

Thanks to the publishing team from Wiley (Mark Hammond, Sarah Hinton and Geoff Farrell), which has given an excellent support throughout this project, which has been very important to accomplish the demanding time schedule. Also thanks to Deerpark for their patience and understanding during the editing phase.

Finally we would like to express our loving thanks to our wives Cristina, Laila and Susanna for the patience and understanding during the holidays, weekends and late nights devoted to the writing of this book.

The authors welcome any comments and suggestions for improvements or changes that could be implemented in possible new editions of this book. The e-mail address for gathering such input is geran_book@hotmail.com.

Malaga, Spain
The editors of the "GSM, GPRS & EDGE Performance" book

Foreword

I have worked in the mobile communications industry longer than I would like to admit. In the early 1970's I started my career as a radio engineer for Motorola. At that time Motorola designed and manufactured low, mid and high tier private land mobile radios. Motorola had few competitors for the mid and high tier product lines (50 to 100 watt radios). However, in the low tier, less than 25 watt radio category, there were numerous contenders, mostly from European manufacturers with a "Nordic Mobile Telephone" heritage.

But times were changing. In the late 1970's the American public got their first taste of mobile communications when Citizen Band (CB) radio became popular ("10-4, good buddy"). It was an unlicensed, short range, "party-line" experience. Those skilled in the art knew that something better was needed. And the American communications industry responded. The Federal Communications Commission and major industry players, like AT&T and Motorola, specified America's first public mobile radio telephone system, AMPS (Advanced Mobile Telephone System). By the mid 1980's AMPS was a proven technology and cellular subscriber growth was constantly exceeding forecasts.

But times were changing. By the early 1990's cellular had become so popular that the first generation analog systems couldn't keep up with demand. New second generation digital systems were developed to address the capacity shortfall. In the U.S. three digital technologies were standardized and deployed: IS-136 (a TDMA technology utilizing the AMPS 30 KHz structure), IS-95 (a 1.25 MHz CDMA carrier scheme), and GSM (the European 200 kHz TDMA standard). This multi-standard wireless environment provided a unique proving ground for the three technologies. While IS-136 and IS-95 engaged in "standards wars," GSM gained a foothold in America. At the same time GSM was achieving global acceptance because it offered a rich selection of capabilities and features that provided real incremental revenues for operators. As more and more countries adopted the technology, GSM experienced tremendous economies of scale for everything from chipsets to handsets, infrastructure and applications.

While the industry continued to experience stellar growth, American manufacturer dominance was challenged by Nordic companies, especially for the GSM technology. They brought to the U.S. innovative, competitively priced products, backed by talented communications professionals with years of experience in designing, manufacturing, engineering and installing cellular equipment and systems throughout the world.

But times were changing. By the late 1990's the Internet was pervasive and the wireless industry looked to mobile data as the growth opportunity. Once again the industry undertook the task of defining new wireless systems – this third generation, 3G, was to be based on packet data. Three new wireless standards emerged; CDMA2000 (evolution of IS-95), EDGE (evolution of GSM for existing spectrum) and W-CDMA (evolution of GSM for new spectrum using a 5 MHz W-CDMA carrier).

The evolution of GSM to 3G is about gradually adding more functionality, possibilities and value to the existing GSM network and business. The evolution begins with an upgrade of the GSM network to 2.5G by introducing GPRS technology. GPRS provides GSM with a packet data air interface and an IP based core network. EDGE is a further evolutionary step of GSM packet data. EDGE can handle about three times more data subscribers than GPRS, or triple the data rate for one end-user. EDGE can be achieved through a very fast and cost effective implementation. The only requirement is to add EDGE-capable transceivers and software.

With the continuation of EDGE standardisation towards GERAN (GSM/EDGE Radio access network) EDGE will achieve a full alignment with W-CDMA. The goal for EDGE is to boost system capacity, both for real-time and best-effort services, and to become perfectly competitive with other 3G technologies.

What emerges with these evolutionary steps from GSM to GPRS, EDGE and W-CDMA is a seamless 3G UMTS (Universal Mobile Telecommunications System) Multi-Radio network, one that maximizes the investments in GSM and GPRS.

It stands to reason that both EDGE and W-CDMA will be mainstream 3G UMTS products from Nordic companies. This book, written by engineers from one of those Nordic companies, is an authoritative treatise on GSM evolution to 3G. The book provides in depth performance analysis of current and future GSM speech and GPRS/EDGE packet data functionality. Furthermore, the concept of a 3G UMTS Multi-Radio network (GSM/EDGE/W-CDMA) is presented in depth as the best solution for wireless operators to evolve their networks towards 3G.

Times change, but some things don't. Nordic companies have been at the forefront of wireless communications for more than a half of a century. They have earned their preeminent position in the industry. I encourage you to listen to what this book has to say.

Mike Bamburak
VP Technology Architecture
& Standards
AT&T

Over the years scientists and dreamers have revolutionised the way we work and live through great inventions. Almost as quickly as news of the inventions spread, soothsayers rise to highlight real or imagined barriers to the success, popularity, or the long-term use of these products and services. As occurred with electricity, the automobile, and the television, soothsayers often misread the long-term impact of these inventions, underestimating qualities that led to their long-term success and adoption by the masses. Ultimately, all three inventions had tremendous social and economic impact on global societies, and the soothsayers were proven to have undervalued the importance of these great inventions over time.

In a slightly different way, the future of EDGE has been understated, underestimated, and undervalued by the latest pundits. Over the last few years, several global wireless organizations, including the GSA, the UWCC and now 3G Americas have stood their ground as advocates for EDGE because of the merits of the technology and its value to operators and customers as a spectrally efficient and cost effective solution for Third Generation (3G) wireless services. 3G Americas is firm in their belief that a comparative review of how EDGE meets three key criteria: performance, cost, and the ease of transformation to 3G, will show that EDGE is indeed a superior technology choice.

With the completion of such a review, the editors of this book, Timo Halonen, Javier Romero and Juan Melero, and have taken the predictions away from the pundits and created a scientifically sound testament to the opportunities presented by EDGE. This long-overdue analysis provides validity and guidelines for the successful deployment of EDGE. Furthermore, the findings arrive at the very best moment for the marketplace: just as 3G operator strategies are beginning to unfold. Recent announcements indicate that many major operators are beginning to transform EDGE's promises into the reality of cost-effective advanced wireless services.

The Reality of EDGE

On October 30, 2001, Cingular Wireless with its vendor partners announced its commitment to become the first operator in the world to deploy EDGE at 850 and 1900 MHz. With over 22 million wireless customers, Cingular is a major player in the global wireless marketplace. The reasons cited by Cingular for its EDGE selection included: capacity and spectral efficiency competitive with any other technology choice (performance); the ability to deploy 3G EDGE in existing spectrum including 850 MHz (cost); a total capital cost of about $18-$19 per Point of Presence (POP) in TDMA markets with plenty of go-forward capacity (cost); ridiculously low cost to upgrade the GSM network (only 10 to 15 percent of the network's cost); the enormous economies of scale and scope offered by the family of GSM technologies, ensuring the availability of equipment and applications at the lowest possible cost; and a transition path through GSM and GPRS achievable seamlessly through the use of the GAIT terminal (GSM-TDMA phone) that will ease transformation and result in customer retention.

Similarly, almost a year before Cingular's announcement, AT&T Wireless Services announced its commitment to EDGE. As of November 2001, reported operator commitments to EDGE in the Americas encompassed hundreds of millions of potential customers. These commitments validate the future of this third generation technology. Cingular's commitment to EDGE in the 850 MHz band sets the stage for an accelerated up-take by operators throughout the Western Hemisphere. Regional U.S. operators and many Latin America operators will find the opportunity to deploy EDGE in 850 MHz especially appealing. Furthermore, these commitments increase the possibility that Europe will recognize that EDGE's capacity and cost qualities make it an important complementary technology to WCDMA. As spectrum shortages inevitability occur in Europe, EDGE will provide an excellent solution for GSM operators as a complement to their WCDMA networks.

Benefits of EDGE

EDGE will benefit operators and customers because it is a cost effective solution for 3G services. Cost efficiency is enabled by the economies of scope and scale demonstrated by the GSM family of technologies, including both TDMA and GSM, which represented nearly 80 percent of the world's digital cellular subscribers in 2001. More than half a billion GSM phones existed by midyear 2001, and within a mere four months that number had risen by 100,000,000 phones to 600,000,000. Bill Clift, Chief Technology Officer of Cingular, noted that the cost differential between GSM and CDMA devices gets fairly significant at $15 to $20 per handset times millions of handsets each year. The economies of scale played a key role in the Cingular decision.

Another major benefit of EDGE cited by operators is that it enables TDMA and GSM carriers to offer 3G services while still realizing lower costs due to higher spectral efficiency and higher data rates. With the implementation of Adaptive Multi-Rate (AMR) Vocoders, and Frequency Hopping, GSM is competitive with CDMA on spectral efficiency, which translates into higher capacity and faster data rates. EDGE offers transmission speeds of 384Kbps – fast enough to support full motion video – and throughput capacity 3-4 times higher than GPRS. Thus, EDGE is fast, EDGE is efficient, and EDGE performs.

Additionally, the opportunity for international roaming with the GSM family of technologies offers yet another major incentive for operators to provide their customers with seamless communications services. Since EDGE and WCDMA will operate on the same GPRS core network, the EDGE/WCDMA customer will be able to travel the world seamlessly, staying connected with one number and one device.

Conclusion

EDGE will contribute to a bright future for 3G services, a vision shared by major analysts and industry groups. The Strategist Group predicts revenue from wireless data will reach $33.5 billion globally by 2004. Frost & Sullivan expects the proportion of operator revenues derived from non-voice services will be in excess of 45 percent by 2006. A UMTS Forum survey has estimated that non-voice revenues may take over voice revenues by 2004, while simple voice revenues will remain a critical revenue component comprising 34% of annual revenues in 2010. The UMTS Forum study also predicts that 3G revenues of $37.4 billion in 2004 will increase to $107 billion by 2006. All in all, predictions may vary, but the consensus is clear that results will be positive for third generation services.

This work offers the reader more than an evolutionary technical strategy of GSM's transition to 3G. It also provides a set of benchmarks for a core evaluation of the merits of EDGE as a central component of the wireless industry's fulfillment of its promise for higher degrees of service and convenience. This process is already being established, as evidenced by the first live EDGE data call completed by Nokia and AT&T Wireless on November 1, 2001. The connection of an EDGE handset with a laptop to the Internet for web browsing and streaming applications marked the first successful completion using EDGE 8-PSK modulation in both directions in the air interface. Indeed, it is another sign that EDGE will flourish in this new billion-dollar marketplace as a leading 3G technology in the Americas due to its performance, cost, interoperability, and ease of transformation. EDGE will outlast the neigh-sayers and in the long-term, EDGE will far exceed expectations. And just as electricity, the automobile, and the television changed our lives, EDGE will change our lives by providing 3G services for the masses.

Chris Pearson
Executive Vice President
3G Americas

Introduction

GERAN: Evolution towards 3G/UMTS

The wireless market has experienced a phenomenal growth since the first second-generation (2G) digital cellular networks, based on global system for mobile communications (GSM) technology, were introduced in the early 1990s. Since then, GSM has become the dominant global 2G radio access standard. Almost 80% of today's new subscriptions take place in one of the more than 400 cellular networks that use GSM technology. This growth has taken place simultaneously with the large experienced expansion of access to the Internet and its related multimedia services.

Cellular operators now face the challenge to evolve their networks to efficiently support the forecasted demand of wireless Internet based multimedia services. In order to do this, they need to perform a rapid technology evolution rolling out new third-generation (3G) radio access technologies, capable of delivering such services in a competitive and cost-effective way. There are multiple recognised 3G technologies, which can be summarised into two main 3G evolution paths. The first one to be conceived and developed, and the more widely supported, is UMTS (universal mobile telecommunications system), and the second one is cdma2000.

The first part of this book will describe the evolution GSM experienced after 5 years of continuous standardisation effort and how the result of such effort has provided GSM networks with a smooth, competitive and cost-efficient evolution path towards UMTS. The first steps of this evolution took place in Release 97 standards, when general packet radio system (GPRS) was introduced to efficiently deliver packet based services over GSM networks. Later, adaptive multi-rate (AMR), included in Release 98, increased the spectral efficiency and quality of speech services to remarkable limits. Enhanced data rates for global evolution (EDGE) was introduced in Release 99 and introduced more efficient modulation, coding and retransmission schemes, boosting the performance of data services. Release 99 also included the support of the UMTS traffic classes hence providing the means for EDGE to support from the very beginning the same set of 3G services UMTS was designed for. This synergy is completed in Release 5 with the development of GERAN, a new radio access network architecture, based on GSM/EDGE radio access technologies, which is fully harmonised with UTRAN (UMTS terrestrial radio access network) through a common connectivity to the UMTS core network, and therefore integrated in the UMTS frame. Today, UMTS standardisation, carried out by the 3rd Generation Partnership Project (3GPP) standardisation body, jointly develops the GERAN and UTRAN as part of the same concept, 3G UMTS multi-radio networks. This GSM evolution towards UMTS is illustrated in Figure 1.

Chapters 1 and 2 will provide a detailed description of the GSM evolution. Chapter 3 will provide an in-depth analysis of the GERAN quality of service (QoS) support in different releases. Finally Chapter 4 will present the different standardised methods to support location services, their architecture and expected accuracy.

Figure 1   GSM evolution towards 3G/UMTS.

The second part of the book will thoroughly cover the performance efficiency of the technologies associated with the GSM evolution. Radio network performance is vital in order to maximise the return on investment from operators. Not only does it provide the ability to increase the capacity of the networks but it improves as well the QoS experienced by the end users. All the performance related dimensions, such as link level voice and data performance, spectral efficiency, maximum and realistic data rates and service support capabilities are studied in depth in this part. Chapter 5 introduces the principles associated with GSM radio network performance analysis. All standardised functionality, such as AMR, GPRS and EDGE, both for voice (Chapter 6) and data (Chapter

7) services, are carefully studied. Non-standardised solutions will be analysed as well in order to provide an overall perspective of the potential GSM/EDGE performance when advanced functionality such as dynamic channel allocation (Chapter 8) or link level enhancement (Chapter 10) techniques are introduced. Chapter 11 analyses the signalling channels performance, since this may limit the overall network performance. A specific chapter on narrowband deployment scenarios (Chapter 9) will provide the guidelines to maximise the performance in these demanding radio conditions.

With the degree of complexity the 3G evolution will bring, planning, optimising and operation of radio networks is a tough challenge operators have to tackle at a minimum cost. Chapter 12 presents new concepts based on automation and control engineering, designed to deal with all these issues, with functionality capable of fully automating the planning, optimisation and operation of radio networks, which not only decrease the operational expenditure (OPEX) of radio networks but maximise as well their performance.

Finally, the third part of the book introduces the principles of the main existing 3G radio technologies, benchmarking their technical capabilities and evolution paths. It covers in detail how GSM/EDGE and wideband code division multiple access (WCDMA) frequency division duplex (FDD) can be efficiently and seamlessly integrated into the UMTS multi-radio network to ensure the achievement of a maximum performance at a minimum cost. Figure 2 illustrates the UMTS multi-radio technology building blocks.

Figure 2 UMTS multi-radio technology building blocks.

The final chapter analyses the current global market dynamics and trends, in order to identify, from a global market perspective, the evolution paths available for different technologies.

Abbreviations

1G	First Generation
2G	Second Generation
3G	Third Generation
4G	Fourth Generation
3GPP	3rd Generation partnership project
3GPP2	3rd Generation partnership project
8-PSK	Octagonal Phase Shift Keying

A

A-GPS	Assisted GPS positioning system
AC	Admission Control
ACELP	Algebraic Code Excitation Linear Prediction
ACK	Acknowledge
ADPCM	Adaptive Differential Pulse Code Modulation
AFP	Automated Frequency Planning
AGCH	Access Grant Channel
ALI	Automatic Location Identification
AMH	Advanced Multilayer Handling
AMPS	American Mobile Phone System
AMR	Adaptive Multi-Rate codec
AoCC	Advice of Charge - Charging
AoCI	Advice of Charge - Information
AP	Access Point
API	Application Programming Interface
ARIB	Association of Radio Industries and Business in Japan

ARQ	Automatic Repeat Request
AT	Absolute Time
ATD	Absolute Time Difference
ATM	Asynchronous Transfer Mode
AuC	Authentication Center

B

BA	BCCH Allocation
BB	Baseband
BEP	Bit Error Probability
BER	Bit Error Rate
BCCH	Broadcast Control Channel
BGIWP	Barring of GPRS Interworking Profile(s)
BIM	Background Interference Matrix
BLER	Block Error Rate
BoD	Bandwidth on Demand
BS	Base station
BSIC	Base Station Identification Code
BSS	Base Station Subsystem
BSSAP	BSS Application Part
BSSGP	Base Station Subsystem GPRS protocol
BSC	Base Station Controller
BTS	Base Transceiver Station

C

CAMEL	Customised Application for Mobile Network Enhanced Logic
CC	Call Control
CCITT	Comité Consultatif International Télégraphique et Téléphonique
CCCH	Common Control Channel
CDMA	Code Division Multiple Access
CEPT	Conference Europeenne des Postes et Telecommunications
CFU	Call Forwarding Unconditional
CFNRc	Call Forwarding on Mobile Subscriber Not Reachable
CGW	Cell Gateway
CI	Cell Identity
C/I	Carrier/Interference
CIR	Carrier/Interference
CLNS	Connectionless Network Service
CM	Communication Management
CN	Core Network
CONS	Connection-Oriented Network Service
CRC	Cyclic Redundancy check
CRS	Cell Resource Server
CS	Circuit Switched
CWTS	China Wireless Telecommunication Standard group in China

D

D-AMPS	Digital AMPS
DCA	Dynamic Channel Allocation
DCR	Dropped Call Rate
DCCH	Dedicated Control Channel
DFCA	Dynamic Frequency and Channel Allocation
DPCCH	Dedicated Physical Control Channel
DPDCH	Dedicated Physical Data Channel
DR	Directed Retry
DRX	Discontinuous Reception
DS-CDMA	Direct Sequence - Code Division Multiple Access
DTAP	Direct Transfer Application Process
DTM	Dual Transfer Mode
DTX	Discontinuous Transmission

E

E112	Emergency Service 911 in Europe
E911	Emergency Service 911 in U.S.
ECDS	Enhanced Circuit Switched Data
EDGE	Enhanced Data Rates for Global Evolution
EFL	Effective Frequency Load
EFR	Enhanced Full Rate (speech coded)
EGPRS	Enhanced General Packet Radio System
EIR	Equipment Identity Register
E-OTD	Enhanced Observed Time Difference
ETSI	European Telecommunications Standards Institute

F

FACCH	Fast Associated Control Channel
FCCH	Frequency Correction Channel
FDD	Frequency Division Duplex
FER	Frame Error Rate
FH	Frequency Hopping
FR	Full Rate
FTP	File Transfer Protocol

G

GAIT	GSM/ANSI Interoperability Team
GERAN	GSM EDGE Radio Access Network
GGSN	Gateway GPRS Support Node
GMLC	Gateway Mobile Location Center
GMSK	Gaussian Minimum Shift Keying
GPRS	General Packet Radio System
GPS	Global Positioning System
GSM	Global System for Mobile Communications
GSN	GPRS Support Node

| GTD | Geometric Time Difference |
| GTP | GPRS Tunnelling Protocol |

H
HC	Handover Control
HLR	Home Location Register
HO	Handover
HPLMN	Home Public Land Mobile Network
HR	Half Rate
HSCSD	High Speed Circuit Switched Data
HSDPA	High Speed Downlink Packet Access
HSN	Hopping Sequence Number
HTTP	Hyper Text Transfer Protocol
HW	Hardware

I
IETF	Internet Engineering Task Force
IFH	Intelligent Frequency Hopping
IM	Interference Matrix
IMEI	International Mobile Equipment Identity
IMGI	International Mobile Group Identity
IMSI	International Mobile Subscriber Identity
IP	Internet Protocol
IPv4	Internet Protocol version 4
IPv6	Internet Protocol version 6
IR	Incremental Redundancy
IS-54	First generation TDMA Radio Interface standard
IS-95	cdmaOne, one of 2nd generation systems, mainly in Americas and Korea
IS-136	US-TDMA, one of 2nd generation systems, mainly in Americas
ISDN	Integrated Services Digital Network
ITU	International Telecommunication Union
ITU-T	Telecommunication standardization sector of ITU
IUO	Intelligent Underlay Overlay
IWMSC	Inter Working MSC

L
LA	Link Adaptation
LAPD	Link Access Protocol for the D channel
LAPDm	Link Access Protocol for the Dm channel
LC	Load Control
LCS	Location Services
LIF	Location Interoperability Forum
LLC	Logical Link Control
LMU	Location Measurement Unit

M

MA	Mobile Allocation
MAC	Medium Access Control
MAIO	Mobile Allocation Index Offset
MAP	Mobile Application Part
MCPA	Multi Carrier Power Amplifier
MCS	Modulation and Coding Scheme
ME	Mobile Equipment
MEHO	Mobile Evaluated Handover
MEO	Medium Earth Orbit
MM	Mobility Management
MoU	Memorandum of Understanding
MS	Mobile Station
MSC	Mobile Services Switching Centre
MRP	Multiple Reuse Pattern

N

NACC	Network Assisted Cell Change
NACK	Negative Acknowledgement
NC	Network Control
NCC	Network (PLMN) Colour Code
NCCR	Network Controlled Cell Re-selection
NMS	Network Management Subsystem
NSAPI	Network layer Service Access Point Identifier
NSS	Network and Switching Subsystem

O

OMC	Operations and Maintenance Center
OSI	Open System Interconnection
OSS	Operation Sub-System
OTD	Observed Time Difference

P

PACCH	Packet Associated Control Channel
PAGCH	Packet Access Grant Channel
PBCCH	Packet Broadcast Control Channel
PC	Power Control
PCCCH	Packet Common Control Channel
PCHM	Packet Channel Manager
PCU	Packet Control Unit
PCH	Paging Channel
PDCP	Packet Data Convergence Protocol
PDP	Policy Decision Point
PDU	Protocol Data Unit
PDTCH	Packet Data Traffic Channel
PEP	Policy Enforcement Point

PFC	Packet Flow Context
PM	Policy Management
PNCH	Packet Data Notification Channel
PRACH	Packet Random Access Channel
PRM	Packet Resource Manager
PS	Packet Switched / Packet Scheduler
PSI	Packet System Information
PSPDN	Packet-Switched Public Data Network
PSTN	Public-Switched Telephone Network
PTM	Point-To-Multipoint
PTM-G	PTM Group call
PTM-M	PTM Multicast
PTM-SC	PTM Service Center
PTP	Point-To-Point

Q

QoS	Quality of Service
QR	Quarter Rate

R

RA	Routing Area
RAB	Radio Access Bearer
RACH	Random Access Channel
RANAP	Radio Access Network Application Part
RAT	Radio Access Technology
RB	Radio Bearer
Rel'97	ETSI release 97
Rel'98	ETSI release 98
Rel'99	ETSI release 99
Rel'4	3GPP release 4
Rel'5	3GPP release 5
Rel'6	3GPP release 6
RLC	Radio Link Control
RM	Resource Manager
RNAS	Radio Access Network Access Server
RNC	Radio Network Controller
RR	Radio Resource
RRBP	Relative Reserved Block Period
RRLP	Radio Resource LCS Protocl
RRM	Radio Resource Management
RSC	Recursive Systematic Convolutional (coding)
RT	Real Time
RTCP	Real Time Control Protocol
RTD	Real Time Difference
RTP	Real Time Transport Protocol
RTSP	Real Time Streaming Protocol

RTT   Round Trip Time
RXLEV   Received Signal Level
RXQUAL   Received Signal Quality

S
SACCH   Slow Associated Control Channel
SAI   Service Area Identifier
SAPI   Service Access Point Indicator
SCH   Syncronisation Channel
SDCCH   Stand alone Dedicated Control Channel
SE   Spectral Efficiency
SGSN   Serving GPRS Support Node
SI   System Information
SIM   GSM Subscriber Identity Module
SIP   Session Initiation Protocol
SM-SC   Short Message Service Center
SMSCB   Short Message Service Cell Broadcast
SMS-GMSC   Short Message Service Gateway MSC
SMS-IWMSC   Short Message Service Inter Working MSC
SMS-MO/PP   Mobile Originated Short Messages
SMS-MT/PP   Mobile Terminated Short Messages
SMS-PP   Point-to-Point Short Message Service
SMG   Special Mobile Group
SMLC   Serving Mobile Location Center
SMSS   Switching and Management Sub-System
SMTP   Simple Mail Transfer Protocol
SNDCP   Sub Network Dependent Convergence Protocol
SNMP   Simple Network Management Protocol
SPSCH   Shared Physical Sub Channel
SS   Supplementary Services
SS7   Signalling System number 7

T
T1P1   North American Telecommunications Industry Association
T1P1.5   Sub Committee of T1P1
TA   Timing Advance
TAI   Timing Advance Index
TBF   Temporary Block Flow
TCH   Traffic Channel
TCP   Transport Control Protocol
TDD   Time Division Duplex
TDMA   Time Division Multiple Access
TFI   Temporary Flow Identity
TIA/EIA   Telecommunications Intrustry Alliance/Electronic Intrusties Alliance
TLLI   Temporary Logical Link Identity
TMSI   Temporary Mobile Subscriber Identity

TOA	Time of Arrival
TRAU	Transcoder/Rate Adapter Unit
TRHO	Traffic Reason Handover
TRX	Transceiver
TSL	Timeslot
TTA	Telecommunication Technology Association in Korea
TTC	Telecommunications Technology Committee in Japan
TU3	Typical Urban 3 km/h
TU50	Typical Urban 50 km/h

U
UDP	User Datagram Protocol
UE	User Equipment
UEP	Unequal Error Protection
UL	Uplink
UMTS	Universal Mobile Telecommunication services
USF	Uplink State Flag
USIM	Universal Subscriber Identity Module
UTD	uniform Theory of Diffraction
UTRAN	UMTS Terrestrial Radio Access Network

V
VAD	Voice Activation Detector
VoIP	Voice over IP
VLR	Visiting Location Register
VPLMN	Visited Public Land Mobile Network

W
WCDMA	Wideband CDMA
WLAN	Wireless LAN
WWW	World Wide Web
X.25	An ITU-T Protocol for Packet Switched Network

Part 1

GERAN Evolution

The standardisation of global system for mobile telecommunications (GSM) has been an ongoing process during the last decade involving more than 100 different wireless operators and manufacturers around the world. The first phases of GSM standards were carried out by the European Telecommunications Standard Institute (ETSI), but today, the 3rd Generation Partnership Project (3GPP) jointly standardises GSM EDGE radio access network (GERAN) and UMTS terrestrial radio access network (UTRAN) as part of the UMTS multi-radio networks concept.

The first part of this book focuses on the standardised technical evolution of GERAN providing GSM networks with a smooth, competitive and cost-efficient evolution path towards UMTS.

Chapter 1 provides an insight about the status of the standardisation up to Rel'4. It covers circuit switched evolution, including high speed circuit switched data (HSCSD) and adaptive multi-rate (AMR) and packet switched general packet radio system (GPRS) definition. Chapter 2 focuses on further evolution, covering Rel'5's new GERAN architecture and beyond. Chapter 3 provides an overview about the quality of service (QoS) architecture of GSM/EDGE for different releases and how the UMTS QoS has been efficiently adopted by GERAN. Finally Chapter 4 covers the location services methods and architectures.

1

GSM/EDGE Standards Evolution (up to Rel'4)

Markus Hakaste, Eero Nikula and Shkumbin Hamiti

The standardisation work of GSM based systems has its roots in the 1980s, when a standardisation body 'Groupe Special Mobile' (GSM)[1] was created within the Conference Europeenne des Postes et Telecommunications (CEPT), whose task was to develop a unique digital radio communication system for Europe, at 900 MHz.

Since the early days of GSM development, the system has experienced extensive modifications in several steps to fulfil the increasing demand from the operators and cellular users. The main part of the basic GSM system development during the last decade until spring 2000 has been conducted in European Telecommunications Standards Institute (ETSI) Special Mobile Group (SMG) and its technical sub-committees, as well as in T1P1, which had the responsibility of the PCS1900 MHz specifications in the US.

Currently, the further evolution of the GSM based systems is handled under the 3rd Generation Partnership Project (3GPP), which is a joint effort of several standardisation organisations around the world to define a global third-generation UMTS (universal mobile communication system) cellular system. The main components of this system are the UMTS terrestrial radio access network (UTRAN) based on wideband code division multiple access (WCDMA) radio

---

[1] The original GSM acronym is 'Groupe Special Mobile'. This changed afterwards to Global System for Mobile communication, which is the current official acronym.

technology and GSM/EDGE radio access network (GERAN) based on global system for mobile communications (GSM)/enhanced data rates for global evolution (EDGE) radio technology.

In the following sections an introduction of the evolution steps of the GSM specifications is presented.

1.1 Standardisation of GSM – phased approach

The need for continuous development of the GSM specifications was anticipated already in the beginning of the specification work, which was as a consequence split into two phases. This phased approach was defined to make sure that specifications supported a consistent set of features and services for multi-vendor operation of the GSM products both on the terminals and networks sides.

The GSM Phase 1 work included most common services to enable as fast as possible deployment of GSM in operating networks still providing a clear technological advance compared with existing analogue networks. The main features in Phase 1 include support for basic telephony, emergency calls, 300–9600 kbps data service, ciphering and authentication as well as supplementary services like call forwarding/barring. Also, short message service (SMS) was also included at this early phase of the specification work, although its commercial success came much later. From a radio performance point of view, frequency hopping, power control and discontinuous transmission were terminal mandatory from GSM Phase 1.

While the GSM Phase 1 networks were being built, GSM Phase 2 was being specified in ETSI SMG. The GSM Phase 2 specifications were frozen in October 1995 and they included a mechanism for cross phase compatibility and error handling to enable evolution of the specifications. Many technical improvements were also introduced including several new supplementary services like line identification services, call waiting, call hold, advice of charge and multi party call. In the speech area a half-rate channel mode codec was introduced to complement the already specified speech codec for GSM full-rate channel mode. In the data side the main improvement included group 3 fax service support.

The first two phases of GSM provided a solid basis for the GSM system evolution towards the third-generation (3G) system requirements and items, which were in the beginning of the work better known as Phase 2 + items. In the core network (CN) side, the evolution led to the introduction of general packet radio system (GPRS) network architecture, especially designed for Internet connectivity. Similarly, the radio access network experienced significant enhancements in both packet and circuit modes to provide higher bit rates and better network capacity, in both packet and circuit switched modes. From the users' perspective, in addition to better speech quality, this evolution consisted of significant enhancements to the capabilities to provide speech and data services. The following list gives an overview of the different work item categories:

- new bearer services and data related improvements like high speed circuit switched (multislot) data (HSCSD), 14.4 kbps (single slot) data, general packet radio service (GPRS) and enhanced data rates for global evolution (EDGE);
- speech related items like enhanced full-rate speech codec (EFR), adaptive multi-rate codec (AMR) with both narrow and wideband options, and tandem free operation (TFO) of speech codecs;
- mobile station (MS) positioning related items like cell identity and timing advance, uplink time of arrival (TOA) and enhanced observed time difference (E-OTD) methods based on measurements within the cellular network, and assisted GPS method based on the GPS (global positioning system) technology;
- frequency band related items like GSM400, 700, 850 MHz, unrestricted GSM multiband operation, release independent support of frequency bands;
- messaging related items like SMS concatenation, extension to alphabet, SMS interworking extensions, forwarding of SMSs and particularly multimedia messaging (MMS);
- new supplementary services like call deflection, calling name presentation, explicit call transfer, user-to-user signalling, completion of calls to busy subscriber and new barring services;
- billing related work items like payphone services, provision for hot billing and support for home area priority;
- service platforms like subscriber identity module (SIM) application toolkit and customised application for mobile network enhanced logic (CAMEL).

Although the phasing is used also when developing 3G standards, grouping into 'Releases' (later in this book referred to as Rel') is a more practical tool used currently in the standardisation to manage changes between different specification versions and introduction of new features. New releases are normally produced every year, and they contain a set of all specifications having added functionality introduced to it compared with previous releases as a result of ongoing standardisation work [1].

GSM/TDMA convergence through EDGE

Standardisation of EDGE was the main cornerstone in order to integrate the two major time division multiple access standards, GSM specified in ETSI and IS-136 specified in Telecommunications Industry Alliance/Electronic Industries Alliance (TIA/EIA) on the same evolution path. At the moment, new frequency bands are not available in the US, so an evolutionary approach to the new services on existing frequency bands is important for the current US operators.

TIA/EIA-136 is a result of the evolution of the US cellular networks. While GSM was digital from the beginning, TIA/EIA-136 has its roots in analogue advanced mobile phone service (AMPS EIA-553), which has been digitised in stages. The first step was the introduction of digital traffic channels in the IS-54 standard version. The IS-136 releases introduced digital signalling channels, improved voice quality with an enhanced voice codec, and standardised dual-band operation. Like in Europe, the next wave in complementing the capabilities of TIA/EIA-136 in the US was targeted in finding suitable 3G high data rate solution.

The start of EDGE development goes back to 1997, when ETSI conducted a feasibility study on improved data rates for GSM evolution. At the same time, the Universal Wireless Communication Consortium's (UWCC) Global TDMA Forum prepared input for TIA's ITU IMT-2000 programme. In 1998 key technical parameters of these two developments were harmonised, forming a basis for a converged time division multiple access (TDMA) standard using 200 kHz carrier and octagonal phase shift keying (8-PSK) modulation based radio interface.

In Europe the work continued under ETSI SMG2's EDGE work items and the first phase EDGE standard, including enhanced circuit switched data (ECSD) based on high speed circuit switched data (HSCSD), and enhanced general packet radio service (EGPRS) based on GPRS, was finalised in spring 2000. In addition, the specifications supported an optional arrangement called EDGE Compact, which enabled the deployment of EDGE in as narrow as 1 MHz frequency bands.

In the US, the work continued in TIA TR45 as further development of EDGE as part of the UWC-136 IMT-2000 proposal. In November 1999, the UWC-136 proposal was approved as a radio interface specification for IMT-2000. Currently, EDGE is part of the GERAN development that is carried out concurrently in 3GPP and UWCC to ensure high synergy between GSM and TDMA systems in the future.

The latest developments in the US market from the two largest IS-136 operators, AT&T and Cingular, define the mainstream evolution path to be likely followed by most IS-136 operators. IS-136 networks will fully migrate to GSM/EDGE/WCDMA UMTS technology making use in some cases of the GAIT (GSM/ANSI-136 Interoperability Team) functionality to ensure a smooth migration [2]. As a result of this, EDGE Compact is not likely to ever be implemented, since the evolution timing and the roaming requirements don't make this option interesting for operators.

GERAN standardisation in 3GPP

In summer 2000, the specification work of GSM radio was moved from ETSI SMG2 to 3GPP, which is a project responsible of the UMTS standards based on the evolved GSM core network. This meant that Rel'99 was the last release of the GSM/EDGE standard that was specified in ETSI SMG and all specifications related to GSM radio access network were moved under 3GPP responsibility with new specification numbers.

The move of the work was motivated by both the need for improved working procedures and technical development closer to 3GPP. The core network part of the GSM specifications was transferred to 3GPP by the time the project work was initiated and the arrangement of standardising the GSM network and radio access parts in two different standardisation bodies, made it too cumbersome to work effectively. More importantly, activities aimed at closer integration of GSM/EDGE radio and WCDMA technologies had led to a decision to adopt the 3GPP specific Iu interface for GERAN. Thereby, this integration achieves a true multi-radio UMTS standard, which is made up of two radio technologies, WCDMA and GSM/EDGE, that can be effectively and seamlessly integrated in order to maximise the efficiency and the quality of service provided to the end users with the introduction of the coming 3G multimedia services.

3GPP organisation

The 3GPP is a joint effort of several standardisation organisations around the world targeting to produce specifications for a global 3G cellular system. These so-called organisational partners (OP) have the authority to develop standards in their region and they currently include: ETSI in Europe, Association of Radio Industries and Business (ARIB) in Japan, Standardisation Committee T1 – Telecommunications (T1) in the US, Telecommunication Technology Association (TTA) in Korea, China Wireless Telecommunication Standard group (CWTS) in China and Telecommunications Technology Committee (TTC) in Japan.

3GPP also have market representation partners (MRP), whose role is to make sure that the standardisation is in line with the market requirements. In practice, the MRPs work together with OPs in the Project Coordination Group (PCG) to guide the overall direction of the 3GPP work. Potential future partners can also follow the work as observers.

More information and up-to-date links about the 3GPP related activities and organisation can be found in http://www.3gpp.org/.

Most of the technical work in 3GPP is carried out by individual members, typically 3G operators and vendors, in technical specification groups (TSGs) and their working groups (WGs). The basic 3GPP work organisation is depicted in Figure 1.1.

Figure 1.1   3GPP organisation and TSG GERAN Working Groups.

Currently there are five TSGs in 3GPP:

- TSG SA – Service and System Aspects, taking care of the overall system architecture and service capabilities, and cross TSG coordination on these areas. Issues related to security, speech and multimedia codecs, network management and charging are also handled in SA.
- TSG CN – Core Network, taking care of the UMTS/GSM core network protocols. This includes the handling of layer 3 call control, mobility management and session management protocols between the mobile terminal and CN, as well as, interworking with external networks.
- TSG T – Terminals, taking care of the aspects related to the terminal interfaces. This includes the responsibility of issues like universal subscriber identity module (USIM), TE performance, service capability protocols, end-to-end service interworking and messaging.
- TSG RAN – Radio Access Network, taking care of the aspects related to UMTS terrestrial radio access network (UTRAN). This includes the handling of specifications for the WCDMA based radio interface towards the WCDMA terminal, RAN specific interfaces Iub and Iur, and CN connectivity over Iu interface.
- TSG GERAN – GSM/EDGE Radio Access Network, taking care of the aspects related to the GSM/EDGE based radio interface including testing for GERAN terminals, GERAN internal interfaces, as well as, external interfaces related to the connections towards the legacy GSM CN.

TSGs have the mandate to approve and modify technical specifications in their area. Although TSGs are reasonably independent in terms of making decisions, liaising between groups is intended to ensure the technical consistency and timely finalisation of the specifications for different releases. For TSG

GERAN, this includes close cooperation particularly with TSG SA on issues related to 3G bearer capabilities, architecture and security, and with TSG RAN on issues related to the common interfaces (Iu, Iur) and protocols (PDCP) under TSG RAN responsibility, as well as, overall optimisation of GSM/EDGE – WCDMA interoperation.

While the TSGs carry the main responsibility for approval of the documents and overall work coordination, the detailed level technical work takes place in WGs. In TSG GERAN there are currently five WGs:

- WG1 – Radio Aspects, specifically having the responsibility over the specifications related to GSM/EDGE layer 1, radio frequency (RF), radio performance and internal Ater specification between Channel Codec Unit (CCU) and Transcoder and Rate Adapter Unit (TRAU).
- WG2 – Protocols Aspects, specifically covering issues related to GERAN layer 2, like radio link control (RLC)/medium access control (MAC) and three radio resource (RR) specifications, A and Gb interfaces towards 2G CN and internal Abis specification between base station controller (BSC) and base transceiver station (BTS).
- WG3 – Base Station Testing and O&M, taking care of all aspects related to conformance testing of GERAN BTSs, as well as, GERAN specific operation and maintenance issues for all GERAN nodes.
- WG4 – Terminal Testing (Radio part), responsible for the creating of specifications for the conformance testing of GERAN terminals in the area of radio interface layer 1 and RLC/MAC.
- WG5 – terminal Testing (Protocol part), responsible of the creating specifications for the conformance testing of GERAN terminals in the area of protocols above RLC/MAC.

1.2 Circuit switched services GSM

The main drivers for the further development of the speech services and codecs are the user demand for better speech quality, and the capacity gains enabling cost savings for network operators. Enhanced full-rate (EFR) codec was the first significant improvement in speech quality when it was introduced in ETSI specifications in 1996, as well as in IS-641 standards in the US during the same year. While the old GSM half-rate (HR) and full-rate (FR) codecs lacked behind the quality of the wireline telephony (32 kbps adaptive differential pulse code modulation (ADPCM)), EFR provided equivalent quality as a normal fixed connection even in typical error conditions. The EFR codec was jointly developed by Nokia and University of Sherbrooke and it has been widely used in operating GSM networks worldwide.

In spite of the commercial success of EFR, the codec left room for further improvements. In particular, the performance of the codec in severe radio channel error conditions could have been better. In addition to this, GSM half-rate codec was not able to provide adequate speech quality, so it was decided to continue the codec specification work with a new codec generation [3].

Adaptive multi-rate codec (AMR)

In October 1997, a new programme was initiated in ETSI aimed at the development and standardisation of AMR for GSM system. Before the standardisation was officially started, a 1-year feasibility study was carried out to validate the AMR concept. The main target of the work was to develop a codec that provided a significant improvement in error robustness, as well as capacity, over EFR.

The actual AMR codec standardisation was carried out as a competitive selection process consisting of several phases. In February 1999, ETSI approved the AMR codec standard, which was based on the codec developed in collaboration between Ericsson, Nokia and Siemens. Two months later 3GPP adopted the AMR codec as the mandatory speech codec for the 3G WCDMA system. Some parts of the AMR codec work, such as voice activity detection and optimised channel coding were finalised and included in the standard later, in June 1999.

The AMR codec contains a set of fixed rate speech and channel codecs, fast in-band signalling and link adaptation. The AMR codec operates both in the full-rate (22.8 kbps) and half-rate (11.4 kbps) GSM channel modes. An important part of AMR is its ability to adapt to radio channel and traffic load conditions and select the optimum channel mode (HR or FR) and codec mode (bit rate trade-off between speech and channel coding) to deliver the best possible combination of speech quality and system capacity.

Even though the AMR codec functionality is not mandatory in Rel'98 compliant terminals, the large benefits this functionality will bring will likely drive the demand for the support of AMR codec in all terminals.

Speech and channel coding

The AMR speech codec utilises the algebraic code excitation linear prediction (ACELP) algorithm employed also in GSM EFR and D-AMPS EFR codecs. The codec is actually a combination of eight speech codecs with bit rates of 12.2, 10.2, 7.95, 7.4, 6.7, 5.9, 5.15 and 4.75 kbps. Table 1.1 in Chapter 2 summarises AMR speech codecs supported in the different releases. AMR narrowband (AMR NB) with Gaussian minimum shift keying (GMSK) modulation is supported already in Rel'98. Each codec mode provides a different distribution of the available bit rate between speech and channel coding. All of the codecs are defined for the full-rate channel mode, while the six lowest ones are defined also for the half-rate channel mode.

Channel coding performs error correction and bad frame detection. The error correction in all the codec modes is based on the recursive systematic convolutional (RSC) coding with puncturing to obtain the required bit rates. Each codec mode utilises a 6-bit cyclic redundancy check (CRC) for detecting bad frames. In order to maximise the commonality with the existing GSM system, all channels use polynomials used for the previous GSM traffic channels.

In-band signalling and link adaptation

In the basic AMR codec operation, shown in Figure 1.2, both the mobile station (MS) and the base transceiver station (BTS) perform channel quality estimation of the received signal. Based on the channel quality measurements, a codec mode command (downlink to the MS) or codec mode request (uplink to the BTS) is sent over the radio interface in in-band messages. The receiving end uses this information to choose the best codec mode for the prevailing channel condition. A codec mode indicator is also sent over the radio to indicate the current mode of operation in the sending side. The basic principle for the codec mode selection is that the mode chosen in the uplink may be different from the one used in the downlink direction, but the channel mode (HR or FR) must be the same.

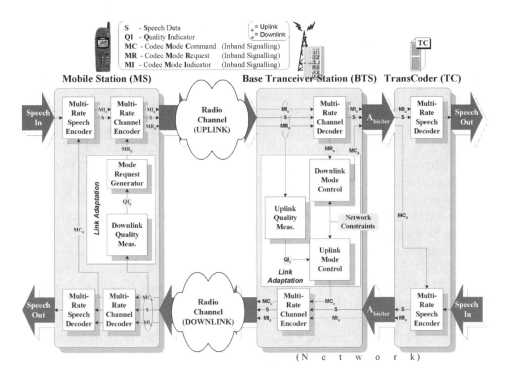

Figure 1.2 Block diagram of the AMR codec system.

The benefit of the in-band method is that it does not require a separate signalling channel for the message transfer. By sending the messages and the indicators together with the speech payload the link adaptation operation can also be made faster leading to improvements in the system performance.

The network controls the uplink and the downlink codec modes and channel modes. The MS must obey the codec mode command from the network, while the network may use any complementing information, in addition to codec mode request, to determine the downlink codec mode. The MS must implement all the codec modes. However, the network can support any combination of them, based on the choice of the operator.

AMR contains also voice activity detection and discontinuous transmission (VAD/DTX). These are used to switch off the encoding and transmission during periods of silence thereby reducing radio interference and extending the battery lifetime.

High speech circuit switched data (HSCSD)

The first phase GSM specifications provided only basic transmission capabilities for the support of data services, with the maximum data rate in these early networks being limited to 9.6 kbps on one timeslot. HSCSD specified in Rel'96, was the first GSM Phase 2 + work item that clearly increased the achievable data rates in the GSM system. The maximum radio interface bit rate of a HSCSD configuration with 14.4 kbps channel coding is 115.2 kbps, i.e. up to eight times the bit rate on the single slot full-rate traffic channel (TCH/F). In practice, the maximum data rate is limited to 64 kbps due to core network and A interface limitations.

The main benefit of the HSCSD feature compared with other data enhancements introduced later is that it is an inexpensive way to implement higher data rates in GSM networks due to relatively small incremental modifications needed to the network equipment. Terminals, however, need to be upgraded to support multislot capabilities. The basic HSCSD terminals with relatively simple implementation (in size and cost compared with single slot terminal) available in the market today can receive on up to four and transmit on up to two timeslots and thus support data rates above 50 kbps. Figure 1.3 depicts the HSCSD network architecture based on the concept of multiple independent TCH/F channels.

In HSCSD, a new functionality is introduced at the network and terminals to provide the functions of combining and splitting the user data into separate n data streams which will then be transferred via n channels at the radio interface, where $n = 1, 2, 3, ..., 8$. Once split, the data streams shall be carried by the n full-rate traffic channels, called HSCSD channels, as if they were independent of each other, for the purpose of data relay and radio interface L1 error control, until the point in the network where they are combined. However, logically the n full-rate traffic channels at the radio interface belong to the same HSCSD

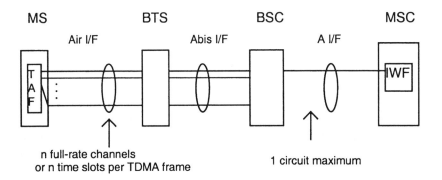

Figure 1.3 Network architecture for supporting HSCSD.

configuration, and therefore they shall be controlled as one radio link by the network for the purpose of cellular operations, e.g. handover.

For both transparent and non-transparent HSCSD connections the call can be established with any number of TCH/F from one up to the maximum number of TCH/F, i.e. the minimum channel requirement is always one TCH/F.

If the wanted air interface user rate requirement cannot be met using a symmetric configuration, an asymmetric configuration can be chosen. The network shall in this case give priority to fulfilling the air interface user rate requirement in the downlink direction.

The MS may request a service level up- or downgrading during the call, if so negotiated in the beginning of the call. This modification of channel requirements and/or wanted air interface user rate is applicable to non-transparent HSCSD connections only.

Radio interface
Two types of HSCSD configurations exist, symmetric and asymmetric. For both types of configurations the channels may be allocated on either consecutive or non-consecutive timeslots taking into account the restrictions defined by the mobile station's multislot classes, described in detail in [2]. An example of the HSCSD operation with two consecutive timeslots is shown in Figure 1.4.

Figure 1.4 Double slot operation in the air interface.

A symmetric HSCSD configuration consists of a co-allocated bi-directional TCH/F channel. An asymmetric HSCSD configuration consists of a co-allocated uni-directional or bi-directional TCH/F channel. A bi-directional channel is a channel on which the data are transferred in both uplink and downlink directions. On uni-directional channels for HSCSD the data are transferred in downlink direction only. The same frequency hopping sequence and training sequence is used for all the channels in the HSCSD configuration.

In symmetric HSCSD configuration individual signal level and quality reporting for each HSCSD channel is applied. For an asymmetric HSCSD configuration individual signal level and quality reporting is used for those channels. The quality measurements reported on the main channel are based on the worst quality measured among the main and the uni-directional downlink timeslots used. In both symmetric and asymmetric HSCSD configuration the neighbouring cell measurement reports are copied on every uplink channel used. See [4] for more detail on signal level and quality reporting.

Transparent data transmission   In transparent data transmission the V.110 data frames on the HSCSD channels carry data sub-stream numbers to retain the order of transmission over GSM, between the split/combine functions. Between these functions channel internal multiframing is also used in order to increase the tolerance against inter channel transmission delays. Depending on the location of the access point to external networks the split/combine functionality is located in the base station subsystem (BSS) or in the IWF on the network side, and at the MS.

Non-transparent data transmission   Non-transparent mode of HSCSD is realised by modifying the radio link protocol (RLP) and L2R functions to support multiple parallel TCH/Fs instead of only one TCH/F (Figure 1.5). In addition the RLP frame numbering is increased to accommodate the enlarged data transmission rate.

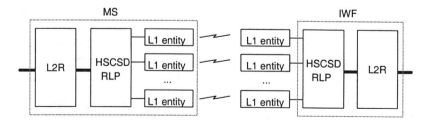

Figure 1.5   The HSCSD concept in non-transparent mode.

1.3 Location services

Different positioning methods found the basis for providing location services (LCS) in the GSM system. In general, these services can be implemented by using vendor/operator specific or standardised solutions. Both of these methods and their combination are likely to be used when LCS enter the operating networks.

The vendor/operator specific solutions utilise the information already available in the cellular network or measured by proprietary units, installed for this purpose. The user location is thus estimated by combining and processing the available information with the help of tools and algorithms implemented for this purpose. In the standardised solution, the basic difference compared with the specific one is that the functionalities and capabilities required to retrieve the information for positioning are integrated into the system itself. The integrated solution affects the specifications and thus standardisation effort is needed, but benefit of this approach is that it enables multi-vendor support, and provides a complete support for user privacy. In this chapter a short overview of the history and drivers behind the standardisation of the location services is given. Chapter 4 describes in more detail the positioning methods, architecture and main technical principles behind them.

LCS standardisation process

The LCS standardisation began in ETSI SMG in 1995 after European operators asked for commercial services and separately in T1P1 in the US after Federal Communications Commission (FCC) established the E911 requirements for the E911 service. The E911 requirement stated that by 1 October 2001 US operators must be able to locate the subscriber calling the emergency number 911.

Although, several operators have applied and received waivers for the E911 requirements and in spite of the different drivers behind the work in the US and Europe, T1P1.5 (T1P1 is a subgroup of T1, a committee of TIA, the North American Telecommunications Industry Association) was given the responsibility to lead the overall standardisation activities in 1997. Four location methods were proposed and taken to the standardisation process:

- cell identity and timing advance
- uplink time of arrival (TOA), and
- enhanced observed time difference (E-OTD) methods based on measurements within the cellular network, as well as
- assisted GPS method based on the GPS technology

In spring 1999, all four methods were included in the GSM specifications Rel'98 and Rel'99, including changes to GSM architecture and introduction of several new network elements for LCS purposes. Later in 2001, TOA method was excluded from the Rel'4 specifications, because no commercial support for the scheme was foreseen due to the complexity in implementing the method in operating networks.

The first specification versions (Rel'98, Rel'99 and Rel'4) include the LCS support for circuit switched connections over the A interface towards the 2G CN only. In Rel'5, the focus is then moved towards providing the same LCS support in PS domain connections over Gb and Iu interfaces. The main reasons for this gradual introduction of different modes were not only the regulatory requirements that in practice were applicable for CS speech connections only, but also the delays in the introduction of GPRS and its interfaces. When the penetration of GPRS users and terminals increases, LCS is expected to be more and more an attractive increment to the services provided to the GPRS subscribers in 'always connected' mode.

1.4 General packet radio system (GPRS)

Introduction of GPRS (Rel'97)

Soon after the first GSM networks became operational in the early 1990s and the use of the GSM data services started, it became evident that the circuit switched bearer services were not particularly well suited for certain types of applications with a bursty nature. The circuit switched connection has a long access time to the network and the call charging is based on the connection time. In packet switched networks, the connections do not reserve resources permanently, but make use of the common pool, which is highly efficient in particular for applications with a bursty nature. The GPRS system will have a very short access time to the network and the call charging could solely be based on an amount of transmitted data.

The GPRS system brings the packet switched bearer services to the existing GSM system. In the GPRS system a user can access the public data networks directly using their standard protocol addresses (IP, X.25), which can be activated when the MS is attached to the GPRS network. The GPRS MS can use between one and eight channels over the air interface depending on the MS capabilities, and those channels are dynamically allocated to a MS when there are packets to be sent or received. In the GPRS network, uplink and downlink channels are reserved separately making it possible to have multislot MSs with various uplink and downlink capabilities. The resource allocation in the GPRS network is dynamic and dependent on demand and resource availability. Packets can also be sent on idle time between speech calls. With the GPRS system it is possible to communicate point-to-point or point-to-multipoint, it also supports the SMS and anonymous access to the network. The theoretical maximum throughput in the GPRS system is 160 kbps per MS using all eight channels without error correction.

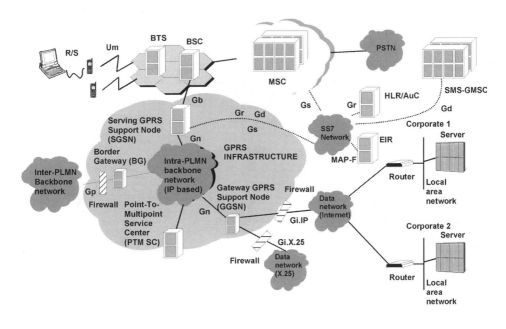

Figure 1.6 Functional view of GPRS.

GPRS network architecture

In Figure 1.6 a functional view of the GPRS network is displayed. GPRS brings a few new network elements to the GSM network. The most important ones are the serving GPRS support node (SGSN) and the gateway GPRS support node (GGSN). Another important new element is the point-to-multipoint service centre (PTM-SC), which is dedicated to the PTM services in the GPRS network. Another new network element is the border gateway (BG), which is mainly needed for security reasons and is situated on the connection to the inter-PLMN backbone network. The inter-PLMN and intra-PLMN backbone networks are also new elements, both Internet protocol (IP) based networks. In addition, there will be a few new gateways in the GPRS system like the charging gateway and the legal interception gateway.

While the current GSM system was originally designed with an emphasis on voice sessions, the main objective of the GPRS is to offer an access to standard data networks such as transport control protocol (TCP)/Internet protocol (IP) and X.25. These other networks consider GPRS just as a normal sub-network (Figure 1.7). A GGSN in the GPRS network behaves as a router and hides the GPRS specific features from the external data network.

The mobile user can have either a static or a dynamic data network address and the data traffic will always use the gateway indicated by this address (Figure 1.8). However, the home network operator could force all the traffic to use a home GGSN, for example for security reasons .

Figure 1.7   GPRS network seen by another data network.

A static address is permanently allocated for one subscriber. As it will point to a gateway of the home network the data packets will always be routed through the home network. Figure 1.8 illustrates the case where a user is in his home network (case 1) and in a visited network (case 2).

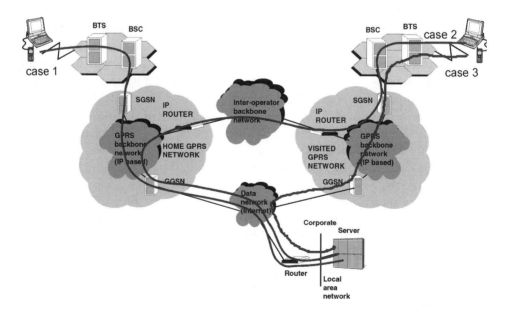

Figure 1.8   Data transfer.

Table 1.1 GPRS mobile classes

Class A mode of operation	Class B mode of operation	Class C mode of operation
Full simultaneous use of packet and circuit mode connections	No simultaneous traffic but automatic sequential service	Alternate use or a GPRS only MS

Each network can also have a pool of available addresses that can be allocated dynamically to the users by the GGSN. This decreases the number of addresses needed by an operator. A dynamic address is allocated to a user only for the time of a connection. To avoid routing the packet through the home network, the dynamic address can be allocated by a GGSN from the visited network (case 3 in Figure 1.8). It can also be allocated by a GGSN from the home network (case 2) to apply, for example, some specific screening.

The incoming and outgoing point-to-point (PTP) traffic can be charged based on the quantity of data transmitted, the protocol used, the length of connection, etc.

GPRS mobiles

A GPRS MS can operate in one of three modes of operation (Table 1.1):

- Class A mode of operation: the MS is attached to both GPRS and other GSM services. The mobile user can make and/or receive calls on the two services simultaneously subject to the quality of service (QoS) requirements, e.g. having a normal GSM voice call and receiving GPRS data packets at the same time.
- Class B mode of operation: the MS is attached to both GPRS and other GSM services, but the MS can only operate one set of services at a time. The MS in idle mode (and packet idle mode) is required to monitor paging channels for both circuit switched and packet switched services. However, the practical behaviour of the MS depends on the mode of network operation. For example, one mode of network operation is defined so that when a MS is engaged in packet data transfer, it will receive paging messages via the packet data channel without degradation of the packet data transfer.
- Class C mode of operation: the MS can only be attached to either the GSM network or the GPRS network. The selection is done manually and there are no simultaneous operations.

In GSM Phase 2 the MS uses one channel for the uplink traffic and one channel for the downlink traffic (Figure 1.9, '1-slot'). In GPRS it is possible to have a multiple slot MS, e.g. a 2-slot MS with two uplink channels and two downlink channels (Figure 1.9, '2-slot'). It is indeed possible to have a GPRS MS with one to eight uplink channels and one to eight downlink channels (Figure 1.9, '3–8-slot'), and the MS can have a different (asymmetric) uplink and downlink capability.

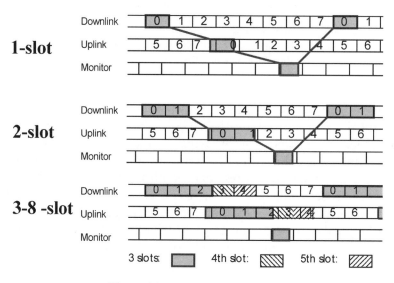

Figure 1.9   GPRS multislot capabilities.

When a MS supports the use of multiple timeslots it belongs to a multislot class as defined in [5]. Multislot class is defined as a combination of several parameters:

- The maximum number of received timeslots that the MS can use per TDMA frame.
- The maximum number of transmit timeslots that the MS can use per TDMA frame.
- Total number of uplink and downlink timeslots that can actually be used by the MS per TDMA frame.
- The time needed for the MS to perform adjacent cell signal level measurement and get ready to transmit.
- The time needed for the MS to get ready to transmit.
- The time needed for the MS to perform adjacent cell signal level measurement and get ready to receive.
- The time needed for the MS to get ready to receive.
- Capability to receive and transmit at the same time. There are two types of MS:

  Type 1: the MS is not required to transmit and receive at the same time.
  Type 2: the MS is required to transmit and receive at the same time.

A combination of these factors has led to a definition of 29 MS classes (in Rel'97 of GSM specifications). For more details on the multislot classes see Annex B of [5].

Base station subsystem (BSS)
The BSS is upgraded with new GPRS protocols for the Gb interface (between the BSS and SGSN) and enhanced layer 2 (RLC/MAC see Section entitled *Medium access control and radio link control layer*) protocols for the air interface. The Gb interface connects the BSS and the SGSN, allowing the exchange of signalling information and user data. Both air and Gb interfaces allow many users to be multiplexed over the same physical resources. The BSS allocates resources to a user upon activity (when data are sent or received) and are reallocated immediately thereafter. The Gb interface link layer is based on frame relay, which is used for signalling and data transmission. The base station subsystem GPRS protocol (BSSGP) provides the radio-related, QoS, and routing information that is required to transmit user data between a BSS and an SGSN.

Mobile services switching centre/home location register (MSC/HLR)
The HLR is upgraded and contains GPRS subscription data. The HLR is accessible from the SGSN via the Gr interface and from the GGSN via the optional Gc interface. For roaming MSs, HLR is in a different PLMN than the current SGSN. All MSs use their HLR in home public land mobile network (HPLMN). The HLR is enhanced with GPRS subscriber information.

The MSC/visiting location register (VLR) can be enhanced for more efficient coordination of GPRS and non-GPRS services and functionality by implementing the Gs interface, which uses the BSSAP + procedures, a subset of BSS application part (BSSAP) procedures. Paging for circuit switched calls can be performed more efficiently via the SGSN. This is the case as well for the combined GPRS and non-GPRS location updates. The Gs interface allows also easy Class B mode of operation MS design. The SMS-GMSC and SMS-IWMSC functions are enhanced for SMS over GPRS. Also the temporary mobile subscriber identity (TMSI) allocation procedure is modified.

SGSN
The SGSN is a main component of the GPRS network, which handles, e.g. the mobility management, and authentication and has register function. The SGSN is connected to the BSC and is the service access point to the GPRS network for the GPRS MS. The SGSN handles the protocol conversion from the IP used in the backbone network to the sub-network dependent convergence protocol (SNDCP) and logical link control (LLC) protocols used between the SGSN and the MS. These protocols handle compression and ciphering. The SGSN also handles the authentication of GPRS mobiles and when the authentication is successful, the SGSN handles the registration of a MS to the GPRS network and takes care of its mobility management. When the MS wants to send (or receive) data to (from) external networks, the SGSN relays the data between the SGSN and relevant GGSN (and vice versa).

GGSN

The GGSN is connected to the external networks like the Internet and the X.25. From the external networks' point of view the GGSN is a router to a sub-network, because the GGSN 'hides' the GPRS infrastructure from the external networks. When the GGSN receives data addressed to a specific user it checks if the address is active. If it is, the GGSN forwards the data to the SGSN serving the MS, but if the address is inactive, the data are discarded. The mobile originated packets are routed to the right network by the GGSN. The GGSN tracks the MS with a precision of SGSN.

GPRS interfaces and reference points

The GPRS system introduces new so-called G-interfaces to the GSM network architecture. It is important to understand the function of every interface and reference point, because this gives an insight to the GPRS system and consequent evolution. Figure 1.10 gives a logical architecture description with the interfaces and reference points of the GSM network with GPRS.

Connections of the GPRS system to the network and switching subsystem (NSS) part of the GSM network are implemented through signalling system number 7 (SS7) network (Gc, Gd, Gf, Gr, Gs), while the other interfaces and reference points are implemented through the intra-PLMN backbone network (Gn), the inter-PLMN backbone network (Gp) or the external networks (Gi).

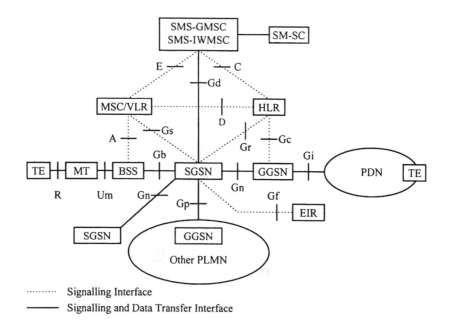

Figure 1.10   Logical architecture.

The different interfaces the GPRS system uses are:

- Gb between a SGSN and a BSS. The Gb interface is the carrier of the GPRS traffic and signalling between the GSM radio network (BSS) and the GPRS part. Frame relay based network services provide flow control for this interface.
- Gc between the GGSN and the HLR. The GGSN may request location information for network requested context activation only via this optional interface. The standard defines also the use of a proxy GSN, which is used as a GPRS tunnelling protocol (GTP) to mobile application part (MAP) protocol converter, thus avoiding implementing MAP in GGSN.
- Gd between the SMS-GMSC and a SGSN, and between SMS-IWMSC and a SGSN. The Gd interface allows more efficient use of the SMS services.
- Gf between a SGSN and the EIR. The Gf gives the SGSN access to equipment information. In the EIR the MSs are divided into three lists: black list for stolen mobiles, grey list for mobiles under observation and white list for other mobiles.
- Gn between two GSNs within the same PLMN. The Gn provides a data and signalling interface in the intra-PLMN backbone. The GTP is used in the Gn (and in the Gp) interface over the IP based backbone network.
- Gp between two GSNs in various PLMNs. The Gp interface provides the same functionality as the Gn interface, but it also provides, with the BG and the Firewall, all the functions needed in the inter-PLMN networking, i.e. security, routing, etc.
- Gr between a SGSN and the HLR. The Gr gives the SGSN access to subscriber information in the HLR, which can be located in a different PLMN than the SGSN.
- Gs between a SGSN and a MSC. The SGSN can send location data to the MSC or receive paging requests from the MSC via this optional interface. The Gs interface will greatly improve effective use of the radio and network resources in the combined GSM/GPRS network. This interface uses BSSAP + protocol.
- Um between a MS and the GPRS fixed network part. The Um is the access interface for the MS to the GPRS network. The MS has a radio interface to the BTS, which is the same interface used by the existing GSM network with some GPRS specific changes.

There are two different reference points in the GPRS network. The Gi is GPRS specific, but the R is common with the circuit switched GSM network. The two reference points in the GPRS are:

- Gi between a GGSN and an external network. The GPRS network is connected to external data networks via this interface. The GPRS system will support a variety of data networks and that is why the Gi is not a standard interface, but merely a reference point.

- R between terminal equipment and mobile termination. This reference point connects terminal equipment to mobile termination, thus allowing, e.g. a laptop-PC to transmit data over the GSM-phone. The physical R interface follows, e.g. the ITU-T V.24/V.28 or the PCMCIA PC-Card standards.

GPRS protocol architecture

The GPRS system introduces a whole new set of protocols for the GSM Phase 2 + network. The interworking between the new network elements is done with new GPRS specific protocols. However, there are a number of existing protocols used at the lower layers of the protocol stacks, namely TCP/user datagram protocol (UDP) IP. Figure 1.11 shows the transmission plane used in the GPRS system.

Physical layer
The physical layer has been separated into two distinct sub-layers, the physical RF layer and the physical link layer.

The *physical RF layer* performs the modulation of the physical waveforms based on the sequence of bits received from the physical link layer. The physical RF layer also demodulates received waveforms into a sequence of bits that are transferred to the physical link layer for interpretation. The GSM physical RF layer is defined in GSM 05 series specifications, which define among other things:

Figure 1.11 Transmission plane.

- The carrier frequency characteristics and GSM radio channel structures [5].
- The modulation of the transmitted waveforms and the raw data rates of GSM channels [6]. Note that in the GPRS Rel'97 radio interface, only the original GSM modulation method, GMSK modulation, with the carrier bit rate of 270.833 kbps is defined. For one single timeslot, this corresponds to the gross data rate of 22.8 kbps.
- The transmitter and receiver characteristics and performance requirements [7].

The *physical link layer* provides services for information transfer over a physical channel between the MS and the network. These functions include data unit framing, data coding, and the detection and correction of physical medium transmission errors. The physical link layer operates above the physical RF layer to provide a physical channel between the MS and the network. The purpose of the physical link layer is to convey information across the GSM radio interface, including RLC/MAC information. The physical link layer supports multiple MSs sharing a single physical channel and provides communication between MSs and the network. In addition it provides the services necessary to maintain communications capability over the physical radio channel between the network and MSs. Radio subsystem link control procedures are specified in [4].

The physical link layer is responsible for:

- Forward error correction (FEC) coding, allowing the detection and correction of transmitted code words and the indication of uncorrectable code words. The coding schemes are described in Section entitled *Physical link layer*.
- Rectangular interleaving of one radio block over four bursts in consecutive TDMA frames, as specified in [8].
- Procedures for detecting physical link congestion.

The physical link layer control functions include:

- Synchronisation procedures, including means for determining and adjusting the MS timing advance to correct for variances in propagation delay [9].
- Monitoring and evaluation procedures for radio link signal quality.
- Cell selection and re-selection procedures.
- Transmitter power control procedures.
- Battery power saving procedures, e.g. discontinuous reception (DRX) procedures.

For more details see Section entitled *Physical link layer*.

RLC/MAC

The RLC/MAC layer provides services for information transfer over the physical layer of the GPRS radio interface. These functions include backward error correction procedures enabled by the selective retransmission of erroneous blocks. The RLC function offers a reliable radio link to the upper layers. The MAC function handles the channel allocation and the multiplexing, i.e. the use of physical layer functions. The RLC/MAC layer together form the open system interconnection (OSI) Layer 2 protocol for the Um interface and uses the services of the physical link layer (see Section entitled *Medium Access Control and Radio Link Control Layer*).

LLC

The logical link control (LLC) layer offers a secure and reliable logical link between the MS and the SGSN to upper layers, and is independent of the lower layers. The LLC layer has two transfer modes, the acknowledged and unacknowledged. The LLC conveys signalling, SMS and SNDCP packets. See [10] for more information.

SNDCP

The sub-network dependent convergence protocol (SNDCP) is a mapping and compression function between the network layer and lower layers. It also performs segmentation, reassembling and multiplexing. The SNDCP protocol is specified in [11].

Signalling has various sets of protocols, which are used with the existing GSM network elements. Internal signalling in the GPRS system is handled by protocols, which carry both data and signalling (LLC, GTP and BSSGP). The GPRS control plane is shown in Figure 1.12.

Figure 1.13 describes how network protocols are multiplexed. NSAPI is the network layer service access point identifier, which is used to identify the PDP context at SNDCP level. Service access point identifier (SAPI) is used to identify the points where LLC provide service to higher layer. SAPIs have different priorities. TLLI is the temporary logical link identity, which unambiguously identifies the logical link between the MS and SGSN. The IP or the X.25 is the packet protocol offered to the subscriber by the GPRS system.

BSSGP

The primary functions of the base station subsystem GPRS protocol (BSSGP) (see [12]) include: in the downlink, the provision by an SGSN to a BSS of radio related information used by the RLC/MAC function; in the uplink, the provision by a BSS to an SGSN of radio related information derived from the RLC/MAC function; and the provision of functionality to enable two physically distinct nodes, an

Figure 1.12 GPRS control plane.

SGSN and a BSS, to operate node management control functions. See [12] for more details. The underlying network service (NS) is responsible for the transport of BSSGP PDUs between a BSS and an SGSN [12].

GTP
The GPRS tunnelling protocol (GTP) is used to tunnel data and signalling between the GSNs. The GTP can have proprietary extensions to allow proprietary features. The relay function relays PDP (packet data protocol) PDUs between the Gb and the Gn interfaces. For details see [13].

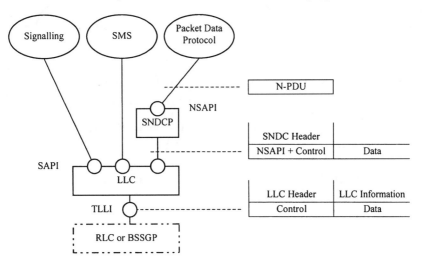

Figure 1.13 Multiplexing of network protocols.

Mobility management

The mobility management in the GPRS network is handled almost the same way as in the existing GSM system. One or more cells form a routing area, which is a subset of one location area. Every routing area is served by one SGSN. The tracking of the location of a MS depends on the mobility management state. When a MS is in a STANDBY state, the location of the MS is known on a routing area level. When the MS is in a READY state, the location of the MS is known on a cell level. Figure 1.14 shows the different mobility management states and transitions between them.

Mobility management states

The GPRS has three various mobility management states. The IDLE state is used when the subscriber (MS) is passive (not GPRS attached). The STANDBY state is used when the subscriber has ended an active phase. A MS is in an active phase

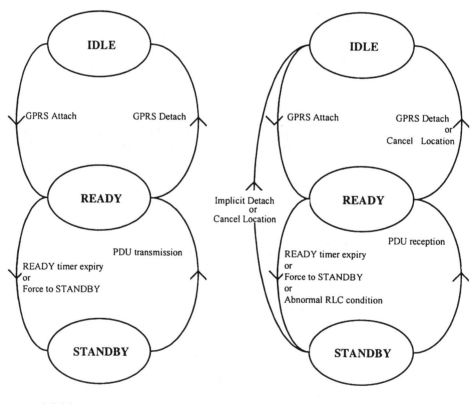

MM State Model of MS   MM State Model of SGSN

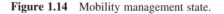

Figure 1.14 Mobility management state.

(READY state) when it is transmitting or has just been transmitting. The change between the states happens upon activity or when a timer expires. A description of the various states is given below.

IDLE state   The subscriber is not reachable by the GPRS network. The MS is only capable of receiving PTM-M data. The network elements hold no valid context for the subscriber and the subscriber is not attached to the mobility management. In order to change state, the MS has to perform a GPRS attach procedure.

STANDBY state   The subscriber is attached to the mobility management and the location of a MS is known on a routing area level. The MS is capable of receiving PTM data and pages for PTP data. The network holds a valid mobility management context for the subscriber. If the MS sends data, the MS moves to READY state. The MS or the network can initiate the GPRS detach procedure to move to IDLE state. After expiry of the MS reachable timer, the network can detach the MS. The MS can use the discontinuous reception (DRX) to save the battery.

READY state   The subscriber is attached to the mobility management and the location of a MS is known on a cell level. The MS is capable of receiving PTM and PTP data. The SGSN can send data to the MS without paging at any time and the MS can send data to the SGSN at any time. The network holds a valid mobility management context for the subscriber. If the READY timer expires, the MS moves to STANDBY state. If the MS performs a GPRS detach procedure, the MS moves to IDLE state and the mobility management context is removed. A MS in READY state does not necessarily have radio resources reserved. The MS can use the DRX to save the battery.

GPRS attach and detach
GPRS attach and GPRS detach are mobility management functions to establish and to terminate a connection to the GPRS network. The SGSN receives the requests and processes them. With the GPRS attach the mobile moves to READY state and the mobility management context is established, the MS is authenticated, the ciphering key is generated, a ciphered link established and the MS is allocated a temporary logical link identity (TLLI). The SGSN fetches the subscriber information from the HLR. After a GPRS attach, the SGSN tracks the location of the MS. The MS can send and receive SMS, but no other data. To transfer other data it has to first activate a PDP context.

When the subscriber wants to terminate a connection to the GPRS network, the GPRS detach is used. The GPRS detach moves the MS to IDLE state and the mobility management context is removed. The MS can be detached from the GPRS also implicitly when the mobile reachable timer expires. The GPRS detach is normally generated by the MS, but can also be generated by the network.

PDP context functions and addresses

The packet data protocol (PDP) context functions are network level functions, which are used to bind a MS to various PDP addresses and after used to unbind the MS from these addresses. The PDP context can also be modified. When a MS is attached to the network, it has to activate all the addresses it wants to use for data traffic with the external networks. Various PDP contexts have to be activated, because these include address, QoS attributes, etc. After the subscriber has finished the use of activated addresses, these have to be deactivated. Also change of address specific attributes have to be performed sometimes. The MS can use these functions when in STANDBY or READY state. The GGSN can also use these functions, but the SGSN is responsible for performing these functions. The MS can select a particular GGSN for the access to certain services and also activate a PDP context anonymously, without using any identification.

Dynamic and static PDP address

The subscriber can use various kinds of PDP addresses. A static PDP address is permanently assigned to a MS by the HPLMN. A HPLMN dynamic PDP address is assigned to a MS by the HPLMN when a MS performs a PDP context activation. A visited public land mobile network (VPLMN) dynamic PDP address is assigned to a MS by the VPLMN when a MS performs a PDP context activation. It is indicated in the subscription if the MS can have a dynamic address or not.

PDP context activation

The PDP context activation can be done by the MS or by the network. The activate PDP context request is sent to the SGSN when a certain address requires activation. As an option, the request can also be made by the GGSN, if packets are received for an address without active context and the MS is GPRS attached. The request contains parameters for the context, like the TLLI, the protocol type, the address type, QoS, requested GGSN, etc.

PDP context modification

The PDP context can be modified by the SGSN with the modify PDP context. Only the parameters' QoS negotiated and radio priority can be changed. The SGSN sends the request to the MS, which either accepts the new QoS by sending a modify PDP context accept to the SGSN or does not accept it by sending a deactivate PDP context request for that context. In GPRS Phase 2 also MS and GGSN can modify PDP context.

PDP context deactivation
The PDP context can be deactivated by the MS or by the network. Every address can be deactivated separately, but when the GPRS detach is performed, the network will automatically remove all the PDP contexts.

Security

The GPRS system uses GSM Phase 2 based security. These security functions include the authentication of the subscriber, the user identity confidentiality and the ciphering of the data traffic between the MS and the SGSN. The authentication of the subscriber is done the same way by the SGSN in the GPRS system as by the MSC/VLR in the Phase 2 GSM network. The TLLI is used to keep the subscriber identity confidential. The correspondence between the international mobile subscriber identity (IMSI) and the TLLI is only known by the MS and the SGSN. The ciphering function used between the MS and the SGSN is not the same as that used in the GSM Phase 2, but an optimised one for the packet switched traffic. Security of the backbone is provided by using a private network, thus avoiding the possibility that an external *hacker* can address it. Each operator must guarantee its physical security.

Location management

The location management procedures are used to handle the changing of a cell and/or a routing area, and the periodic routing area updates. If a MS stays a long time at the same place, the network has to receive an indication that the MS is still reachable. This is the reason why periodic routing area updates are made. All the MSs attached to the GPRS will perform a periodic routing area update. Only the Class B mode of operation mobiles which are engaged in a circuit switched communication cannot. The MS performs a cell update when it changes cell within a routing area in READY mode. When the MS changes cell between the different routing areas, it performs a routing area update. There are two types of routing area updates, the intra-SGSN routing area update and the inter-SGSN routing area update. A SGSN can manage many routing areas and if the new routing area belongs to the management of a new SGSN, the inter-SGSN routing area update is used. If the new routing area belongs to the management of the same SGSN than the old one, the intra-SGSN routing area update is used. The old SGSN forwards user packets to the new SGSN, until it receives a cancel location from the HLR.

GPRS radio interface

This section gives an overview of the GPRS Rel'97 radio interface.

GPRS packet data logical channels

This section describes the logical channels of the GPRS radio interface. The reason for defining a new set of control channels for the GPRS packet data, which is partly parallel to that of the circuit switched control channels, is to be able to flexibly allocate more signalling capacity for the packet data traffic without sacrificing the quality of the speech traffic.

The packet data logical channels are mapped on to the physical channels that are dedicated to packet data. The physical channel dedicated to packet data traffic is called a packet data channel (PDCH).

The packet data logical channels can be grouped into different categories as shown in Figure 1.15. These categories (packet common control channels, packet broadcast control channels, packet traffic channels and packet dedicated control channels) are introduced in more detail in the following list [17].

Packet Common Control Channel (PCCCH)   PCCCH comprises logical channels for common control signalling used for packet data:

- Packet random access channel (PRACH) – uplink only

  PRACH is used by the MS to initiate uplink transfer to send data or signalling information.

- Packet paging channel (PPCH) – downlink only

  PPCH is used to page a MS prior to downlink packet transfer.

- Packet access grant channel (PAGCH) – downlink only

  PAGCH is used in the packet transfer establishment phase to send resource assignment to a MS prior to packet transfer.

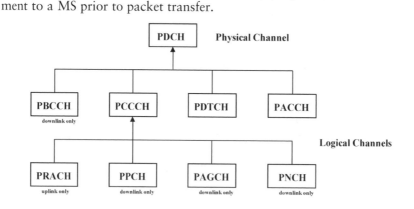

Figure 1.15   GPRS logical channels.

- Packet notification channel (PNCH) – downlink only

  PNCH is used to send a PTM-M (point to multipoint – multicast) notification to a group of MSs prior to a PTM-M packet transfer. (Note that the PTM-M service is not specified in GPRS Rel'97.)

Packet broadcast control channel (PBCCH) – downlink only PBCCH broadcasts packet data specific system information. If PBCCH is not allocated, the packet data specific system information is broadcast on BCCH.

Packet traffic channels

- Packet data traffic channel (PDTCH)

  PDTCH is a channel allocated for data transfer. It is temporarily dedicated to one MS. In the multislot operation, one MS may use multiple PDTCHs in parallel for individual packet transfer.

  All packet data traffic channels are uni-directional, either uplink (PDTCH/U), for a mobile originated packet transfer or downlink (PDTCH/D), for a mobile terminated packet transfer.

Packet dedicated control channels

- Packet associated control channel (PACCH)

  PACCH conveys signalling information related to a given MS. The signalling information includes, e.g. acknowledgements and power control information. PACCH carries also resource assignment and reassignment messages, comprising the assignment of a capacity for PDTCH(s) and for further occurrences of PACCH. The PACCH shares resources with PDTCHs, which are currently assigned to one MS. Additionally, a MS that is currently involved in packet transfer, can be paged for circuit switched services on PACCH.

- Packet timing advance control channel, uplink (PTCCH/U)

  PTCCH/U is used to transmit random access burst to allow estimation of the timing advance for one MS in packet transfer mode.

- Packet timing advance control channel, downlink (PTCCH/D)

  PTCCH/D is used to transmit timing advance information updates to several MSs. One PTCCH/D is paired with several PTCCH/Us.

Mapping of packet data logical channels into physical channels

Different packet data logical channels can occur on the same physical channel (i.e. PDCH). The sharing of the physical channel is based on blocks of four consecutive bursts, except for PTCCH. The mapping in frequency of PDCH on to the physical channel is defined in [5].

On PRACH and PTCCH/U, access bursts are used. On all other packet data logical channels, radio blocks comprising four normal bursts are used. The only exceptions are some messages on uplink PACCH which comprise four consecutive access bursts (to increase robustness). The following list shows some more details how the packet data logical channels are mapped on to physical channels.

Packet common control channel (PCCCH)   At a given time, the logical channels of the PCCCH are mapped on different physical resources than the logical channels of the CCCH. The PCCCH does not have to be allocated permanently in the cell. Whenever the PCCCH is not allocated, the CCCH shall be used to initiate a packet transfer. The PCCCH, when it exists, is mapped on one or several physical channels according to a 52-multiframe. In that case the PCCCH, PBCCH and PDTCH share the same physical channels (PDCHs). The existence and location of the PCCCH shall be broadcast on the cell.

- Packet random access channel (PRACH)

  The PRACH is mapped on one or several physical channels. The physical channels on which the PRACH is mapped are derived by the MS from information broadcast on the PBCCH or BCCH.
  PRACH is determined by the uplink state flag (USF) being marked as free, which is broadcast continuously on the corresponding downlink. Additionally, a predefined fixed part of the multiframe structure for PDCH can be used as PRACH only and the information about the mapping on the physical channel is broadcast on PBCCH. During those time periods a MS does not have to monitor the USF that is simultaneously broadcast on the downlink.

- Packet paging channel (PPCH)

  The PPCH is mapped on one or several physical channels. The exact mapping on each physical channel follows a predefined rule, as it is done for the paging channel (PCH).
  The physical channel on which the PPCH is mapped, as well as the rule that is followed on the physical channels, are derived by the MS from information broadcast on the PBCCH.

- Packet access grant channel (PAGCH)

  The PAGCH is mapped on one or several physical channels. The exact mapping on each physical channel follows a predefined rule.
  The physical channels on which the PAGCH is mapped, as well as the rule that is followed on the physical channels, are derived by the MS from information broadcast on the PBCCH.

- Packet notification channel (PNCH)

  The PNCH is mapped on one or several blocks on PCCCH. The exact mapping follows a predefined rule. The mapping is derived by the MS from information broadcast on the PBCCH.

Packet broadcast control channel (PBCCH)   The PBCCH shall be mapped on one or several physical channels. The exact mapping on each physical channel follows a predefined rule, as it is done for the BCCH.

The existence of the PCCCH, and consequently the existence of the PBCCH, is indicated on the BCCH.

Packet timing advance control channel (PTCCH)   Two defined frames of multiframe are used to carry PTCCH. The exact mapping of PTCCH/U sub-channels and PTCCH/D are defined in [5].

On PTCCH/U, access bursts are used. On PTCCH/D, four normal bursts comprising a radio block are used.

Packet traffic channels

- Packet data traffic channel (PDTCH)

  One PDTCH is mapped on to one physical channel. Up to eight PDTCHs, with different timeslots but with the same frequency parameters, may be allocated to one MS at the same time.

- Packet associated control channel (PACCH)

  PACCH is dynamically allocated on a block basis on the same physical channel that is carrying PDTCHs. However, one block PACCH allocation is used on the physical channel that is carrying only PCCCH when the MS is polled to acknowledge the initial assignment message.

  PACCH is of a bi-directional nature, i.e. it can dynamically be allocated both on the uplink and on the downlink regardless of whether the corresponding PDTCH assignment is for uplink or downlink.

Different packet data logical channels can be multiplexed in one direction (either on the downlink or uplink) on the same physical channel (i.e. PDCH). The reader is advised to see details in [5]. The type of message that is indicated in the radio block header allows differentiation between the logical channels. Additionally, the MS identity allows differentiation between PDTCHs and PACCHs assigned to different MSs.

Radio interface (Um)

Radio resource management principles *[17]* A cell supporting GPRS may allocate resources on one or several physical channels in order to support the GPRS traffic. Those physical channels (i.e. PDCHs), shared by the GPRS MSs, are taken from the common pool of physical channels available in the cell. The allocation of physical channels to circuit switched services and GPRS can vary dynamically. Common control signalling required by GPRS in the initial phase of the packet transfer is conveyed on PCCCH (when allocated), or on CCCH. This allows capacity to be allocated specifically to GPRS in the cell only when a packet is to be transferred.

At least one PDCH, acting as a primary PDCH, accommodates packet common control channels that carry all the necessary control signalling for initiating packet transfer (i.e. PCCCH), whenever that signalling is not carried by the existing CCCH, as well as user data and dedicated signalling (i.e. PDTCH and PACCH). Other PDCHs, acting as secondary PDCHs, are used for user data transfer and for dedicated signalling.

The GPRS does not require permanently allocated PDCHs. The allocation of capacity for GPRS can be based on the needs for actual packet transfers ('capacity on demand' principle). The operator can as well decide to dedicate permanently or temporarily some physical resources (i.e. PDCHs) for GPRS traffic.

When the PDCHs are congested due to the GPRS traffic load and more resources are available in the cell, the network can allocate more physical channels as PDCHs. However, the existence of PDCH(s) does not imply the existence of PCCCH. When no PCCCH is allocated in a cell, all GPRS attached MSs camp on the CCCH.

In response to a packet channel request sent on CCCH from the MS that wants to transmit GPRS packets, the network can assign resources on PDCH(s) for the uplink transfer. After the transfer, the MS returns to CCCH.

When PCCCH is allocated in a cell, all GPRS attached MSs camp on it. PCCCH can be allocated either as the result of the increased demand for packet data transfers or whenever there are enough available physical channels in a cell (to increase the QoS). The information about PCCCH is broadcast on BCCH. When the PCCCH capacity is inadequate, it is possible to allocate additional PCCCH resources on one or several PDCHs. If the network releases the last PCCCH, the MS performs cell re-selection.

The number of allocated PDCHs in a cell can be increased or decreased according to demand.

Multiframe structure for PDCH The mapping in time of the logical channels is defined by a multiframe structure. The multiframe structure for PDCH consists of 52 TDMA frames, divided into 12 blocks (of 4 frames), 2 idle frames and 2 frames used for the PTCCH according to Figure 1.16.

52 TDMA Frames

X = Idle frame
T = Frame used for PTCCH
B0 - B11 = Radio blocks

Figure 1.16 Multiframe structure for PDCH. [17]

The mapping of logical channels on to the radio blocks is defined in this section by means of the ordered list of blocks (B0, B6, B3, B9, B1, B7, B4, B10, B2, B8, B5, B11).

The PDCH that contains PCCCH (if any) is indicated on BCCH. That PDCH is the only one that contains PBCCH blocks. On the downlink of this PDCH, the first block (B0) in the ordered list of blocks is used as PBCCH. If required, up to three more blocks on the same PDCH can be used as additional PBCCH. Any additional PDCH containing PCCCH is indicated on PBCCH.

On any PDCH with PCCCH (with or without PBCCH), up to the next 12 blocks in the ordered list of blocks are used for PAGCH, PNCH, PDTCH or PACCH in the downlink. The remaining blocks in the ordered list are used for PPCH, PAGCH, PNCH, PDTCH or PACCH in the downlink. In all cases, the actual usage of the blocks is indicated by the message type. On an uplink PDCH that contains PCCCH, all blocks in the multiframe can be used as PRACH, PDTCH or PACCH. Optionally, the first blocks in the ordered list of blocks can only be used as PRACH. The MS may choose to either ignore the USF (consider it as FREE) or use the USF to determine the PRACH in the same way as for the other blocks.

The mapping of channels on multiframes is controlled by several parameters broadcast on PBCCH. On a PDCH that does not contain PCCCH, all blocks can be used as PDTCH or PACCH. The actual usage is indicated by the message type. Two frames are used for PTCCH (see [5]) and the two idle frames as well as the PTCCH frames can be used by the MS for signal measurements and base station identification code (BSIC) identification.

A MS attached to GPRS shall not be required to monitor BCCH if a PBCCH exists. All system information relevant for GPRS and some information relevant for circuit switched services (e.g. the access classes) shall in this case be broadcast on PBCCH. In order to facilitate the MS operation, the network is required to transmit certain types of packet system information (PSI) messages in specific multiframes and specific PBCCH blocks within the multiframes. The exact scheduling is defined in [5]. If no PCCCH is allocated, the MS camps on the CCCH and receives all system information on BCCH. In this case, any necessary GPRS specific system information has to be broadcast on BCCH.

Table 1.2 Correspondence between RR operating modes and MM states [17]

Radio resource BSS	Packet transfer mode	Measurement report reception	No state	No state
Radio resource MS	Packet transfer mode	Packet idle mode		Packet idle mode
Mobility management NSS and MS	Ready			Standby

Radio resource operating modes [17] Radio resource (RR) management procedures are characterised by two different RR operating modes, packet idle mode and packet transfer mode.

In *packet idle mode* no temporary block flow (TBF) exists. Upper layers can require the transfer of a LLC PDU, which implicitly, may trigger the establishment of TBF and transition to packet transfer mode.

In packet idle mode, the MS listens to the PBCCH and to the paging sub-channel for the paging group the MS belongs to. If PCCCH is not present in the cell, the MS listens to the BCCH and to the relevant paging sub-channels.

In *packet transfer mode*, the mobile station is allocated radio resource providing a TBF on one or more physical channels. Continuous transfer of one or more LLC PDUs is possible. Concurrent TBFs may be established in opposite directions. Transfer of LLC PDUs in RLC acknowledged or RLC unacknowledged mode is provided.

When selecting a new cell, the MS leaves the packet transfer mode, enters the packet idle mode where it switches to the new cell, reads the system information and may then resume to packet transfer mode in the new cell.

The mobility management states are defined in [14]. Table 1.2 provides the correspondence between radio resource states and mobility management states.

Each state is protected by a timer. The timers run in the MS and the network. Packet transfer mode is guarded by RLC protocol timers.

The GPRS RR procedures and RR operating modes are defined in detail in [15].

Physical link layer Different radio block structures are defined for the data transfer and for the control message transfer. The radio block consists of MAC header, RLC data block or RLC/MAC control block (Figure 1.17). It is always carried by four normal bursts. For detailed definition of radio block structure, see [16].

The MAC header (8 bits) contains control fields, which are different for uplink and downlink directions. The RLC header is of variable length and contains control fields, which are as well different for uplink and downlink directions. The RLC data field contains octets from one or more LLC packet data units (PDUs). The block check sequence (BCS) is used for error detection.

Radio Block

MAC header	RLC header	RLC data	

RLC data block   (part of physical link

Radio Block

MAC header	RLC/MAC Control Message	

RLC/MAC control block   (part of physical link

Figure 1.17 Radio block structures. [17]

The RLC/MAC control message field contains one RLC/MAC control message.

Channel coding Four coding schemes, CS-1 to CS-4, are defined for the packet data traffic channels. For all other packet control channels than packet random access channel (PRACH) and packet timing advance control channel on uplink (PTCCH/U), coding scheme CS-1 is always used. For access bursts on PRACH, two coding schemes are specified. All coding schemes are mandatory for the MS. Only CS-1 is mandatory for the network.

Channel coding for PDTCH Four different coding schemes, CS-1 to CS-4, are defined for the radio blocks carrying RLC data blocks. The block structures of the coding schemes are shown in Figs. 18 and 19.

The first step of the coding procedure is to add a BCS for error detection. For the CS-1–CS-3, the second step consists of pre-coding USF (except for CS-1), adding four tail bits and a half-rate convolutional coding for error correction that is punctured to give the desired coding rate. For the CS-4, there is no coding for error correction. The details of the codes are shown in Table 1.3, including the length of each field, the number of coded bits (after adding tail bits and convolutional coding), the number of punctured bits and the data rate (including the RLC header and RLC information).

The CS-1 is the same coding scheme as specified for the slow associated control channel (SACCH) in GSM 05.03. It consists of a half-rate convolutional code for forward error correction and a 40-bit FIRE code for the BCS. The CS-2 and CS-3 are punctured versions of the same half-rate convolutional code as CS-1. The CS-4 has no forward error correction. The CS-2–CS-4 use the same 16-bit CRC for the BCS. The CRC is calculated over the whole uncoded RLC data block including the MAC header.

The USF has eight states, which are represented by a binary 3-bit field in the MAC header. For the CS-1, the whole radio block is convolutionally coded and

Table 1.3 Coding parameters for the coding schemes

Scheme	Code rate	USF	Pre-coded USF	Radio block excl. USF and BCS	BCS	Tail	Data rate kb/s	Data rate kb/s excl. RLC/ MAC headers
CS-1	1/2	3	3	181	40	4	9.05	8
CS-2	≈ 2/3	3	6	268	16	4	13.4	12
CS-3	≈ 3/4	3	6	312	16	4	15.6	14.4
CS-4	1	3	12	428	16	–	21.4	20

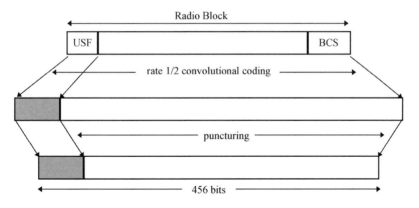

Figure 1.18   Radio block structure for CS-1 to CS-3.

Figure 1.19   Radio block structure for CS-4. [17]

the USF needs to be decoded as part of the data. All other coding schemes generate the same 12-bit code for the USF. The USF can be decoded either as a block code or as part of the data.

In order to simplify the decoding, the stealing bits (1-bit flag in either side of the training sequence code, as defined in [8]) of the block are used to indicate the actual coding scheme.

Channel coding for PACCH, PBCCH, PAGCH, PPCH, PNCH and PTCCH   The channel coding for the PACCH, PBCCH, PAGCH, PPCH, PNCH and the downlink PTCCH is the same as the coding scheme CS-1. The coding scheme used for the uplink PTCCH is the same as for the PRACH.

Channel coding for the PRACH   Two types of packet access bursts may be transmitted on the PRACH: an 8 information bits access burst or an 11 information bits access burst called the extended packet access burst. The reason for two different packet access bursts is that the 8-bit access burst cannot carry the additional information (USF) in the dynamic allocation mode. The mobile supports both access bursts. The channel coding for both burst formats is indicated as follows.

- Coding of the 8 data bit packet access burst:

  The channel coding used for the burst carrying the 8 data bit packet access uplink message is identical to the coding of the access burst as defined for the random access channel in [8].

- Coding of the 11 data bit packet access burst:

  The channel coding for the 11-bit access burst is the punctured version of the same coding as used for the 8-bit access burst.

Cell re-selection   Initial PLMN selection and cell selection are done when the mobile station power is switched on. In GPRS packet idle *cell re-selection* is performed autonomously by the MS. In GPRS packet transfer mode the cell re-selection mechanism depends on the network operational mode. There are three network control modes: NC0, NC1 and NC2. In NC0 the MS performs the cell re-selection autonomously as in idle mode. In NC1 the MS also performs the cell re-selection autonomously but, in addition, it sends periodically measurement reports to the network. The measurement reports are analogous to the ones sent every 0.48 s in voice and CS data connections. In NC2 the MS sends periodical measurement reports and, additionally, the network sends cell re-selection commands to the MS, so the network has control of the cell re-selection process.

When the MS is attached to the circuit switched core network (to MSC) and at the same time it is in packet transfer mode, the cell is determined by the network according to the handover procedures (handovers have precedence over GPRS cell re-selection).

In GPRS, two additional cell re-selection criteria, C31 and C32, are provided as a complement to the other GSM cell re-selection criteria. Cell re-selection criteria are specified in [4].

Also a network controlled cell re-selection can be used in GPRS. The network can order mobile stations to send measurement reports to the network and to suspend its normal cell re-selection, and instead to accept decisions from the network. This applies to both packet idle mode and packet transfer mode. The MS measures the received RF signal strength on the BCCH frequencies of the serving cell and the neighbour cells as indicated in the BA-GPRS list, and calculates the received level average (RLA) for each frequency, as defined in [4]. In addition, the MS verifies the BSIC of the cells. Only channels with the same BSIC as broadcast together with BA-GPRS on PBCCH are considered for re-selection.

The PBCCH broadcasts GPRS specific cell re-selection parameters for serving and neighbour cells, including the BA-GPRS list. A BA-GPRS identifies the neighbour cells, including BSIC, that need to be considered for the GPRS cell re-selection.

Timing advance The timing advance procedure is used to derive the correct value for timing advance that the MS has to use for the uplink transmission of radio blocks. The timing advance procedure comprises two parts:

- initial timing advance estimation
- continuous timing advance update

The *initial timing advance estimation* is based on the single access burst carrying the packet channel request. The packet uplink assignment or packet downlink assignment then carries the estimated timing advance value to the MS. The MS uses this value for the uplink transmissions until the continuous timing advance update provides a new value (a few special cases for the initial timing advance are explained in [17]).

In the packet transfer mode, the MS uses the *continuous timing advance update* procedure. The continuous timing advance update procedure is carried on the PTCCH allocated to the MS. For the uplink packet transfer, within the packet uplink assignment, the MS is assigned the timing advance index (TAI) and the PTCCH. For the downlink packet transfer, within the packet downlink assignment, the MS is assigned the TAI and the PTCCH. The TAI specifies the PTCCH sub-channel used by the MS. On the uplink, the MS sends the assigned PTCCH access burst, which is used by the network to derive the timing advance.

The network analyses the received access burst and determines new timing advance values for all MSs performing the continuous timing advance update procedure on that PDCH. The new timing advance values are sent via a downlink signalling message (TA message) on PTCCH/D. The network can send timing advance information also in packet timing advance/power control and packet uplink Ack/Nack messages on PACCH. The mapping of the uplink access bursts and downlink TA messages on groups of eight 52-multiframes is shown in [17].

The BTS updates the timing advance values in the next TA message following the access burst.

Power control procedure Power control is used in order to improve the spectrum efficiency and to reduce the power consumption in the MS. For the uplink, the MS follows a flexible power control algorithm, which the network can optimise through a set of parameters. It can be used for both open loop and closed loop power control. For the downlink, the power control is performed in the BTS. There is no need to specify the actual algorithms, but information about the downlink performance is needed. Therefore, the MSs transfer channel quality reports to the BTS. For the detailed specification of the power control see [4].

The MS shall calculate the RF output power value, P_{CH}, to be used on each individual uplink PDCH assigned to the MS [17]:

$$P_{CH}= \min((\Gamma_0 - \Gamma_{CH} - \alpha^*(C+48),\ PMAX)$$

where

Γ_{CH} = a MS and channel specific power control parameter. It is sent to the MS in any resource assigning message. The network can, at any time during a packet transfer, send new Γ_{CH} values to the MS on the downlink PACCH

Γ_0 = frequency band dependent constant

$\alpha \in [0, 1]$ = a system parameter. Its default value is broadcast on the PBCCH. MS and channel specific values can be sent to the MS together with Γ_{CH}

C = received signal level at the MS (see [17] for the derivation of this value)

PMAX = maximum allowed output power in the cell

All power values are expressed in dBm.

P_{CH} is not used to determine the output power when accessing the cell on PRACH or RACH, in which case PMAX is used.

The BTS uses constant power on those PDCH radio blocks which contain PBCCH or which may contain PPCH. This power may be lower than the output power used on BCCH. The difference is broadcast on PBCCH. On the other PDCH radio blocks, downlink power control can be used. Thus, a procedure can be implemented in the network to control the power of the downlink transmission based on the channel quality reports. The network has to ensure that the output power is sufficient for the MS for which the RLC block is intended as well as the MS(s) for which the USF is intended, and that for each MS in packet transfer mode, at least one downlink RLC block per multi-frame is transmitted with an output power that is sufficient for that MS, on a block monitored by that MS.

The MS has to monitor periodically the downlink Rx signal level and quality from its serving cell.

In order to derive the value of C, the MS measures periodically the received signal strength. In the packet idle mode, the MS measures the signal strength of the PCCCH or, if PCCCH does not exist, the BCCH. In the packet transfer mode, the MS measures the signal strength on BCCH. The same measurements as for cell re-selection are used. Alternatively, if indicated by a broadcast para-meter, the MS measures the signal strength on one of the PDCHs where the MS receives PACCH. This method is suitable in the case BCCH is in another frequency band than the one used by PDCHs. It requires that constant output power is used on all downlink PDCH blocks.

MS measurements   The MS measures the signal strength of each radio block monitored by the MS. The C value is achieved by filtering the signal strength with a running average filter. The filtering is usually continuous between the packet modes. The different filter parameters for the packet modes are broadcast on PBCCH or, if PBCCH does not exist, on BCCH. The variance of the received signal level within each block is also calculated. The filtered value SIGN_VAR is included in the channel quality report.

The channel quality is measured as the interference signal level during the idle frames of the multiframe, when the serving cell is not transmitting. In the packet transfer mode, the MS measures the interference signal strength of all eight channels (slots) on the same carrier as the assigned PDCHs. In the packet idle mode, the MS measures the interference signal strength on certain channels which are indicated on the PBCCH or, if PBCCH does not exist, on BCCH. Some of the idle frames and PTCCH frames are used for the channel quality measurements, while the others are required for BSIC identification and the timing advance procedure (see [4] for details).

The MS may not be capable of measuring all eight channels when allocated some configurations of channels. The MS has to measure as many channels as its allocation allows taking into account its multislot capability. The interference, γ_{CH}, is derived by filtering the measured interference in a running average filter.

In packet transfer mode the MS transfers the eight γ_{CH} values and the RXQUAL, SIGN_VAR and C values to the network in the channel quality report included in the PACKET DOWNLINK ACK/NACK message.

The BSS has to monitor the uplink Rx signal level and quality on each uplink PDCH, active as well as inactive. The BSS also has to measure the Rx signal level and the quality of a specific MS packet transfer.

Scheduling the MS activities during the PTCCH and idle frames. The MS uses the PTCCH and idle frames of the PDCH multiframe for the following tasks:

- neighbour cell level measurement and BSIC identification for cell re-selection
- continuous timing advance procedures
- interference measurements

The scheduling of these tasks is not specified in detail. During the frames when the MS receives TA messages it can also make interference measurements. Additionally, during the frames when the MS transmits access bursts it may also be possible to make measurements on some channels.

The MS tries to schedule the BSIC identification as efficiently as possible, using the remaining PTCCH frames and the idle frames and also considering the requirements for interference measurements.

Discontinuous reception (DRX). The MS has to support discontinuous reception (DRX; sleep mode) in packet idle mode. The DRX is independent from mobility management (MM) states ready and standby. The negotiation of the DRX parameters is per MS. The DRX parameters and the overview of DRX operation are explained in [17].

Medium access control and radio link control layer The medium access control (MAC) and radio link control (RLC) layer operates above the physical link layer in the protocol architecture. The MAC function defines the procedures that enable multiple MSs to share a common transmission medium, which may consist of several physical channels. The RLC function defines the procedures for a selective retransmission of unsuccessfully delivered RLC data blocks. MAC/RLC is a complex protocol and for details the reader is referred to [16]. This section gives only an introduction of the main functions of the RLC/ MAC layer.

The basic principle of data transfer is illustrated in Figure 1.20. The network protocol data units (N-PDU), that correspond to full IP packets, are compressed and segmented into the sub-network protocol data units (SN-PDU) by the sub-network dependent convergence protocol (SNDCP). IP packet compression is optional. The SN-PDUs are encapsulated into one or several LLC frames. The size of the data part the LLC frames is a parameter between 140 and 1520. LLC frames are segmented into RLC data blocks. At the RLC/MAC layer, a selective automatic repeat request (ARQ) protocol (including block numbering) between the MS and the network provides retransmission of erroneous RLC data blocks. When a complete LLC frame is successfully transferred across the RLC layer, it is forwarded to the LLC layer.

There are two important concepts that constitute the core of RLC/MAC operation: temporary block flow and temporary flow identifier.

Temporary block flow [17] A temporary block flow (TBF) is a temporal connection between the MS and the network to support the uni-directional transfer of LLC PDUs on packet data physical channels. A TBF can use radio resources on one or more PDCHs and comprise a number of RLC/MAC blocks

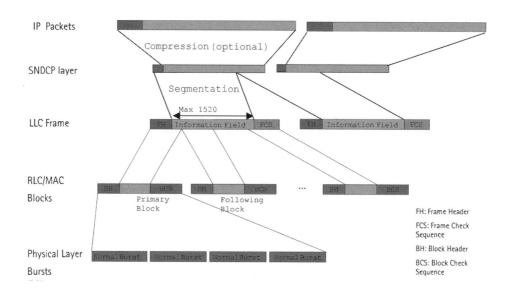

Figure 1.20   Transmission and reception data flow.

carrying one or more LLC PDUs. A TBF is temporary and is maintained only for the duration of the data transfer.

Temporary flow identity [17] Each TBF is assigned a temporary flow identity (TFI) by the network. The assigned TFI is unique among concurrent TBFs in each direction and is used instead of the MS identity in the RLC/MAC layer. The same TFI value may be used concurrently for TBFs in opposite directions. The TFI is assigned in a resource assignment message that precedes the transfer of LLC frames belonging to one TBF to/from the MS. The same TFI is included in every RLC header belonging to a particular TBF as well as in the control messages associated with the LLC frame transfer (e.g. acknowledgements) in order to address the peer RLC entities.

Uplink radio blocks multiplexing In uplink the packet switched users are multiplexed in the same PDCH by the MAC layer. There are three access modes supported by RLC/MAC protocol: dynamic allocation, extended dynamic allocation and fixed allocation.

Dynamic allocation. In order to make the multiplexing of mobile stations on the same PDCH possible, the uplink state flag (USF) is introduced. During the establishment of the uplink TBF, an USF is assigned to each mobile, that will be used by the network to indicate which terminal is allowed to transmit in the following uplink radio block.

With this method each downlink block has to be decoded by all the MSs assigned to the PDCH in order to obtain the USF and the TFI identifying the owner of the data or control information. In addition, even if no downlink information is available, a dummy radio block must be sent just to transmit the USF when required.

The USF is a 3-bit flag, so each PDCH can manage up to eight MSs, except on the packet control channel (PCCH), where the value '111' (USF = Free) indicates that the corresponding uplink radio block is reserved for the packet random access channel (PRACH) [17]. There are only 16 TA indexes for the continuous timing advance procedure to keep the MS synchronised with the TDMA frame at the BTS. If the maximum number of TBF multiplexed in UL is 7, the maximum number of TBF in DL is limited to 9 in the extreme case that the 16 TBF (7 in UL and 9 in DL) multiplexed on the same PDCH belongs to different MSs.

A special channel coding is used for the USF added to the different coding schemes that makes it quite reliable (even more than CS-1) and prevents uplink radio block losses. For multislot classes, the mobile must decode the USF in all PDCHs separately.

Whenever the MS detects an assigned USF value on an assigned PDCH, the MS shall transmit either a single RLC/MAC block or a sequence of four RLC/MAC blocks on the same PDCH. The number of RLC/MAC blocks to transmit is controlled by the USF_GRANULARITY parameter characterising the uplink TBF [17].

The fact that the MS has to receive a USF on each timeslot (TSL), introduces limitations on the maximum number of TSLs supported in the UL direction for certain MS classes. In most practical cases, with the simplest (E)GPRS radio implementation, the maximum number of TSLs in UL is limited to two. This can be seen in Figure 1.21 with a 3 + 2 MS allocation. Note that this limitation is very dependent on MS class.

Extended dynamic allocation. This procedure allows a higher number of TSLs in UL and eliminates the need of receiving USF on each TSL (which may create too many dummy blocks in DL). In this case, when a USF is received in a particular TSL, the MS is allowed to transmit on the same TSL and all the subsequent TSLs. Let's consider the allocation in Figure 1.22 where a 2 + 3 MS allocation is considered. When a USF is received on TSL 1 the MS will transmit on TSL 1 and 2. If simple dynamic allocation would be used, the MS would transmit only on TSL 1.

Fixed allocation. Using this procedure, the network communicates with the MS an UL transmission turn pattern by using a specific control message. In this case USF is not utilised and therefore DL communication is point to point. This later characteristic makes it more suitable for link optimisation techniques like DL power control and smart antennas. Once a MS has requested a resource assignment for the uplink, the network responds with a bitmap (ALLOCA-

tb: guard time between receiving and transmitting
ra: BCCH frequencies measurements
R: MS receiving
T: MS transmitting

Figure 1.21   Dynamic UL allocation example.

tb: guard time between receiving and transmitting
ra: BCCH frequencies measurements
R: MS receiving
T: MS transmitting

Figure 1.22   Extended dynamic UL allocation example.

TION_BITMAP) indicating timeslots and radio blocks where the MS is allowed to transmit after a frame number contained in the parameter *TBF starting time*.

Time is slotted in *allocation periods* with a configurable duration that will be the basic time unit to assign resources for an uplink TBF. If more than one allocation is needed for the MS initial request, the network will send new bitmaps in packet ACK/NACK messages at the end of the current allocation period.

Modes for RLC/MAC operation   In addition there are two distinctive modes of operation of RLC/MAC layer:

* unacknowledged operation
* acknowledged operation

Acknowledged mode for RLC/MAC operation. The transfer of RLC data blocks in the acknowledged RLC/MAC mode is controlled by a selective ARQ mechanism coupled with the numbering of the RLC data blocks within one temporary block flow. The sending side (the MS or the network) transmits blocks within a window and the receiving side sends packet uplink Ack/Nack or packet downlink Ack/Nack message when needed. Every such message acknowledges all correctly received RLC data blocks up to an indicated block sequence number (BSN). Additionally, the bitmap is used to selectively request erroneously received RLC data blocks for retrans-mission.

Unacknowledged mode for RLC/MAC operation. The transfer of RLC data blocks in the unacknowledged RLC/MAC mode is controlled by the numbering of the RLC data blocks within one temporary block flow and does not include any retransmissions. The receiving side extracts user data from the received RLC data blocks and attempts to preserve the user information length by repla-cing missing RLC data blocks by dummy information bits.

Mobile originated packet transfer   Uplink access. A MS initiates a packet transfer by making a packet channel request on PRACH or RACH depending on the availability of packet control channels in the network. The network responds on PAGCH (if packet control channels are supported by the network) or AGCH, respectively. It is possible to use one or two phase packet access methods (see Figure 1.23).

In the one phase access, the network will respond to the packet channel request by sending a packet uplink assignment message and reserving the resources on PDCH(s) for uplink transfer of a number of radio blocks. The reservation is done according to the requested resources indicated in the packet channel request. On RACH, there are only two cause values available for denot-ing GPRS, which can be used to request limited resources or two phase access. On PRACH, the packet channel request may contain more adequate informa-tion about the requested resources and, consequently, uplink resources on one or more PDCHs can be assigned by using the packet uplink assignment message.

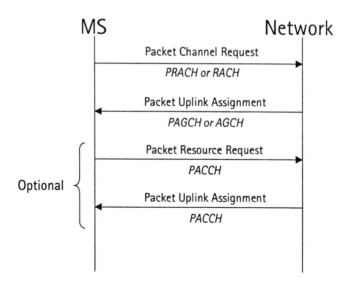

Figure 1.23   Access and allocation for the one or two phase packet access, uplink packet transfer.

In the two phase access, the packet channel request is responded to with the packet uplink assignment, which reserves the uplink resources for transmitting the packet resource request. A two phase access can be initiated by the network or a MS. The network can order the MS to send a packet resource request message by setting a parameter in packet uplink assignment message. The MS can require two phase access in the packet channel request message. In this case, the network may order the MS to send a packet resource request or continue with a one phase access procedure.

The packet resource request message carries the complete description of the requested resources for the uplink transfer. The MS can indicate the preferred medium access method to be used during the TBF. The network responds with the packet uplink assignment reserving resources for the uplink transfer and defining the actual parameters for data transfer (e.g. medium access mode). If there is no response to the packet channel request within the predefined time period, the MS makes a retry after a random time.

On PRACH, a two-step approach is used including a long-term and a short-term estimation of the persistence [17]. The optimal persistence of the mobile stations is calculated at the network side.

The actual persistence values depend on:

- the priority of the packet to be transmitted;
- the amount of traffic within higher priority classes;
- the amount of traffic within the own priority class.

Occasionally, more packet channel requests can be received than can be served. To handle this, a packet queuing notification is transmitted to the sender of the packet channel request. The notification includes information that the packet channel request message is correctly received and packet uplink assignment may be transmitted later. If the timing advance information becomes inaccurate for a MS, the network can send packet polling request to trigger the MS to send four random access bursts. This can be used to estimate the new timing advance before issuing the packet uplink assignment [17].

In case of dynamic or extended dynamic allocation, the packet uplink assignment message includes the list of PDCHs and the corresponding USF value per PDCH. A unique TFI is allocated and is thereafter included in each RLC data and control block related to that temporary block flow. The MS monitors the USFs on the allocated PDCHs and transmits radio blocks when the USF value indicates to do so. In case of fixed allocation, the packet uplink assignment message is used to communicate a detailed fixed uplink resource allocation to the MS. The fixed allocation consists of a start frame, slot assignment, and block assignment bitmap representing the assigned blocks per timeslot [17]. The MS will wait until the start frame and then transmit radio blocks on those blocks indicated in the block assignment bitmap. The fixed allocation does not include the USF and the MS is free to transmit on the uplink without monitoring the downlink for the USF. If the current allocation is not sufficient, the MS may request additional resources in one of the assigned uplink blocks. A unique TFI is allocated and is thereafter included in each RLC data and control block related to that temporary block flow. Because each radio block includes an identifier (TFI), all received radio blocks are correctly associated with a particular LLC frame and a particular MS.

Contention resolution Contention resolution is an important part of RLC/MAC protocol operation, especially because one channel allocation can be used to transfer a number of LLC frames. There are two basic access possibilities, one phase and two phase access as seen in Figure 1.23. The two phase access eliminates the possibility that two MSs can perceive the same channel allocation as their own. Basically the second phase of access will uniquely identify the MS by TLLI and that TLLI will be included in packet uplink assignment therefore ruling out the possibility for a mistake.

Mobile terminated packet transfer *Packet paging.* The network initiates a packet transfer to a MS that is in the standby state by sending one or more packet paging request messages on the downlink (PPCH or PCH). The MS responds to one packet paging request message by initiating a mobile originated packet transfer. This mobile originated packet transfer allows the MS to send a packet paging response message containing an arbitrary LLC frame. The message sequence described in Figure 1.24 is conveyed either on PCCCH or on CCCH. After the packet paging response is sent by the MS and received by the network, the mobility management state of the MS is ready.

Figure 1.24 Paging message sequence for paging, downlink packet transfer.

The network can then assign some radio resources to the MS and perform the downlink data transfer.

Downlink packet transfer. The transmission of a packet to a MS in the ready state is initiated by the network using a packet downlink assignment message. In case there is an uplink packet transfer in progress, the packet downlink assignment message is transmitted on PACCH. Otherwise it is transmitted in the PCCCH, in case there is one allocated in the cell. If that is not the case it will be transmitted in the CCCH. The packet downlink assignment message includes the list of PDCH(s) that will be used for downlink transfer.

The network sends the RLC/MAC blocks belonging to one temporary block flow on downlink on the assigned downlink channels. Multiplexing the RLC/MAC blocks addressed for different MSs on the same PDCH downlink is enabled with the TFI identifier, included in each RLC/MAC block [17].

The sending of the packet downlink Ack/Nack message is obtained through the periodical network initiated polling of the MS. The MS sends the packet downlink Ack/Nack message in a reserved radio block which is allocated together with polling.

Release of the resources. The release of the resources is initiated by the network by terminating the downlink transfer and polling the MS for a final packet downlink Ack/Nack message.

Simultaneous uplink and downlink packet transfer. During the ongoing uplink temporary block flow, the MS continuously monitors one downlink PDCH for possible occurrences of packet downlink assignment or packet timeslot reconfigure messages on PACCH. The MS is therefore reachable for downlink packet transfers that can then be conveyed simultaneously on the PDCH(s) that respects the MS multislot capability.

If the MS wants to send packets to the network during the ongoing downlink temporary block flow, it can be indicated in the acknowledgement that is sent from the MS. By doing so, no explicit packet channel requests have to be sent to the network. Furthermore, the network already has the knowledge of which PDCH(s) that particular MS is currently using so that the uplink resources can be assigned on the PDCH(s) that respect the MS multislot capability.

1.5 EDGE Rel'99

Enhanced data rates for GSM evolution (EDGE) is a major enhancement to the GSM data rates. GSM networks have already offered advanced data services, like circuit switched 9.6 kbps data service and SMS, for some time. High speed circuit switched data (HSCSD), with multislot capability and the simultaneous introduction of 14.4 kbps per timeslot data, and GPRS are both major improvements, increasing the available data rates from 9.6 kbps up to 64 kbps (HSCSD) and 160 kbps (GPRS).

EDGE is specified in a way that will enhance the throughput per timeslot for both HSCSD and GPRS. The enhancement of HSCSD is called ECSD (enhanced circuit switched data), whereas the enhancement of GPRS is called EGPRS (enhanced general packet radio service). In ECSD, the maximum data rate will not increase from 64 kbps due to the restrictions in the A-interface, but the data rate per timeslot will triple. Similarly, in EGPRS, the data rate per timeslot will triple and the peak throughput, with all eight timeslots in the radio interface, will reach 473 kbps.

8-PSK modulation in GSM/EDGE standard

The 'enhancement' behind tripling the data rates is the introduction of the 8-PSK (octagonal phase shift keying) modulation in addition to the existing GMSK (Gaussian minimum shift keying). An 8-PSK signal is able to carry 3 bits per modulated symbol over the radio path, while a GMSK signal carries only 1 bit per symbol (Table 1.4). The carrier symbol rate (270.833 kbps) of standard GSM is kept the same for 8-PSK and the same pulse shape as used in GMSK is applied to 8-PSK. The increase in data throughput does not come for free, the price being paid in the decreased sensitivity of the 8-PSK signal. This affects, e.g. the radio

Table 1.4   Key physical layer parameters for GSM/EDGE

	8-PSK	GMSK
Symbol rate	270.833 ksps	270.833 ksps
Number of bits/symbol	3 bits/symbol	1 bit/symbol
Payload/burst	342 bits	114 bits
Gross rate/timeslot	68.4 kbps	22.8 kbps

network planning, and the highest data rates can only be provided with limited coverage. The GMSK spectrum mask was the starting point for the spectrum mask of the 8-PSK signal but along the standardisation process, the 8-PSK spectrum mask was relaxed with a few dB in the 400 kHz offset from the centre frequency [18]. This was found to be a good compromise between the linearity requirements of the 8-PSK signal and the overall radio network performance.

It was understood already at the selection phase of the new modulation method that a linear modulation would have quite different characteristics than the GMSK as a constant envelope modulation. A practical problem is how to define the receiver performance requirements for the physical channels with 8-PSK modulation, in particular how to incorporate the phenomena resulting from the non-idealities in the transmitter and receiver. This problem is solved in the specification by using the concept of error vector of magnitude (EVM). The EVM effectively measures how far an observed signal sample is from the ideal signal constellation point (Figure 1.25). Thus the EVM takes into account deviation in both I- and Q-axes, not explicitly distinguishing whether the signal impairment is due to the phase noise or the amplitude distortion. For the 8-PSK signal, [18] sets specific EVM percentage requirements for the transmitter and receiver performance.

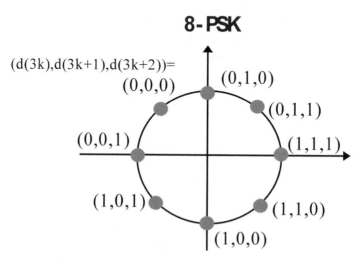

Figure 1.25   8-PSK signal constellation.

A further peculiarity, introduced by bringing the second modulation method, 8-PSK, to GSM standard, is the need in downlink to blindly recognise the transmitted modulation method in the mobile station receiver. This is due to the characteristics of the EGPRS link quality control where the used modulation and coding scheme (MCS) is adjusted according to the channel conditions to the most suitable one, and in downlink, no prior information is sent to the receiver but the receiver is able to find out the used MCS based on (1) the blind modulation identification, and (2) the decoding of the RLC/MAC header field which contains indication of the coding and puncturing scheme. The modulation identification is based on the different phase rotation characteristics in the GMSK and 8-PSK training sequences. In the GMSK training sequence, the symbol-by-symbol phase rotation is p/2 whereas in the 8-PSK training sequence, the rotation is 3p/8. Otherwise, the set of 8-PSK training sequences has identical information content (the same 26-bit sequence) as the GMSK training sequences.

Enhanced general packet radio service (EGPRS)

EGPRS is built on top of GPRS, which is the packet switched data service of GSM. Examples of typical high bit rate packet services include fast file transfer, Internet service access, web browsing and remote email.

EGPRS has a major impact on the RF and the physical layer of the radio interface as well as on RLC/MAC protocol, but the changes to other protocols and protocol layers are minor. The EDGE RF specification and the definition of the burst structures are common to both EGPRS and ECSD. One large conceptual modification in EGPRS, compared with GPRS, is the link quality control (LQC) which, in EGPRS, also supports incremental redundancy (type II hybrid ARQ), in addition to GPRS-type link adaptation mode (type I hybrid ARQ). The LQC includes nine different modulation and coding schemes (MCS-1–MCS-9) as well as related signalling and other procedures for link adaptation (switching between different MCSs).

Table 1.5 shows the EGPRS modulation and coding schemes (MCS) and their data throughputs. New GMSK coding schemes (MCS-1–MCS-4), different from GPRS GMSK coding schemes (CS-1–CS-4), are needed due to the incremental redundancy support. Figure 1.26 illustrates the coding and puncturing principle of the header and payload part of an MCS-9 radio block. In EGPRS radio blocks, the stealing bits (SB) indicate the header format. The RLC/MAC header is strongly coded (close to rate 1/3 code in 8-PSK modes and close to rate 1/2 code in GMSK modes) to allow reliable decoding of the header in incremental redundancy (IR) operation. There is also a separate header check sequence (HCS) for the error detection in the header part. The RLC data payload is encoded along with the final block indicator (FBI) and extension bit (E) fields. Also, the block check sequence (BCS) and tail bits (TB)

Table 1.5   EGPRS modulation and coding schemes

Modulation and coding scheme	Code rate	Modulation	Data rate/ timeslot (kbps)	Family
MCS-9	1.0	8-PSK	59.2	A
MCS-8	0.92		54.4	A
MCS-7	0.76		44.8	B
MCS-6	0.49		29.6	A
MCS-5	0.37		22.4	B
MCS-4	1.0	GMSK	17.6	C
MCS-3	0.80		14.8	A
MCS-2	0.66		11.2	B
MCS-1	0.53		8.8	C

are added before encoding. In the illustrated case (MCS-9), two RLC data blocks are mapped on one radio block. Note that in the case of MCS-8 and MCS-9 a RLC data block is interleaved over two bursts only. The RLC/MAC header is though interleaved over the whole radio block (i.e. four bursts).

EGPRS modulation and coding schemes are organised in families according to their RLC data block sizes. For example, the MCS-9 carries 1184 payload bits in two RLC data blocks (592 payload bits in each of them) during one EGPRS radio block of four consequent bursts. The MCS-6 carries 592 payload bits in one RLC data block, and the MCS-3 carries 296 payload bits within the radio block. The number of payload bits in MCS-6 and MCS-3 being submul-

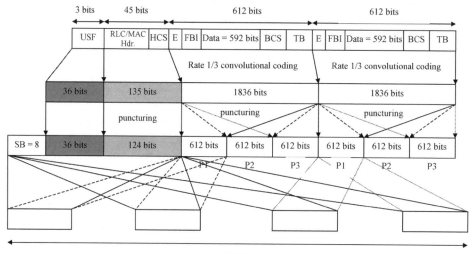

Figure 1.26   EGPRS coding and puncturing example (MCS-9: uncoded 8-PSK, two RLC blocks per 20 ms). [17]

tiples of the higher coding schemes in the same family (family A) enables effective retransmission of the negatively acknowledged RLC data blocks with the lower coding schemes within the family if needed. A typical case where this functionality is useful is the sudden change in the channel conditions.

The transfer of RLC data blocks in the acknowledged RLC/MAC mode can be controlled by a selective type I ARQ mechanism, or by type II hybrid ARQ (IR) mechanism, coupled with the numbering of the RLC data blocks within one temporary block flow. The sending side (the MS or the network) transmits blocks within a window and the receiving side sends packet uplink Ack/Nack or packet downlink Ack/Nack message when needed.

The initial MCS for an RLC block is selected according to the link quality. For the retransmissions, the same or another MCS from the same family of MCSs can be selected. For example, if MCS-7 is selected for the first transmission of an RLC block, any MCS of the family B can be used for the retransmissions. The selection of MCS is controlled by the network.

In the EGPRS type II hybrid ARQ scheme, the information is first sent with one of the initial code rates (i.e. rate 1/3 encoded data are punctured with puncturing scheme (PS) 1 of the selected MCS). If the RLC data block is received in error, additional coded bits (i.e. the output of the rate 1/3 encoded data which is punctured with PS 2 of the prevailing MCS) are sent and decoded together with the already received codewords until decoding succeeds. If all the codewords (different punctured versions of the encoded data block) have been sent, the first codeword (which is punctured with PS 1) is sent. Alternatively, it is possible to use incremental redundancy modes called MCS-5–7 and MCS-6–9, in which the initial transmissions are sent with either MCS-5 or MCS-6 (respectively) and the retransmissions are sent with MCS-7 or MCS-9 (respectively). In the EGPRS type I ARQ, the operation is similar to that of the EGPRS type II hybrid ARQ, except that the decoding of an RLC data block is solely based on the prevailing transmission (i.e. erroneous blocks are not stored). The motivation for supporting the incremental redundancy mode is the improved data throughput and the robustness for varying channel conditions and measurement errors. Incremental redundancy is defined to be mandatory for the MS and optional for the network.

Another RLC/MAC layer modification in Rel'99 due to EGPRS is the increased RLC window size. In GPRS with four-burst radio blocks (20 ms) the RLC window size of 64 is defined. With the higher coding schemes of EGPRS, MCS-7 to MCS-9, there are two RLC blocks per 20 ms radio block, which makes the RLC window size 64 too small and RLC protocol subject to stalling. Therefore, the EGPRS RLC window size was increased to 128 and a compression method was defined for the acknowledgement bitmap of the RLC blocks (see details in GSM 04.60).

Bit error probability (BEP) measurements
In Rel'99, also a generic enhancement to the channel quality measurements was introduced. Earlier, the measure to be used for the channel quality indication was the RX_QUAL which is an estimate of the pseudo bit error rate, i.e. computed by comparing the received bit sequence (before decoding) with the encoded version of the decoded bit sequence and counting the number of bit errors. It was shown in [19] that the alternative measure – bit error probability (BEP) – reflects better the channel quality in varying channel conditions – including frequency hopping and varying mobile speed. BEP is estimated burst by burst, for example from the soft output of the receiver. The detailed realisation of the measurement itself is left for the equipment vendors whereas the method of reporting the BEP measurements and the accuracy requirements are standardised. The measurements to be reported by the MS for the network are MEAN_BEP and CV_BEP where the MEAN_BEP is the average of the block wise mean BEP during the reporting period and the CV_BEP is the average of the block wise coefficient of variation (CV) (CV = Std(BEP)/Mean(BEP)) during the reporting period.

Adjustable filtering length for the EGPRS channel quality measurements
A further standardised means of optimising the EGPRS performance is the filtering length of the link quality control measurements, which can be varied [20]. The network and the MS can negotiate with each other – using the radio link control signalling – the optimum length of the filter which is used to smooth the BEP measurements that the MS sends to the network. For example, typically the filter length should be shorter for the MS with higher speed, so that the received channel quality information at the network would be as correct as possible. On the other hand, for slowly moving mobiles the longer averaging of the measurements improves the accuracy.

Enhanced circuit switched data (ECSD)

ECSD uses current HSCSD as a basis. The user data rates are not increased compared with HSCSD (up to 64 kbps), but these rates can be achieved with smaller numbers of timeslots and simpler MS implementation. Data rates to be provided with ECSD, although limited to 64 kbps, are still sufficient for providing various transparent and non-transparent services. ECSD enables interworking with audio modems at higher data rates than in current GSM networks, interworking with ISDN at various data rates and various video based services ranging from still image transfer to videoconferencing services.

Higher data rates are defined for both transparent and non-transparent services. Radio interface rates (and user rates per timeslot) for ECSD are shown in Table 1.6. In the definition of the ECSD, only a few new traffic

Table 1.6 ECSD data rates

Coding scheme	Code rate	Modulation	Gross rate (kbps)	Radio interface rate (kbps)	User rates (kbps)
TCH/28.8 (NT/T)	0.419	8-PSK	69.2	29.0	28.8
TCH/32.0 (T)	0.462	8-PSK	69.2	32.0	32.0
TCH/43.2 (NT)	0.629	8-PSK	69.2	43.5	43.2

channels are defined. The control channels – both the common, dedicated and associated control channels – are the same as for the other circuit switched services (like, e.g. for the 9.6 kbps circuit switched data, HSCSD) except the fast associated control channel for the ECSD (FACCH/E) which uses GMSK modulation and is interleaved rectangularly on the four consequent TDMA frames. Further, the detection of the FACCH/E frame is based on the similar principles as the identification of the 8-PSK and GMSK modulated data blocks in EGPRS, i.e. the modulation method is blindly detected in the receiver and if a block of four GMSK bursts is received in the middle of the 8-PSK data frames the receiver knows that a FACCH/E frame was received.

The 28.8 kbps data rate is available for transparent and non-transparent services for both single and multislot configurations. The 32.0 kbps service is available only in multislot configuration (two-slot) and can be used for offering 64 kbps transparent service, while 43.2 kbps is available for non-transparent services only.

The ECSD architecture is largely based on HSCSD transmission and signalling. This ensures a minimum impact on existing specifications. The basic principle is to use the same transcoder and rate adaptation unit (TRAU) frame formats and multiple 16 kbps sub-channels in the network side. For example, 28.8 kbps service provided with one radio interface timeslot is supported with two 16 kbps sub-channels and 14.4 kbps TRAU frames on the network side.

Figure 1.27 shows the network architecture for providing 57.6 kbps non-transparent service (a) and 56/64 kbps transparent service (b).

ECSD 57.6 kbps non-transparent service is provided using two timeslots of 28.8 kbps as shown in Figure 1.27a. The same frame formats as in HSCSD are used end-to-end between the MS/TAF and MSC/IWF. The BTS is responsible for the de-/multiplexing of data between two radio timeslots and four Abis sub-channels.

The network architecture for supporting bit transparent 56/64 kbps service is shown in Figure 1.27b. It is provided using two air interface timeslots of 32 kbps. The rate adaptation functions are located in the BTS.

Higher data rates can be offered with limited coverage and, therefore, the link adaptation mechanism becomes essential. Switching between 8-PSK and GMSK based channels is done with the standard intra-cell handover procedure. Signalling for link adaptation is based on the existing HSCSD signalling mechanism.

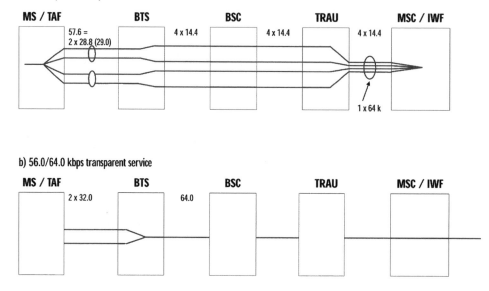

Figure 1.27   Network architecture.

Class A dual transfer mode (DTM)

With a Class A mobile station a subscriber can simultaneously connect to the circuit switched core network (to mobile switching centre (MSC)) and to the packet switched core network (to serving GPRS support node (SGSN)), through A and Gb interfaces, respectively. Until recently, the implementation of a Class A MS has been considered impractical and costly – due to the uncoordination of the radio resources for the CS and PS connections which implies duplicating many key parts of a MS like the transceiver. To enable practical low cost implementation of the Class A mobile, the Class A dual transfer mode (DTM) was defined. This functionality was deemed necessary in the Rel'99 specification – at the same schedule with the first UTRAN standard release (Rel'99) and with the interworking between GSM and UMTS (e.g. inter-system handovers). This package of specifications ensures that an operator can provide the same type of service – e.g. simultaneous voice and data – in both radio access networks, GSM and UTRAN.

The DTM concept includes:

- The single timeslot method: TCH/H + PDTCH/H. In this method, AMR (adaptive multi-rate speech codec) voice frames are carried in the half-rate circuit switched channel (TCH/H) and the user data in the half-rate packet switched channel (PDTCH/H). The significance of the single timeslot method is that it enables implementation of the Class A DTM feature even in single timeslot MSs. Another advantage is that in case of a handover, the availability of the radio resources in the target cell can be easier guaranteed.

- The multiple timeslot method: TCH + one or more adjacent PDTCH. In this method, one full timeslot is reserved for the AMR voice frames and the adjacent timeslots are used for the user data transfer.

A further significant difference between the single and multiple timeslot methods is that in the former one, the PDTCH/H resource is always dedicated for the DTM user (dedicated mode of (E)GPRS MAC protocol) whereas in the latter one, the PDTCH resources can be shared (shared mode of (E)GPRS MAC protocol). More details of Class A DTM can be found, e.g. in [21,22].

EDGE compact

Compact is a particular mode – or variant – of EGPRS which is designed for the deployment in the narrow frequency band allocations – the minimum requirement being the deployment in less than 1 MHz of spectrum. It is known that the capacity (spectral efficiency) of the GSM network is limited by the BCCH reuse, and as available bandwidth decreases its relative impact will be higher. This is due to the nature of the broadcast control channel concept in the GSM standard which requires continuous transmission with constant transmit power in the beacon frequency (BCCH frequency). This causes the BCCH frequency reuse to be relatively large to ensure reliable reception of the system information in the broadcast and common control channels.

The key component of the compact mode is the new control channel structure which is based on the discontinuous time rotating BCCH, which enables significantly lower reuse factors for the BCCH channel and thus operation in the narrow spectrum allocation. The characteristics of the compact mode are:

- It is a stand-alone high-speed packet data system (EGPRS variant with alternative control channel structure).
- It can be deployed in only 600 kHz of spectrum (+ guardband) by using three carriers in a 1/3 reuse pattern.
- It requires inter base station time synchronisation.
- It uses discontinuous transmission on the control carriers, and a new logical control channel combination (compact packet broadcast control channel (CPBCCH) based on a standard 52-multiframe). It uses timeslot mapping of control channels in a rotating fashion, which makes neighbour channel measurements more feasible during traffic mode.

More details on compact mode can be found in [23].

GPRS and EGPRS enhancements in Rel'4

In GERAN Rel'4, a few new features are added to the GPRS and EGPRS protocols and signalling to improve the efficiency of the data transfer with certain
typical traffic characteristics and the seamlessness of the service in the cell change.
These features, the delayed temporary block flow (TBF) release and the network
assisted cell change (NACC), are introduced briefly in the following sections.

Delayed TBF release

The early GPRS protocols are inefficient when dealing with the bursty TCP/IP
traffic. The GPRS radio interface protocols are initially designed to free the
unused radio resources as soon as possible. With bursty IP traffic this may
lead to the frequent set-up and release of the radio resources (more specifically
TBFs) which results in an increased signalling load and inefficiency in the data
transfer as the release and set-up of the TBFs takes some time and radio interface
resources. Unnecessary TBF set-ups and releases can be avoided by delaying the
release of the TBF. During the inactive period the connection is maintained by
sending periodically dummy LLC frames in the downlink [20].

Network assisted cell change

The GPRS cell change was not initially designed for the services that would
require seamless cell change operation, such as real-time services. Later on,
these requirements gained more importance and a new type of cell change,
network assisted cell change (NACC), was defined. In the NACC, the network
can send neighbour cell system information to a MS. The MS can then use this
system information when making the initial access to a new cell after the cell
change. This way, the MS does not need to spend some time in the new cell
receiving the system information, and the typical break in the ongoing service
decreases from seconds to a few hundreds of milliseconds [24].

References

[1] Kurronen K., 'Evolution of GSM towards the Third Generation Requirements'.
[2] Pirhonen R., Rautava T., Penttinen J., 'TDMA Convergence for Packet Data Services', IEEE
    Personal Commun., Vol. 6, No. 3, June 1999, pp. 68–73.
[3] K. Järvinen, 'Standardisation of the Adaptive Multi-rate Codec', Proc. X European Signal
    Processing Conference (EUSIPCO Tampere, Finland, 4– 8 (2000) 2000.
[4] 3GPP TS 05.08 V8.6.0 (2000-09), 3rd Generation Partnership Project; Technical Specifica-
    tion Group GERAN; Digital Cellular Telecommunications System (Phase 2 +); Radio
    Subsystem Link Control (Release 1999).
[5] 3GPP TS 05.02 V8.10.0 (2001-08), 3rd Generation Partnership Project; Technical Specifi-
    cation Group GSM/EGGE Radio Access Network; Digital Cellular Telecommunications
    System (Phase 2 +); Multiplexing and Multiple Access on the Radio Path (Release 1999).

[6] 3GPP TS 05.04, Modulation, 3rd Generation Partnership Project; Technical Specification Group GERAN.

[7] GSM 05.05 Version 8.8.0 Release 1999: 'Digital Cellular Telecommunications System (Phase 2 +)'; Radio Transmission and Reception.

[8] GSM 05.03 Version 8.11.0 Release 1999: 'Digital Cellular Telecommunications System (Phase 2 +)'; Channel Coding.

[9] 3GPP TS 05.10, Radio Subsystem Synchronisation, 3rd Generation Partnership Project; Technical Specification Group GERAN.

[10] 3GPP TS 04.64, LLC Specification GPRS, 3rd Generation Partnership Project; Technical Specification Group GERAN.

[11] 3GPP TS 04.65, GPRS Subnetwork Dependent Convergence Protocol (SNDCP), 3rd Generation Partnership Project; Technical Specification Group GERAN.

[12] 3GPP TS 48.018, General Packet Radio Service (GPRS); Base Station System (BSS) – Serving GPRS Support Node (SGSN); BSS GPRS Protocol, 3rd Generation Partnership Project; Technical Specification Group GERAN.

[13] 3GPP TS 09.60, GPRS Tunnelling Protocol (GTP) across the Gn and Gp Interface, 3rd Generation Partnership Project; Technical Specification Group GERAN.

[14] 3GPP TS 03.60, GPRS Stage 2, 3rd Generation Partnership Project; Technical Specification Group GERAN.

[15] 3GPP TS 04.07, Mobile Radio Interface Signalling Layer 3 – General Aspects, 3rd Generation Partnership Project; Technical Specification Group GERAN.

[16] GSM 04.60 V8.1.0 (1999-11), European Standard (Telecommunications Series), Digital Cellular Telecommunications System (Phase 2 +); General Packet Radio Service (GPRS); Mobile Station (MS)–Base Station System (BSS) Interface; Radio Link Control/Medium Access Control (RLC/MAC) Protocol (GSM 04.60 Version 8.1.0 Release 1999).

[17] 3GPP TS 03.64, General Packet Radio Service (GPRS); Overall Description of the GPRS Radio Interface; Stage 2, 3rd Generation Partnership Project; Technical Specification Group GERAN.

[18] 3GPP TS 45.005, Radio transmission and reception 3rd Generation Partnership Project; Technical Specification Group GERAN.,

[19] Tdoc SMG2, 464/99, Link Quality Control Measurements for EGPRS, April 12–16, 1999, Dublin, Ireland.

[20] 3GPP TS 44.060, General Packet Radio Service (GPRS); Mobile Station (MS)–Base Station System (BSS) Interface; Radio Link Control/Medium Access Control (RLC/MAC) Protocol, Radio Resource Control Protocol, 3rd Generation Partnership Project; Technical Specification Group GERAN.

[21] 3GPP TS 43.055, Dual Transfer Mode, 3rd Generation Partnership Project; Technical Specification Group GERAN.

[22] Pecen M., Howell A., 'Simultaneous Voice and Data Operation for GPRS/EDGE: Class A Dual Transfer Mode'. IEEE Personal Communications, April 2001, pp. 14–29.

[23] ETSI SMG2 Tdoc 1549/99, Concept Proposal for EGPRS-136, Revision 1.6.

[24] 3GPP TSG GERAN #3 GP-010361 Network Assisted Cell Change; Concept Document.

2

Evolution of GERAN Standardisation (Rel'5 and beyond)

Eero Nikula, Shkumbin Hamiti and Markus Hakaste

After the introduction of the Gb interface in Rel'97, all the consequent GSM standard releases (Rel'98, Rel'99, Rel'4) have been based on the assumption that there are two separate core network interfaces in GSM/GPRS – namely the A interface between the base station subsystem (BSS) and the mobile switching centre (MSC) for the circuit switched services, and the Gb interface between the BSS and the serving GPRS support node (SGSN) for the packet switched services. At the same time, the universal mobile telecommunications service (UMTS) Rel'99 has adopted much of the functionalities of the GSM/GPRS core network, modified the functional split between the radio access network (RAN) and the core network (CN) and evolved the CN further. In the UMTS CN, the interface between the RAN and the CN is called Iu – with two component interfaces, the Iu-cs for the circuit switched services and the Iu-ps for the packet switched services.

Other key points in the introduction to the GSM/EDGE Rel'5 development are the service continuity request of the operators who are deploying both GSM/EDGE and wideband code division multiple access (WCDMA) networks, and the dual mode nature of the handsets supporting these two different radio access technologies. It would be clearly beneficial for the whole GSM/WCDMA community – operators, service providers, subscribers, equipment manufacturers to name a few – if the similar set of services could be provided through both radio access technologies, GSM/EDGE and WCDMA. Against this assumption, there are two main alternatives for the GSM/EDGE to evolve: either to develop further the functionality and the capabilities of the A and Gb interfaces to match the same requirements as set for the UMTS, or adopt the UMTS Iu interface and modify the GSM/EDGE BSS functionality accordingly. With the adoption of the UMTS Iu interface and the UMTS quality of service (QoS) architecture, GERAN and UTRAN can be efficiently integrated under a single UMTS multi-radio network.

In the 3rd Generation Partnership Project (3GPP) GSM/EDGE radio access network (GERAN) Rel'5, the adoption of the Iu interface was selected as a way to go. Although this selection probably causes larger changes to the standard in the short term, it is a more future-proof solution – allowing the same services and CN solutions to be applied to both GERAN and UTRAN fluently. Section 2.1 presents an overview of the GERAN Rel'5 features with Iu alignment as a major topic. Section 2.2 explains basics of the new architecture for GSM/EDGE radio access networks that are currently being specified in 3GPP Rel'5 and Section 2.3 looks beyond Rel'5 – seeking for topics that might be of interest in the future evolution of the GERAN.

2.1 GERAN Rel'5 features

In the 3GPP Rel'5 overall, the largest new functionality is the Internet multimedia subsystem (IMS) domain of the CN (Figure 2.1). The IMS provides a set of new services that are not available in the circuit switched (CS) and packet switched (PS) domains of the Rel'4 CN. Some examples of the new IMS services are presence, multimedia and mixed media interactive communication [1].

IMS can be considered as a new architecture that provides a platform to launch a flexible set of new Internet based multimedia services. This will have a major relevance in the third-generation (3G) service provision and the subsequent operators' revenue generation. With the support of the Iu interface the GERAN network can be directly connected to IMS, seamlessly delivering these services.

From the GERAN perspective, the support for the IMS services implies the following enhancements:

- Introduction of the Iu interface.
- Definition of the header adaptation mechanism for the real-time protocol (RTP)/user datagram protocol (UDP)/Internet protocol (IP) traffic. The mechanism can be either header compression or header removal or both.

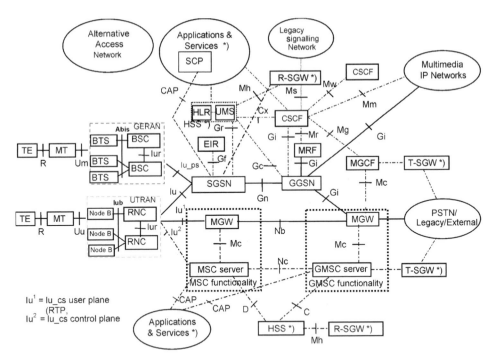

Figure 2.1   Internet Multimedia Subsystem domain of the core network.

The Iu interface for GERAN and the support for the header adaptation are standardised in Rel'5. The other major GERAN enhancements in Rel'5 are:

- wideband adaptive multi-rate (AMR) speech for enhanced speech quality
- half-rate octagonal phase shift keying (8-PSK) speech for improved speech capacity
- fast power control for speech
- location service enhancements for Gb and Iu interfaces
- inter-BSC and base station controller (BSC)/radio network controller (RNC) network assisted cell change (NACC)

The following subsections highlight the main characteristics of the new Rel'5 features, refer to the other chapters of this book for more details and given references for the more detailed descriptions.

Iu interface for GERAN and the new functional split

Adopting the Iu interface and aligning the GERAN architecture with the UTRAN architecture is the major change to the GSM/EDGE Rel'5 standard. Not too many modifications are needed for the Iu-cs and Iu-ps interfaces themselves – they can be more or less used as such as they are in the existing specifications – but the majority of the modifications lies on the higher layers of the

radio interface protocols, especially in the radio resource control (RRC) protocol. The RRC protocol happens to be the central point where the requirements
from the two different worlds – namely the legacy GSM/EDGE physical layer
and radio link control (RLC)/medium access control (MAC), and the expectations from the UMTS core network – conflict with each other.

Figure 2.2 shows the GERAN Rel'5 architecture. The legacy interfaces of
GSM/EDGE – A and Gb – are intentionally included in the figure as they
continue to coexist in the GERAN standard, and for many years to come
they will coexist in the live networks in the field.

Generally speaking, the GERAN Iu alignment work includes:

- Adoption of the Iu interface. Both Iu-cs and Iu-ps interfaces are supported.
- Definition of the Iur-g interface between two GERAN networks. The Iur-g is
  a subset of the UTRAN Iur interface – enhanced with GERAN specific
  requirements.
- Aligning the GERAN user plane and control plane protocol stacks with
  UTRAN. This implies that in the user plane the sub network dependent
  convergence protocol (SNDCP) and the logical link control (LLC) protocols
  are replaced with the packet data convergence protocol (PDCP). In the
  control plane, the radio related functions will be moved from the CN to
  the RAN and defined as part of the new RRC protocol.
- Adoption of the PDCP protocol from UTRAN. No major changes are
  needed to the UTRAN PDCP protocol for supporting GERAN, too.
- Definition of the GERAN RRC protocol. This constitutes the major part of
  the needed changes to align the GERAN architecture with UTRAN. The
  GSM radio resource (RR) protocol is a large one and the new functional
  split between the GERAN and CN results as a large amount of modifications.

Figure 2.2   Architecture of the GSM/EDGE radio access network with Iu interface support.

- The RLC and MAC protocols are not much affected by the new functional split as they are below the PDCP and RRC protocols in the user and control plane stacks, respectively. However, some enhancements are still needed like:

  - Dedicated MAC mode (dedicated channels in general) for the new user plane protocol stack.
  - Replacement of the link access protocol for the Dm channel (LAPDm) protocol with the RLC and MAC functions.
  - Multiple simultaneous temporary block flows (TBFs) for a mobile station. This can partly be seen as a part of the functionality alignment with the UTRAN as multiple TBFs enable the support of the multiple parallel traffic flows with different QoS requirements.

Section 2.2 gives a deeper description of the new GERAN architecture and protocols. In the following paragraphs, the other new GERAN Rel'5 features – independent of the Iu alignment – are shortly visited.

Header adaptation of the IP data streams

The header adaptation function is needed for the efficient transfer of the IP data streams – whether real time (e.g. RTP/UDP/IP) or non-real time (e.g. TCP/IP) – over the radio interface. The header adaptation is a function in the non-transparent mode of the PDCP protocol.

In the header compression, the transport and network level headers (e.g. RTP/UDP/IP) are compressed in such a way that the decompressed headers are semantically identical to the original uncompressed headers. Header compression is suited for standard Internet applications that are not designed to work only with GERAN and especially for multimedia applications, and therefore the scheme will be used with generic real-time multimedia bearers.

The header compression algorithms are standardissed in the IETF (Internet Engineering Task Force) ROHC (Robust Header Compression) WG (Working Group) but the application of the IETF ROHC scheme to the 3GPP radio access technologies is jointly defined by the IETF and 3GPP.

Speech capacity and quality enhancements

GERAN Rel'5 contains a number of new features that are aimed for the capacity increase and the quality improvement of the speech services.

Wideband AMR speech
Wideband AMR (adaptive multi-rate) codec is a new speech codec in the 3GPP specifications. Earlier, GSM has had the FR (full rate), EFR (enhanced full rate) and AMR (adaptive multi-rate) speech codecs. AMR codec was standardised in

GSM Rel'98 and also adopted in the UTRAN Rel'99 as a mandatory speech codec. The AMR Rel'98 codec can be referred also as a AMR-NB (AMR narrowband) – to distinguish it from the new AMR-WB (AMR wideband) codec.

The significance of the wideband AMR lies in the speech coding bandwidth. The bandwidth for the AMR-WB varies between 50 and 7000 Hz whereas the bandwidth of the narrowband speech (e.g. AMR-NB) varies between 150 and 3500 Hz. This difference in the speech coding bandwidth relates directly to the speech quality difference. The AMR-WB speech increases the naturalness and the intelligibility of the speech, and makes the AMR-WB codec suitable for applications with high quality audio parts [2,3].

The application of the AMR-WB codec in the radio interface is based on the similar principles as defined for the AMR-NB. The AMR-WB has nine different codec rates, ranging from 6.60 to 23.85 kbps. In the GSM/EDGE single timeslot configuration, all these codec rates can be supported in the 8-PSK speech channels, and the codec rates from 6.60 to 19.85 kbps can be supported in the Gaussian minimum shift keying (GMSK) speech channels. The link adaptation mechanism is similar to the one defined for the AMR-NB and based on the inband signalling bits that are transmitted along the speech frames. The major difference in the deployment scenarios is that for preserving the superior quality of the wideband speech, no transcoders are allowed in the transmission path of the speech from one terminal to another.

The ITU-T codec subgroup has selected the 3GPP AMR-WB codec as its recommended wideband speech codec. This implies that the AMR-WB codec is the first truly global speech codec standard.

Fast power control for speech
The basic power control of the GSM/EDGE operates on the 480 ms basis using the regular slow associated control channel (SACCH) frames for the reporting of the channel quality measurements and sending the power control (PC) commands. In Rel'99, faster power control with 20 ms basis was defined for the enhanced circuit switched data (ECSD). The ECSD fast power control uses inband signalling for carrying the measurements and the PC commands and thus it is specific to the ECSD channels. In GERAN Rel'5, a third power control mechanism, called enhanced power control (EPC), is defined. The EPC uses a modified SACCH channel for carrying the measurements and the PC commands. The modified SACCH channel allows these messages to be carried on the single SACCH burst, resulting as a 1 per 120 ms frequency [18].

Half-rate 8-PSK speech
As in the GSM/EDGE Rel'99, the 8-PSK modulation was applied only to enhance the capabilities of the data traffic channels (EGPRS and ECSD), in

Rel'5 the 8-PSK modulation is applied also to the speech services. Earlier in this chapter, the wideband AMR was already mentioned as one application area of the 8-PSK speech. In addition to that, also the AMR-NB (Rel'98 AMR codec) can be enhanced with the use of the 8-PSK modulation. After extensive studies, it was decided to standardise only the 8-PSK half-rate channels for the AMR-NB codec [4].

As the full-rate and half-rate GMSK channels are already in the standard, the significance of the half-rate 8-PSK speech channels is due to the better link level performance of the half-rate 8-PSK channels, compared with the half-rate GMSK channels. The improvement is largest for the highest codec rates that can be supported by the GMSK channels (7.95 and 7.4 kbps) as there is more room for the channel coding with 8-PSK channels. Further, the two highest modes of the AMR-NB (12.2 and 10.2 kbps) can be supported with half-rate 8-PSK channels but not half-rate GMSK channels.

Table 2.1 summarises all speech codecs supported in the GERAN specifications and their basic characteristics.

Table 2.1   Speech codecs supported in GERAN

Speech codec		Bits per speech frame	Average code rate GMSK channel		8-PSK channel	
			FR	HR	FR	HR
GSM FR		260	0.57	x	x	x
GSM HR		112	x	0.49	x	x
EFR		244	0.54	x	x	x
AMR NB	12.2 kbit/s	244	0.54	x	x	0.36
	10.2 kbit/s	204	0.45	x	x	0.30
	7.95 kbit/s	159	0.35	0.70	x	0.23
	7.4 kbit/s	148	0.32	0.65	x	0.22
	6.7 kbit/s	134	0.29	0.59	x	0.20
	5.9 kbit/s	118	0.26	0.52	x	0.17
	5.15 kbit/s	103	0.23	0.45	x	0.15
	4.75 kbit/s	95	0.21	0.42	x	0.14
AMR WB	23.85 kbit/s	477	x	x	0.35	0.70
	23.05 kbit/s	461	x	x	0.34	0.67
	19.85 kbit/s	397	0.87	x	0.29	0.58
	18.25 kbit/s	365	0.80	x	0.27	0.53
	15.85 kbit/s	317	0.70	x	0.23	0.46
	14.25 kbit/s	285	0.63	x	0.21	0.42
	12.65 kbit/s	253	0.55	x	0.18	0.37
	8.85 kbit/s	177	0.39	x	0.13	0.26
	6.60 kbit/s	132	0.29	x	0.10	0.19

Location service enhancements for Gb and Iu interfaces

The support for positioning a mobile station (referred to in the standard as location services (LCS)) is already defined in Rel'98 of the GSM standard. Three positioning mechanisms are supported for LCS: timing advance (TA), enhanced observed time difference (E-OTD), and global positioning system (GPS). Rel'98 defines the LCS only for the CS domain. In Rel'4, the first step was taken to define the LCS service support also for the PS domain. This solution assumes that in case of the positioning request the mobile station will suspend its GPRS connection, perform the needed location procedures using the existing LCS solution in the CS domain and resume the GPRS connection.

In Rel'5, the LCS service for the PS domain develops in two parallel tracks: one solution for the A/Gb mode and another solution for the Iu mode. In both of these modes, the same positioning mechanisms (TA, E-OTD and GPS) as in earlier releases are supported. The differences lie on the mode-specific protocol and signalling solutions that are tailored separately for the A/Gb and Iu modes. The enhancement compared with the Rel'4 solution for the PS domain LCS service is that it is not needed to suspend and resume the packet data connection for determining the location of the mobile but the location procedures are performed parallel to the connection.

Readers with more detailed interest on the LCS specifications are recommended to take a look at [5].

Inter-BSC and BSC/RNC NACC (network assisted cell change)

In Rel'4 of the GERAN specifications, the NACC procedures cover only cell re-selection to other GSM/EDGE cells within the same BSC where the BSC has system information available for the target cell. This limits the value of the NACC as inter-BSC cell changes and cell changes between GSM/EDGE and UTRAN cells in some network configurations are of frequent occurrence.

The inter-BSC and BSC-RNC NACC is an extension of the Rel'4 NACC feature. For the fast re-selection of cells, the mobile station (MS) needs to have knowledge of certain system information for the target cell before performing the cell change. If the cells belong to different BSCs/RNCs, the required system information has to be transported between BSCs/RNCs when updated to be available for the MSs when re-selecting cells [6].

High multislot classes for type 1 mobiles

There are two main types of mobile stations in the GERAN standard:

- type 1 mobiles: not required to transmit and receive at the same time;
- type 2 mobiles: required to be able to transmit and receive at the same time.

Due to the need to perform neighbour cell measurements between the transmit (Tx) and receive (Rx) timeslots the highest type 1 multislot class is in practice number 12 (4 Rx timeslots, 4 Tx timeslots, sum of Rx and Tx time slots active at one time = 5). To overcome this restriction in the total number of active timeslots (five in the example), the UL multiframe of those BTSs in the network that would support high multislot classes is shifted. This shift is transparent for any MS (legacy MS or new MS supporting high multislot classes) because this shift is just a timing advance minimum value. If the UL multiframe is shifted 100 ms, then there is 100 ms available for the mobile station to switch from one frequency to another, and another 100 ms window for making the actual neighbour cell measurement, even for the 6 Rx timeslot + 1 Tx timeslot configuration.

This example UL multiframe shift (100 ms) restricts the maximum cell size to 18 km. However, this restriction should not become a problem as this solution can be deployed in selected areas, on a cell-per-cell basis, depending on the size of the cells, the expected data traffic and BTS capabilities [7].

2.2 GERAN *architecture*

General

This section explains basics of new architecture for GSM/EDGE radio access networks that are currently being specified in 3GPP Rel'5. Description in this chapter assumes that the reader is familiar with basic protocols and interfaces used in pre-Rel'5 GSM/GPRS. In addition the UMTS architecture as specified in [8] would provide useful insights for fully understanding the concepts presented here.

The evolution of GSM standards has been described in Chapter 1 and the beginning of Chapter 2 as well. Here, let's briefly recall the goals of GERAN Rel'5 specification. In Rel'5 the functional split between the GSM/EDGE radio access network (GERAN) and the core network (CN) will be harmonised with the functionality split that has been specified between UTRAN and the CN. As a consequence we find that this harmonisation has three important impacts in the overall system:

- It will enable GERAN to connect to the same 3G CN as UTRAN creating first steps towards efficient resource optimisations in multi-radio networks.
- It will enable GERAN to provide the same set of services as UTRAN, making the radio technology invisible to the end user, while allowing operators to efficiently manage the available spectrum.
- Existing GERAN radio protocols need to undergo significant modifications, and this will obviously increase the complexity or radio interface protocols.

In the following sections we can see details of the architecture. The reader should note that the specification of GERAN Rel'5 was ongoing at the time this book was published. For more accurate information the reader is advised to check the 3GPP specifications mentioned in the bibliography.

Architecture and interfaces

The basic principles that are used for the design of GERAN architecture for Rel'5 are:

- Separation of the radio related and non-radio related functionalities between the CN and the RAN. This principle is important in order to enable development of the platform that would allow for provisioning of services independent of the access type. For example an operator could run one CN and offer services to the end users by utilising different radio access technologies, e.g. WCDMA, GSM/EDGE radio access networks or wireless local area network (LAN).
- Support of services for pre-Rel'5 terminals must be ensured. Obviously it is important to evolve the GSM/EDGE radio access in a backwards compatible way.
- Maximise commonalties between GERAN and UTRAN but maintain the backward compatibility for GERAN. The key alignment point between the GERAN and UTRAN is the common interface towards the 3G core network (Iu interface). Consequently, GERAN should provide the same set of services as UTRAN, e.g. support the same radio access bearers and QoS mechanisms as in UTRAN.
- Standardisation of the GERAN Rel'5 should support a true multi-vendor environment.

In accordance with those principles the GERAN architecture for Rel'5 has been designed in 3GPP and is shown in the Figure 2.2.

The following sections describe the details of the interfaces that are part of GERAN Rel'5 architecture.

Legacy interfaces

The protocol model for legacy interfaces is based on the Rel'4 of GSM/EDGE and consequently the functionality split between the radio access and the core network is unchanged. The two interfaces towards core network are:

Gb interface. The Gb interface is used in GSM/GPRS to connect the GSM serving GPRS support node (SGSN) and base station subsystem (BSS). This interface is needed in GERAN for supporting pre-Rel'5 terminals. There are two peer protocols that have been identified across Gb interface:

- The base station subsystem GPRS protocol (BSSGP) (see [9]). The primary functions of the BSSGP include: in the downlink, the provision by an SGSN to a BSS of radio related information used by the RLC/MAC function; in the uplink, the provision by a BSS to an SGSN of radio related information derived from the RLC/MAC function; and the provision of functionality to enable two physically distinct nodes, an SGSN and a BSS, to operate node management control functions. See [9] for more details.
- The underlying network service (NS). The NS is responsible for the transport of BSSGP protocol data units (PDUs) between a BSS and an SGSN (see [10]).

A interface. Traditional interface in GSM that is used to connect BSS and MSC will be supported by GERAN. This interface is needed to support pre-Rel'5 terminals, as well as to provide new enhanced services like, e.g. wideband telephony (with wideband adaptive multi-rate – WB-AMR – speech codec) and support for 8-PSK modulated speech channels. There are a number of available references that describe the protocol and procedures over the A interface, therefore these description are omitted from this chapter.

New interfaces

The general protocol model for new interfaces is based on the protocol model used in UMTS and is shown in Figure 2.3. The layers and planes are logically independent of each other which offers additional flexibility. For example, if needed, transport network protocol layers, may be changed in the future with virtually no impact on the radio network layer.

GERAN Rel'5 will connect to the UMTS CN using Iu interface. The Iu interface is specified at the boundary between the CN and the RAN. The Iu interface

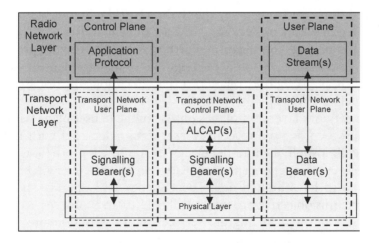

Figure 2.3   General UMTS protocol model.

is used in Rel'99 to connect UTRAN and CN, and the basic idea is to adopt this interface for connecting GERAN to the UMTS CN. There are two flavours of Iu interface:

- *Iu-cs interface*: the interface between the MSC and its BSS and/or RNS is specified in the 25.41x-series of UMTS technical specifications and is called Iu-cs. The Iu-cs interface is used to carry information concerning BSS management, call handling and mobility management.
- *Iu-ps interface*: the BSS-UMTS-SGSN interface is called Iu-ps interface and is used to carry information concerning packet data transmission and mobility management. The Iu-ps interface is defined in the 25.41x-series of UMTS technical specifications.

The protocols over Iu interface are divided into two structures:

- *User plane protocols.* These are the protocols implementing the actual radio access bearer service, i.e. carrying user data through the access stratum.
- *Control plane protocols.* These are the protocols for controlling the radio access bearers and the connection between the UE and the network from different aspects (including requesting the service, controlling different transmission resources, handover and streamlining, etc.). Also a mechanism for transparent transfer of NAS messages is included.

The protocol stacks across Iu-cs and Iu-ps interfaces as specified in 3GPP TSs are shown in Figures 2.4 and 2.5. From these figures it is seen that on the radio network layer there is no difference between Iu-cs and Iu-ps interfaces.

The radio network signalling (control plane) over Iu consists of the radio access network application part (RANAP). The RANAP protocol consists of mechanisms to handle all procedures between the CN and RAN. It is also capable of conveying messages transparently between the CN and the MS without interpretation or processing by the RAN. Over the Iu interface the RANAP protocol is, e.g. used for provisioning of a set of general RAN procedures from the CN such as paging – notification. Further, RANAP provides procedures for the transfer of transparent non-access signalling, management of radio access bearers as well as procedures related to the relocation function. The RANAP protocol is defined in [11].

The radio network layer user plane consists of the Iu user plane (UP) protocol. The Iu UP protocol is used to convey user data associated with radio access bearers. The philosophy behind the design of the Iu UP protocol is to keep the radio network layer independent of the CN domain (circuit switched or packet switched) and to have limited or no dependency with the transport network layer. Meeting this objective provides the flexibility to evolve services regardless of the CN domain and to migrate services across CN domains. The Iu UP protocol is therefore defined with modes of operation that can be activated

Figure 2.4 Protocol stack across Iu-cs interface.

on a radio access bearer (RAB) basis rather than on a CN domain basis or (tele) service basis. The Iu UP mode of operation determines if and which set of features shall be provided to meet, e.g. the RAB QoS requirements. The Iu UP protocol is defined in [12].

The difference between Iu-cs and Iu-ps in terms of protocol stack is visible in the transport network layer. In case of Iu-cs, the choice of signalling bearer for RANAP over the Iu interface is based on broadband signalling system No.7 consisting of well-standardised components. In case of Iu-ps the standard allows operators to choose one out of two standardised protocols to suite transport of SCCP messages.

So far, we have seen that GERAN of Rel'5 could have four interfaces towards the CN. This could lead to a quite large number of possible combinations of interfaces towards the CN thus increasing dramatically the complexity of both RAN and the MS, therefore some simplifications are necessary. The standard specifies two modes of operations for the MS:

- *A/Gb mode*, e.g. for pre-Rel'5 terminals or for Rel'5 terminals when connected to a GERAN with no Iu interface towards the CN;
- *Iu mode* (i.e. Iu-cs and Iu-ps), e.g. for Rel'5 terminals when connected to a GERAN with Iu interfaces towards the CN.

Figure 2.5   Protocol stack across Iu-ps interface.

The mobile station shall operate only on one of the modes. This in practice means that if the MS is operating in A/Gb mode then it is utilising services from the legacy GSM CN, while services from the UMTS CN are available when the MS is operating in Iu mode.

Radio access network interfaces

BSS–MS interface (Um)

Adoption of the Iu interface in GERAN Rel'5 requires significant modification on the radio interface protocol stack. These modifications are required primarily due to the significant difference in functionality split between GERAN and core networks depending on the interface that is used: A/Gb or Iu. This difference, together with the necessity to fulfil the backwards compatibility principle mentioned in the beginning of this chapter, leads to a rather complex protocol constellation over the air interface. Overall description of GERAN Rel'5 is given in [13] and the reader is referred to this specification for more details. As an illustration, Figure 2.6 shows the user plane protocol stack for PS domain (Figure 2.4 in [13]).

Figure 2.6 User plane protocol stack for PS domain [13].

As it can be seen from Figure 2.6, the protocol stack over the air interface (Um) consists of common protocol layers (common to Gb and Iu-ps) as well as protocol layers specific to each interface. Similar figures for the control plane of the PS domain, as well as the user and control plane of the CS domain are available in [13]. The radio interface protocol architecture when GERAN connects to A or Gb is the same as defined in earlier releases. The sections below describe the protocol architecture specific to the Iu interface support.

Radio interface protocol architecture   When designing protocols across the Um interface the usual layering principle has been followed:

- the physical layer (L1)
- the data link layer (L2), and
- the network layer (L3)

Further, the principle of user and control plane separation is applied. The radio interface protocol architecture is shown in Figure 2.7.

The RLC, MAC protocol and physical layer carry data for both control and user plane. Layer 2 consists of RLC and MAC protocols as well as PDCP. The PDCP protocol is used only on the user plane.

Layer 3 of the control plane is split into two sublayers, RRC and duplication avoidance functionality. The RRC protocol utilises services offered by RLC/MAC for transfer of data, except for data related to the operation on the broadcast control channel (BCCH) and common control channel (CCCH), where the data link layer is used. The duplication avoidance functionality provides the

Figure 2.7   Radio interface protocol architecture [13].

access stratum services to higher layers such as mobility management (MM) and call control (CC).

Physical layer (PHY) protocol   The physical layer of GERAN interfaces the MAC and the RRC protocols. Within a carrier, a series of timeslots of consecutive TDMA frames forms a basic physical channel. *Basic physical channels* are organised into multiframe. Three types of multiframe are defined:

- 26-multiframe formed from 26 TDMA frames (120 ms) and is used for traffic and associated control channels.
- 52-multiframe formed from 52 TDMA frames (240 ms) and used for traffic, associated control and common control channels.
- 51-multiframe comprising 51 TDMA frames (235.4 ms). This multiframe supports broadcast and common control. It always carries synchronisation and frequency correction bursts. It also supports short message service (SMS).

A *basic physical subchannel* (BPSCH) is defined as a basic physical channel or a part of a basic physical channel. Some BPSCHSs are reserved by the network for control and broadcast (51-multiframe), others are assigned to dedicated connections with MSs called dedicated BPSCH (DBPSCH), or are assigned to a shared usage between several MSs and called shared BPSCH (SBPSCH).

A DBPSCH always follows the 26-multiframe structure. It can be either full rate (DBPSCH/F) or half rate (DBPSCH/H). Among the 26 timeslots of a DBPSCH/F, one is always reserved for control and one is always idle. That leaves 24 timeslots available for traffic every 120 ms. Each timeslot is thus equivalent to 5 ms of traffic on DBPSCH/F. Similarly, each timeslot is equivalent to 10 ms of traffic on DBPSCH/H.

A SBPSCH always follows the 52-multiframe structure. It can be either full rate (SBPSCH/F) or half rate (SBPSCH/H). A SBPSCH/H is available in MAC dual transfer mode (DTM) state only. Among the 52 timeslots of a SBPSCH/F, two are always reserved for timing advance procedures and two are always idle. That leaves 48 timeslots available for traffic every 240 ms. As on DBPSCH, each timeslot is thus equivalent to 5 ms of traffic on SBPSCH/F. Similarly, each timeslot is equivalent to 10 ms of traffic on SBPSCH/H.

The physical layer offers logical channels and the associated transmission services. Logical channels are divided in two categories:

• traffic channels
• control channels

Traffic channels are intended to carry either encoded speech or user data while control channels carry signalling or synchronisation data. Logical channels are multiplexed either in a fixed predefined manner, or dynamically by the MAC on physical subchannels. Traffic channels of type TCH are intended to carry either speech or user data on dedicated basic physical subchannels (for one user only). TCH can be either full rate (TCH/F) or half rate (TCH/H), and is GMSK modulated. Traffic channels of type O-TCH (octal traffic channel) are intended to carry speech on dedicated basic physical subchannels (for one user only). O-TCH can be either full rate (O-TCH/F) or half rate (O-TCH/H) and is 8-PSK modulated. Traffic channels of type E-TCH (enhanced traffic channel) are intended to carry user data on dedicated basic physical subchannels (for one user only). E-TCH is full rate and always 8-PSK modulated only. Packet data traffic channels are intended to carry user data on either SBPSCH or DBPSCH. The packet data traffic channel (PDTCH) is temporarily dedicated to one MS but one MS may use multiple PDTCHs in parallel. PDTCH allows several MSs to be multiplexed on the same SBPSCH. PDTCH also allows several traffic classes from the same MS to be multiplexed on the same PSCH (shared or dedicated). PDTCHs can be either full rate (PDTCH/F) or half rate (PDTCH/H), and can be either GMSK or 8-PSK modulated.

Control channels are intended to carry signalling or synchronisation data. Three categories of control channel are defined: broadcast, common and dedicated control channels.

Broadcast channels consist of frequency correction channels (FCCH) which carry information for frequency correction of the MS, synchronisation channels (SCH) which carry information for frame synchronisation of the MS and identification of a base transceiver station, broadcast control channel (BCCH) used for broadcasting general information on a base transceiver station and packet broadcast control channel (PBCCH) used for broadcasting parameters valuable to the MS to access the network for packet transmission operation.

Common control type channels, known when combined as a common control channel (CCCH) consist of paging channel (PCH) used to page mobiles, random access channel (RACH), used to request allocation of a dedicated control channel, access grant channel (AGCH) used to allocate a basic physical subchannel and notification channel used to notify MSs of voice group and voice broadcast calls. Packet common control channels (PCCCH) are used to support packet switched operations and consist of packet paging channel (PPCH), packet random access channel (PRACH), packet access grant channel (PAGCH) and packet notification channel (PNCH). PCCCH is optional for the network and if a PCCCH is not allocated, the information for packet switched operation is transmitted on the CCCH. If a PCCCH is allocated, it may transmit information for circuit switched operation.

There are a number of dedicated control channels to support DBPSCH. They are all derived from three types of control channel:

- Slow associated control channel (SACCH) is used mainly for the transmission of the radio measurement data needed for radio resource management algorithms. In the uplink, the MS sends measurements reports to the BTS, while in the downlink, the BTS sends commands to the MS. SACCH is also used for SMS transfer during a call.
- Fast associated control channel (FACCH) which is always associated with one TCH, and occurs on DBPSCH only.
- Stand-alone dedicated control channel (SDCCH).

Similarly, control channels for support of SBPSCH are:

- The packet associated control channel (PACCH): the PACCH is bi-directional. For description purposes PACCH/U is used for the uplink and PACCH/D for the downlink.
- Packet timing advance control channel uplink (PTCCH/U): used to transmit random access bursts to allow estimation of the timing advance for one MS in packet transfer mode.

- Packet timing advance control channel downlink (PTCCH/D): used to transmit timing advance updates for several MS. One PTCCH/D is paired with several PTCCH/Us.

For more information on the physical layer see 3GPP TS 45.x series.

Medium access control (MAC) protocol   The MAC sublayer allows the transmission over the physical layer of upper layer PDUs in dedicated or shared mode. A MAC mode is associated with a basic physical subchannel for use by one or more MSs (dedicated or shared mode, respectively). The MAC layer handles the access to and multiplexing on to the physical subchannels of mobile stations and traffic flows.

MAC protocol for Rel'5 is heavily based on the MAC protocol of Rel'4 as specified in [14]. Main modifications in Rel'5 are related to the support of multiple temporary block flows, ciphering, support for radio bearers and support for RRC signalling. The functions of MAC include [13]:

- Configuring the mapping between logical channels and basic physical subchannels. The MAC is responsible for configuring the mapping of logical channel(s) on to the appropriate basic physical subchannel(s).
- Defining logical channels to be used for each radio bearer service.
- Assignment, reconfiguration and release of shared radio resources for a temporary block flow (TBF). The MAC layer may handle the assignment of radio resources needed for a TBF including needs from both the control and user plane. The MAC layer may reconfigure radio resources of a TBF.
- MS measurement reporting and control of the reporting. The MAC layer is responsible for sending information that controls the MS measurement reporting when using PBCCH or PACCH channels. The MAC layer also performs the reporting of the measurements from the MS to the network using PACCH.
- Broadcasting/listening of/to PBCCH and PCCCH. The MAC layer broadcasts/listens (to) the PBCCH of the serving cell for the sending/decoding of packet system information messages. The MAC layer listens to the BCCH of neighbouring cells for neighbour cell measurements. The MAC layer also sends paging information on the PCCCH or monitors the paging occasions according to the discontinuous reception (DRX) cycle.
- Timing advance control. The MAC layer controls the operation of timing advance on shared basic physical subchannels.
- Ciphering. The MAC is responsible for ciphering user data blocks when RLC operates in a transparent mode.
- Identification of different traffic flows of one or more MSs on the basic physical subchannels. Inband identification is needed to address a flow to a MS in the downlink or identify a flow from a MS in the uplink.

- Multiplexing/demultiplexing of higher layer PDUs. This may include priority handling between data flows of one or more MSs, e.g. by attributes of
  radio bearer services.
- Multiplexing/demultiplexing user and control plane data to/from the physical layer for PDTCHs. The MAC layer is responsible for multiplexing/
  demultiplexing RLC data blocks carried on PDTCH and RLC/MAC control
  blocks carried on PACCH.
- Scheduling of RLC/MAC data and control PDUs delivered to the physical
  layer on shared basic physical subchannels. This includes uplink state flag
  (USF) and relative reserved block period (RRBP) field monitoring for uplink
  transfer and sharing radio resources on the downlink.
- Splitting/recombining. This includes splitting/recombining of the RLC/
  MAC PDU flow belonging to one or more TBF(s) on to/from several shared
  logical channels. This function does not apply for RLC/MAC control
  blocks.

Figure 2.8 describes the MAC state model for GERAN in Iu mode. The model
shows the MAC state for the MAC control entity of a MS and not for individual
radio bearers (see [13]).

There are four states in the MAC state machine. The MAC control entity of a
MS is in MAC-IDLE state when there are no dedicated or shared basic physical
subchannels. RRC can be in RRC-Cell_Shared state, RRC-GRA_PCH state or
RRC-Idle mode (see sections below). In this state, the MS camps on the PCCCH
or on the CCCH. The MAC control entity is in MAC-SHARED state when at
least one TBF is ongoing but no dedicated basic physical subchannels have been
allocated. The MAC control entity of a MS is in MAC-DTM state when it has
one or more dedicated basic physical subchannel(s) and one or more shared
basic physical subchannel(s) allocated. The MAC control entity will be in MAC-

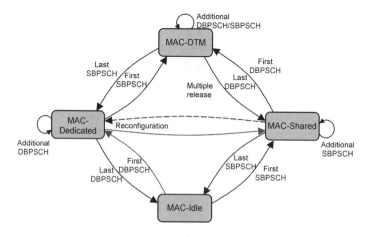

Figure 2.8   MAC protocol state model [13].

DEDICATED state when a dedicated basic physical subchannel is used and no shared basic physical subchannel is established. In MAC-DEDICATED state MAC has no control functionality.

Radio link control (RLC) protocol   The RLC protocol allows for the data transfer in transparent, acknowledged or unacknowledged modes, and in addition notifies unrecoverable errors to the upper layer.

When in transparent mode, RLC has no functionality and does not alter the data units of the upper layer.

In non-transparent mode, the RLC is responsible for ciphering RLC PDUs in order to prevent any unauthorised acquisition of data. RLC non-transparent mode is built based on the (E)GPRS if the radio access bearer (RAB) is mapped on to a packet data traffic channel (PDTCH). Alternatively, the RAB may be mapped on to an (E-)TCH (enhanced traffic channel). Signalling is performed on a packet associated control channel (PACCH) when PDTCH is used, and on fast associated control channel (FACCH) and slow associated control channel (SACCH) when (E-)TCH is used. The mapping of a RAB on to the proper traffic channel is dependent on the MAC mode in use and the targeted traffic class, as described in the next chapter. In acknowledged mode, backward error correction (BEC) procedures are provided that allow error-free transmission of RLC PDUs through two possible selective automatic repeat request (ARQ) mechanisms: type I ARQ and hybrid type II ARQ (incremental redundancy). In unacknowledged mode, no BEC procedure is available.

For more details see [14].

Packet data convergence protocol (PDCP)   The PDCP layer is operational only on the user plane of the radio interface protocol stack. This protocol layer is common to UTRAN and GERAN and is specified in [15]. The PDCP allows for the transfer of user data using services provided by the RLC, and header adaptation of the redundant network PDU control information (IP headers) in order to make the transport over the radio interface spectrum efficient. One important function of PDCP is header compression functionality which consists of compressing transport and network level headers (e.g. RTP/UDP/IP) so that de-compressed headers are semantically identical to the original un-compressed headers. Header compression is suited for generic Internet applications (i.e. not specific to GERAN), especially multimedia applications.

Radio resource control (RRC)   The RRC layer handles the control plane signalling of Layer 3 between the MSs and GERAN. The RRC performs the following functions (see [13]):

- Broadcast of information provided by the non-access stratum (core network): the RRC layer performs system information broadcasting from the network to all MSs. The system information is normally repeated on a regular basis. The RRC layer performs the scheduling, segmentation and repetition when such broadcasting is carried on a broadcast control channel (BCCH). This function supports broadcast of higher layer (above RRC) information. This information may be cell specific or not. As an example RRC may broadcast CN location service area information related to some specific cells.

- Broadcast of information related to the access stratum: the RRC layer performs system information broadcasting from the network to all MSs. The system information is normally repeated on a regular basis. The RRC layer performs the scheduling, segmentation and repetition when such broadcasting is carried on BCCH. This function supports broadcast of typically cell-specific information.

- Establishment, re-establishment, maintenance and release of an RRC connection between the MS and GERAN: the establishment of an RRC connection is initiated by a request from higher layers at the MS side to establish the first signalling connection for the MS. The establishment of an RRC connection includes an optional cell re-selection, an admission control, and a layer 2 signalling link establishment. The release of an RRC connection can be initiated by a request from higher layers to release the last signalling connection for the MS or by the RRC layer itself in case of RRC connection failure. In case of connection loss, the MS requests re-establishment of the RRC connection. In case of RRC connection failure, RRC releases resources associated with the RRC connection.

- Establishment, reconfiguration and release of radio bearers: the RRC layer can, on request from higher layers, perform the establishment, reconfiguration and release of radio bearers in the user plane. A number of radio bearers can be established to a MS at the same time. At establishment and reconfiguration, the RRC layer performs admission control and selects parameters describing the radio bearer processing in layer 2 and layer 1, based on information from higher layers.

- Assignment, reconfiguration and release of radio resources for the RRC connection: depending on the RRC and MAC states, the RRC layer may handle the assignment of radio resources needed for the RRC connection including needs from both the control and user plane. The RRC layer may reconfigure radio resources during an established RRC connection. RRC signals to the MS to indicate resource allocations for purposes of inter-system handovers.

- RRC connection mobility functions: the RRC layer performs evaluation, decision and execution related to RRC connection mobility during an established RRC connection, such as handover, preparation of handover to UTRAN or other systems, cell re-selection and cell/GERAN registration area (GRA) update procedures, based on, e.g. measurements done by the MS.

- Release of signalling connections: the RRC layer provides the necessary functions for the GERAN or the MS to request the release of a signalling connection.

- Paging/notification: the RRC layer can broadcast paging information from the network to selected MSs on CCCH. Higher layers on the network side can request paging and notification. The RRC layer can also initiate paging during an established RRC connection.

- Listening to the BCCH and common control channels (CCCH). The RRC layer listens to the BCCH of the serving cell for the decoding of system information messages and to the BCCH of neighbouring cells for neighbour cell measurements; the RRC layer also monitors the paging occasions according to the DRX cycle and receives paging information on the CCCH.

- Routing of higher layer PDUs: this function performs at the MS side routing of higher layer PDUs to the correct higher layer entity, at the GERAN side to the correct RANAP entity.

- Control of requested QoS: this function shall ensure that the QoS requested for the radio bearers can be met. This includes the allocation of a sufficient number of radio resources.

- MS measurement reporting and control of the reporting: the measurements performed by the MS are controlled by the RRC layer including both GSM/EDGE air interface and other systems. The RRC layer is responsible for sending information that controls the MS measurement reporting when using BCCH or SACCH channels. The RRC layer also performs the reporting of the measurements from the MS to the network using SACCH.

- Power control: the RRC layer controls parameters for normal power control, enhanced power control and fast power control.

- Control of ciphering: the RRC layer provides procedures for setting of ciphering (on/off) between the MS and GERAN.

- Integrity protection: this function controls integrity protection and performs integrity protection of those RRC messages that are considered sensitive and/or contain sensitive information.

- Support for location services. Signalling between MS and GERAN to support positioning of a MS.

- Timing advance control. The RRC controls the operation of timing advance on dedicated basic physical subchannels.

The RRC protocol for GERAN Iu mode is designed by combining necessary procedures from the GSM RR sublayer specified in [16] and UTRAN RRC protocol specified in [17]. The basic idea with combining those two protocols is to align the GERAN Iu mode with UTRAN to a maximum extent while maintaining backwards compatibility by keeping critical procedures common to A/Gb and Iu modes. Further, the RRC protocol for Iu mode requires capabilities that are in line with UMTS bearer concept and should provide smooth interaction with Iu signalling, without causing modification to the Iu interface. New procedures that are created for GERAN Iu mode are based on the UTRAN RRC protocol and include the following procedures: RRC connection management procedures, RRC connection mobility procedures, radio bearer control procedures, signalling flow procedures, security mode control and delivery of non-access stratum messages. Other necessary procedures are modification of existing GSM RR procedures, e.g. system information, handover procedure, paging, etc.

The result, on the protocol state machine level, is shown in Figure 2.9.

The state machine on the right applies to the GERAN A/Gb mode (denoted as RR modes), while the state machine on the left is applicable to the GERAN Iu mode. Let's look into the details of the Iu mode state machine.

Two modes of operation are defined for the MS, *RRC idle mode* and *RRC connected mode*. After power on, the MS stays in RRC idle mode until it transmits a request to establish an *RRC connection*. In RRC idle mode the MS is identified by non-access stratum identities such as international mobile subscriber identity (IMSI), temporary mobile subscriber identity (TMSI) and packet TMSI (P-TMSI). While in RRC idle mode the MS has no relation to GERAN, only to the CN. For example the MS has already attached to the network, performed necessary procedures, and released the RRC connection, so the MS is already known to the CN. For data transfer, a signalling connection has to be established. The RRC connected mode is entered when the RRC connection is established at which point the MS is assigned a GERAN radio

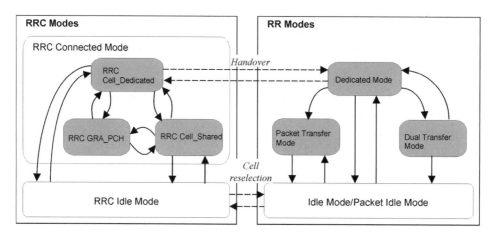

Figure 2.9   RRC and RR protocol state machine [13].

network temporary identity (G-RNTI) to be used as MS identity. The MS is identified within a GERAN with the G-RNTI. The MS leaves the RRC connected mode and returns to RRC idle mode when the RRC connection is released or at RRC connection failure.

RRC connection is established by using RRC connection management procedures that are specified based on the corresponding UTRAN procedure described in [17]. After a successful establishment of the RRC connection the MS enters the RRC connected mode and will be in one of the states shown in Figure 2.9. Three states are defined in RRC connected mode: RRC-Cell_Shared, RRC-Cell_Dedicated and RRC-GRA_PCH. In RRC-Cell_Shared state, no dedicated basic physical subchannel is allocated to the MS and the position of the MS is known by GERAN on the cell level. In RRC-Cell_Dedicated state, the MS is assigned one or more dedicated basic physical subchannels in the uplink and downlink, which it can use anytime. Furthermore, the MS may be assigned one or more shared basic physical subchannels. The position of the MS is known by GERAN on a cell level. In RRC-GRA_PCH state, no basic physical subchannel is allocated to the MS. The location of the MS is known at the GRA level. Introduction of GRA is believed to decrease the MS power consumption and signalling load within GERAN. The GRA concept has been adopted for GERAN in Iu mode. The behaviour of the MS while in the RRC-GRA_PCH state is limited to monitoring the paging occasions according to the DRX cycle and receive paging information on the (P)CCCH. The MS will listen to the (P)BCCH control channel of the serving cell for the decoding of system information messages and initiate a GRA updating procedure upon GRA change. If the network wants to initiate any activity, it shall make a paging request on the (P)CCCH logical channel within the GRA where the MS is. GRA updating is initiated by the MS, which, upon the detection of the new GRA, sends the network the registration area update information to the new cell. Any activity causes a transition to either the RRC-Cell_Shared or the RRC-Cell_Dedicated state, depending on the activity.

Here it is important to note one essential element in the radio interface protocol stack. Due to legacy reasons the radio resource management tasks are split between RRC layer and RLC/MAC layer. There is a complex interaction between those two layers when it comes to the radio resource assignment, maintenance and release. This interaction is shown in Table 2.2 but as a simplification one could note the main rule that the RRC layer is responsible for assignment, maintenance and release of *dedicated* channels, while the MAC layer is responsible for those of the *shared* channels. It is worth noting that even if the MAC layer is responsible for the management of the shared channel, the RRC layer will ensure the QoS for a particular radio bearer and therefore configure the MAC layer correspondingly.

Table 2.2 Interaction between RRC and MAC layers

Currently allocated channel(s)		Allocation of new resources		Current control plane states	
TCH or PDTCH on DBPSCH	PDTCH on SBPSCH	TCH or PDTCH on DBPSCH	PDTCH on SBPSCH	MAC state	RRC state
1 (or more)	–	RRC	RRC	Dedicated	Cell_Dedicated
1 (or more)	1 (or more)		MAC	DTM	
	1 (or more)			Shared	Cell_Shared
–	–	NA		Idle	
–	–		NA		GRA_PCH
–	–	NA			RRC Idle Mode

BSS–BSS and BSS–RNS interface (Iur-g)

The interface between GERAN BSSs or between GERAN BSS and UTRAN RNS is called the Iur-g interface. The support of the Iur-g interface in GERAN allows for registration areas to span across several BSS areas. Similarly, an interface between GERAN BSS and UTRAN RNS would allow for registration areas that are made up of both GERAN and UTRAN cells, typically in the same geographical location. The benefits of introducing registration areas are visible in the fact that it will reduce the amount of signalling in the network since the MS would make fewer updates. In case of registration areas that consist of both UTRAN and GERAN cells the introduction of Iur-g will reduce the amount of signalling on the radio interface.

Procedures over the inter BSS Iur-g and BSS–RNS Iur-g are exactly the same. The difference is in the parameters that are used within specific messages on that interface.

The Iur-g interface has been designed according to the following principles:

- This interface shall be open.
- It is assumed that the Iur-g is based on the existing Iur specification (3GPP TS 25.42x series) and uses a subset of the messages and procedures defined for the RNSAP. This interface shall support the exchange of signalling information between a BSS and an RNC. The Iur-g does not support user data.
- From a logical standpoint, this interface is a point-to-point interface between two BSSs or between one BSS and one RNC within a PLMN. From a physical point of view, the interface could share Iu or other transmission resources.
- The Iur-g interface is optional as the Iur in UTRAN today. The presence of the Iur-g is transparent to the UE/MS: the GERAN specifications shall ensure that all mobiles function correctly irrespective of the presence or absence of the Iur-g interface. This 'transparency principle' can be used to

allow infrastructure manufacturers to implement this interface independently of other features. This implies that the UE/MS should behave like a UTRAN UE in UTRAN and a GERAN MS in GERAN, and UTRAN legacy mobiles will also be supported.

Following the principles above the protocol stack across the Iur-g interface has been designed based on the UTRAN Iur interface and is shown in Figure 2.10.

When comparing this with the protocol stack specified across the Iur interface in UTRAN we find that in Iur-g there is no user plane. One reason for this choice is that procedures that are required on the Iur-g interface are fulfilled using the subset of radio network subsystem application part (RNSAP) basic mobility procedures and RNSAP global procedures and are run over the control plane only. The list of procedures that need support over Iur-g are related to the MS operation on RRC-GRA_PCH state. Those are paging, cell update and GRA update procedures. The list of elementary procedures in RANSAP that support those are: uplink signalling transfer, downlink signalling transfer, relocation commit, paging and error indication.

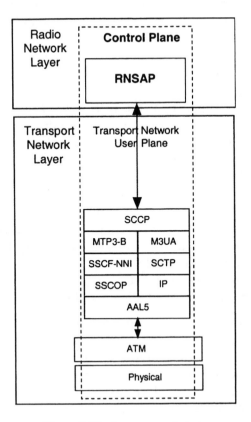

Figure 2.10 Iur-g protocol stack.

2.3 Future standardisation items

Apart from the previous Section 2.1, this section deals with the further evolution of the GERAN standard beyond Rel'5. A part of the contents in this section is based on the current discussions in the standardisation and the scheduling of the standardisation work between the different releases of the standard whereas another part of the contents in this section is more speculative and based on the authors' views. The maturity and the acceptability level of the future work areas will be made clear in the text which follows.

Optimised support for IMS services

The definition of the IMS services in 3GPP takes place in Rel'5 and '6. From the GERAN perspective, there are no fundamental reasons why GERAN could not provide radio bearers for all IMS services. The same reasoning applies for the UTRAN as well. Another question is then whether GERAN can provide the optimum radio bearer for any one IMS service. Some examples of the IMS services that might require some particular optimisation of the radio bearer are basic voice (voice over IP), video telephony and audio/video streaming. It may be beneficial from the radio interface performance point of view to opti- mise radio bearers for these kind of services. Some examples of potential opti- misation are unequal error protection and variable interleaving length.

Taking a broader look at the GERAN physical layer capabilities – not looking only for the optimised support for the IMS services – it can be claimed that building the support for new services (e.g. new codecs) in GERAN is cumber- some and it takes a lot of time and effort in standardisation to agree on the new channel codings for each new service. Defining a similar kind of flexible layer 1 solution for GERAN as there is in UTRAN today may be worth exploring. Larger flexibility in the physical layer could make the introduction of the new services in GERAN easier. On the other hand, this flexibility does not come for free, the price being paid is some penalty in the performance of the radio bearers as the increased flexibility would require a few additional headers or indicators to be carried over the radio interface.

Based on the discussion in the earlier paragraphs, there are two basic alter- natives for the optimisation of the GERAN radio bearers for the IMS services:

- Define a configurable physical layer concept – using similar principles as in UTRAN.
- Decide on optimisation of the radio bearers on a case-by-case basis when needed. This solution would avoid the generic overhead resulting from the additional headers that are subject to the additional flexibility, but on the other hand, would not lead to a flexible platform for the introduction of the new services.

Capacity, quality and data rate enhancements

There is a generic pressure to strive for better radio network capacity (spectral efficiency) and better service quality (speech quality, quality of data services) in GSM/EDGE radio access. Here it is important to note that only part of these improvements needs to be standardised. There is already a lot of flexibility and room within the current GSM/EDGE standard to develop proprietary solutions for improving the radio network capacity and service quality. For adding such features in the GERAN standard that enhance the radio network capacity it is required to prove the feasibility of such an improvement in terms of overall radio network performance and the equipment complexity. As the GSM/EDGE reference performance may not always be directly based on the standard, but may comprise some proprietary well known and acknowledged techniques, it is often difficult to justify the standardisation of new capacity improvements. One example of this kind are some transmit diversity schemes which need to be compared with some non-standardised diversity techniques like delay diversity or antenna hopping.

For the speech quality, the wideband AMR (AMR-WB) speech codec has already brought the 3GPP cellular speech to a very high level – both in terms of absolute speech quality and relative to the current landline speech. Time will tell whether even better quality speech codecs are needed in the future for cellular applications. Likely the challenge in cellular speech quality relates to the IP-based speech services. Even though the same speech codecs like AMR-WB could be used in rich calls and other IP-based multimedia calls there is still work to be done in ensuring the overall end-to-end speech quality without sacrificing the spectral efficiency. As many of the protocols or algorithms for the IP based services (like session initiation protocol (SIP), robust header compression (ROHC)) are developed in the IETF it is very important that the 3GPP and IETF are cooperating efficiently.

The data rate evolution in GSM/EDGE radio access took a big step in Rel'99 as EDGE with 8-PSK modulation was defined. In terms of the theoretical peak data rate, eight timeslots with 8-PSK modulation provides 473.6 kbps (8×59.2 kbps). Future standardisation items will likely explore the capability of GSM/EDGE radio technology to provide even larger data rates.

References

[1] 3GPP TR 22.941, IP Based Multimedia Services Framework.
[2] J. Rotola-Pukkila, J. Vainio, H. Mikkola, K. Järvinen, B. Bessette, R. Lefebvre, R. Salami and M. Jelinek, "AMR Wideband codec – leap in mobile communication voice quality", Eurospeech 2001, September 3–7, Aalborg, Denmark.

[3] B. Bessette, R. Lefebvre, R. Salami and M. Jelinek, J. Rotola-Pukkila, J. Vainio, H. Mikkola, K. Järvinen, "Techniques for High-Quality ACELP Coding of Wideband Speech", Eurospeech 2001, September 3–7, Aalborg, Denmark.

[4] Contribution to 3GPP TSG GERAN AdHoc, February 12–16, 2001, Sophia Antipolis, France: "On GERAN Speech Capacity with Different Quality Criteria".

[5] 3GPP TS 43.059, Functional Stage 2 Description of Location Services in GERAN (Rel'5).

[6] 3GPP TR R3-012181, Study on Network Assisted Inter BSC/RNC Cell Change for REL-5.

[7] 3GPP Tdoc GP-011755, High Multislot Classes for Type 1 Mobiles, 27–31 August 2001.

[8] 3GPP TS 23.110, UMTS Access Stratum; Services and Functions, 3rd Generation Partnership Project; Technical Specification Group GERAN.

[9] 3GPP TS 48.018, General Packet Radio Service (GPRS); Base Station System (BSS)–Serving GPRS Support Node (SGSN); BSS GPRS Protocol, 3rd Generation Partnership Project; Technical Specification Group GERAN.

[10] 3GPP TS 48.016, General Packet Radio Service (GPRS); Base Station System (BSS)–Serving GPRS Support Node (SGSN) Interface; Network Service, 3rd Generation Partnership Project; Technical Specification Group GERAN.

[11] 3GPP TS 25.413, UTRAN Iu Interface RANAP Signalling, 3rd Generation Partnership Project; Technical Specification Group GERAN.

[12] 3GPP TS 25.415, UTRAN Iu Interface User Plane Protocols, Radio Resource Control Protocol, 3rd Generation Partnership Project; Technical Specification Group GERAN.

[13] 3GPP TS 43.051 v5.3.0, GSM/EDGE Radio Access Network (GERAN) Overall Description; Stage 2, 3rd Generation Partnership Project; Technical Specification Group GERAN.

[14] 3GPP TS 44.060, General Packet Radio Service (GPRS); Mobile Station (MS)–Base Station System (BSS) Interface; Radio Link Control/Medium Access Control (RLC/MAC) Protocol, Radio Resource Control Protocol, 3rd Generation Partnership Project; Technical Specification Group GERAN.

[15] 3GPP TS 25.323, Description of the Packet Data Convergence Protocol (PDCP), Radio Resource Control Protocol, 3rd Generation Partnership Project; Technical Specification Group GERAN.

[16] 3GPP TS 44.018, Mobile Radio Interface Layer 3 Specification, Radio Resource Control Protocol, 3rd Generation Partnership Project; Technical Specification Group GERAN.

[17] 3GPP TS 25.331, Radio Resource Control (RRC) Protocol Specification, 3rd Generation Partnership Project; Technical Specification Group GERAN.

[18] 3GPP TS 45.008, Radio subsystem link control, 3rd Generation Partnership Project; Technical Specification Group GERAN.

3

GERAN QoS Evolution towards UMTS

Renaud Cuny, Gerardo Gómez and Héctor Montes

Third-generation (3G) mobile networks are expected to offer a wide range of new services (e.g. multimedia) that will have different requirements in terms of quality of service (QoS). Internet web browsing applications requirements (low loss, medium delay) are for instance far from voice services requirements (low jitter, low delay). The QoS management in universal mobile telecommunications services (UMTS) networks is therefore a key element that must enable different types of services to coexist.

The evolution from circuit switched to packet switched networks is clearly the main motivation to define a complete new mobile QoS architecture and solution. Another major transformation is that new generation networks will extensively use Internet technologies in many of their operations (e.g. transport, traffic management and policy), in order to maximise network utilisation. As an example, the differentiated services scheme from the Internet Engineering Task Force (IETF) may be implemented in some access and core parts of UMTS networks to handle traffic at the user plane. The integration of UMTS technologies and Internet technologies is certainly a challenge for 3G network designers, especially from the QoS perspective.

The mobile network convergence and compatibility towards Internet protocol (IP) will also need special tuning and optimisation. It is not enough to provide gateways to the Internet in order to make effective use of it. For instance, transport control protocol (TCP) traffic needs to be handled in a special way over radio links to allow a good throughput. Typically, the mobile terminal and server/proxy TCP stacks require a different configuration than the one in use in fixed networks. Also, some radio network elements may support clever mechanisms to manage the likely congestion cases or to speed up TCP streams. TCP optimisations over wireless are studied extensively in the IETF and standard recommendations already available, see Chapter 7.

One important feature in the UMTS QoS architecture is the traffic differentiation. The 3rd Generation Partnership Project (3GPP) specifies four different traffic classes (conversational, streaming, interactive and background). In addition, other parameters like maximum bit rate or delay can also be precise in order to make refined differentiation inside each traffic class. It should be noted that traffic differentiation is meaningful only up to a certain level that is determined essentially by the end user perception.

It is important to keep in mind that customers do not always make their selections based on technical facts. QoS designers and service providers need to take this characteristic into account in order to provide flexible and attractive enough solutions that adapt to various needs. The ultimate goal is to offer the services that customers are ready to pay for.

Policy is another facet of QoS. Policy is a matter of allocating resources according to business decisions. QoS policy tools shall allow the operators to easily define and enforce traffic differentiation based on the customer subscription and profile. In order to control the network in an effective way, the policy rules need to be consistent over the entire path.

As we can see, the end-to-end QoS concept itself contains many different aspects. It is not intended to describe in detail all topics evoked above in the following sections but rather concentrate on those directly related to GSM EDGE radio access network (GERAN) QoS architecture and provision.

It is important to highlight that the GERAN QoS model will be based on the complete UMTS QoS architecture. Hence, GERAN provides the same services to the UMTS core network (CN) as the UMTS terrestrial radio access network (UTRAN). A new architecture separated in user and control plane is adopted by GERAN Rel'5 in order to achieve these services, as described in Chapter 2.

This is realised by the adoption of all UMTS radio access bearer classes (conversational, streaming, interactive and background) and the harmonisation of QoS management between GERAN and UTRAN, which will ensure the feasibility to achieve full system integration between both 3G radio access network technologies.

3.1 The 3GPP QoS architecture

QoS architecture evolution

The evolution of the QoS management architecture for mobile networks began with Rel'97, where general packet radio system (GPRS) was introduced. The main features of Rel'97 are the introduction of the Gb interface, which connects the second-generation serving GPRS support node (2G-SGSN) and the base station subsystem (BSS), as well as the concept of GPRS bearer service, which is the base of the GPRS QoS management framework. The GPRS bearer service is based on the concept of policy decision point (PDP) context, which is, in essence, a logical connection set up between user equipment (UE) and gateway GPRS support node (GGSN) to carry all the IP traffic to and from UE. Although the PDP context was introduced by the GPRS architecture, the UMTS standard has further developed the concept to support a variety of new requirements. The first GPRS standard (Rel'97) already allows one mobile station (MS) to have multiple simultaneous PDP contexts, but they all need a dedicated PDP address. This drawback is solved in UMTS (and in GPRS Rel'99), as it enables the usage of several PDP contexts per PDP address, each having a QoS profile of its own. The first PDP context opened for a PDP address is called the primary or default PDP context. The subsequent contexts opened for the same PDP address are *secondary* contexts.

The QoS requirements expressed through the QoS profile in Rel'97 are defined in terms of the following attributes: precedence, delay, reliability and throughput classes (see Table 3.1). The service precedence indicates the relative importance of maintaining the service commitments under abnormal conditions (e.g. which packets are discarded in the event of problems such as limited resources or network congestion). The delay class determines the per-packet GPRS network transit delay. The combination of the transmission modes of the different network protocol layers supports the reliability class performance requirements. User data throughput is specified in terms of a set of throughput classes that characterise the expected bandwidth required for such traffic.

Rel'99 [1] includes some features allowing a better control of the QoS provision in the radio domain, like sending the QoS requirements from the 2G-SGSN to the BSS by means of the *BSS packet flow context* (PFC). A given BSS PFC can be shared by one or more activated PDP contexts with identical or similar negotiated QoS requirements. The data transfer related to PDP contexts that share the same PFC constitute one aggregated packet flow, whose QoS requirements are called *aggregate BSS QoS profile*. The aggregate BSS QoS profile defines the QoS that must be provided by the BSS for a given packet flow between the MS and the SGSN, i.e. for the Um and Gb interfaces combined.

Table 3.1   QoS attributes in Rel'97

QoS attribute	Description
Precedence class	There are three different service precedence levels, which indicate the priority of maintaining the service. High precedence level guarantees service ahead of all other precedence levels. Normal precedence level guarantees service ahead of low priority users. Low precedence level receives service after the high and normal priority commitments have been fulfilled.
Delay class	Delay attribute refers to end-to-end transfer delay through the GPRS system. There are three predictive delay classes and one best-effort class.
Reliability class	There are five different reliability classes. Data reliability is defined in terms of residual error rates for the following cases: • Probability of data loss • Probability of data delivered out of sequence • Probability of duplicate data delivery, and • Probability of corrupted data
Throughput class	User data throughput is specified in terms of a set of throughput classes that characterise the expected bandwidth required for a PDP context. The throughput is defined by both peak and mean classes: • The *peak throughput* specifies the maximum rate at which data are expected to be transferred across the network for an individual PDP context. There is no guarantee that this peak rate can be achieved or sustained for any time period, this depends upon the MS capability and available radio resources. The peak throughput is independent of the delay class that determines the per-packet GPRS network transit delay. • The *mean throughput* specifies the average rate at which data are expected to be transferred across the GPRS network during the remaining lifetime of an activated PDP context. A best-effort mean throughput class may be negotiated, and means that throughput shall be made available to the MS on a per need and availability basis.

Rel'99 introduces the UMTS QoS traffic classes and attributes (described in Table 3.2), and GPRS QoS at session management level integrated with them. The set of UMTS QoS attributes known as QoS profile is composed of the following attributes: *traffic class, guaranteed* and *maximum bit rate, transfer delay, delivery order, delivery of erroneous Service Data Units (SDUs), residual bit error rate (BER), SDU error ratio, maximum SDU size, SDU format information, traffic handling priority* and *allocation/retention priority*.

In Rel'5, GERAN and UTRAN QoS management frameworks are fully harmonised, since GERAN adopts the UMTS architecture with the 3G-SGSN being connected to both types of radio access networks through the Iu interface. The Iu interface towards the packet switched (PS) domain of the CN is called Iu-ps, and the Iu interface towards the circuit switched (CS) domain is called Iu-cs. This chapter is focusing on the Iu-ps interface though the QoS architecture is similar for both interfaces.

Table 3.2   QoS attributes in Rel'99

QoS attribute	Description
Traffic class	Type of application for which the radio access bearer service is optimised ('conversational', 'streaming', 'interactive', 'background').
Delivery order	Indicates whether the UMTS bearer shall provide in-sequence SDU delivery or not.
Maximum SDU size (octets)	Defines the maximum allowed SDU size.
SDU format information (bits)	List of possible exact sizes of SDUs. If unequal error protection shall be used by a RAB service, SDU format information defines the exact subflow format of the SDU payload.
Delivery of erroneous SDUs	Indicates whether SDUs with detected errors shall be delivered or not. In case of unequal error protection, the attribute is set per subflow.
Residual BER	Specifies the undetected bit error ratio for each subflow in the delivered SDUs. For equal error protection, only one value is needed. If no error detection is requested for a subflow, residual bit error ratio indicates the bit error ratio in that subflow of the delivered SDUs.
SDU error ratio	Defines the fraction of SDUs lost or detected as erroneous. SDU error ratio is defined only for conforming traffic. In case of unequal error protection, SDU error ratio is set per subflow and represents the error ratio in each subflow. SDU error ratio is only set for subflows for which error detection is requested.
Transfer delay (ms)	Indicates maximum delay for 95th percentile of the distribution of delay for all delivered SDUs during the lifetime of a bearer service, where delay for an SDU is defined as the time from a request to transfer an SDU at one SAP to its delivery at the other SAP.
Maximum bit rate (kbps)	Maximum number of bits delivered at a SAP time within a period of time divided by the duration of the period.
Guaranteed bit rate (kbps)	Guaranteed number of bits delivered at a SAP within a period of time (provided that there are data to deliver), divided by the duration of the period.
Traffic handling priority	Specifies the relative importance for handling of all SDUs belonging to the radio access bearer compared with the SDUs of other bearers.
Allocation/retention priority	Specifies the relative importance compared with other radio access bearers for allocation and retention of the radio access bearer. The allocation/retention priority attribute is a subscription parameter which is not negotiated from the mobile terminal.

The harmonisation of GERAN and UTRAN QoS management frameworks implies that the same QoS parameters are also employed in lower level protocols. The Rel'5 of GERAN will support both interfaces. The harmonisation of the QoS management for both radio technologies is very convenient from service support outlook, as services will be seamlessly provided to end users regardless of the radio access technology used. It will be as well convenient for the operator network management, as an integrated management tool can be used for both technologies.

Furthermore, Rel'5 introduces the IP bearer service [2], which may include the support of DiffServ [3] edge function and the resource reservation protocol (RSVP) function [4]. In the GGSN, the UMTS QoS parameters are mapped into IP QoS parameters. In the UE, the QoS requirements determined from the application are mapped to either the PDP context parameters or IP layer parameters (e.g. DiffServ CodePoint and RSVP flowspec.). As shown in Figure 3.1, the IP bearer service consists of the UMTS bearer service and the external bearer service. External bearer service is out of scope of 3GPP standards.

The UMTS bearer service provides the UMTS QoS and consists of two parts, the radio access bearer service and the CN bearer service. Both services reflect the optimised way to realise the UMTS bearer service over the respective cellular network topology taking into account such aspects as, e.g. mobility and mobile subscriber profiles.

The radio access bearer service provides confidential transport of signalling and user data between UE and CN Iu edge node (SGSN) with the QoS adequate to the negotiated UMTS bearer service or with the default QoS for signalling. This service is based on the characteristics of the radio interface and is maintained for a moving MS. Radio access bearer (RAB) services through the Iu interface provide higher granularity and flexibility than BSS PFC through the Gb interface, since a per-flow management is performed instead of aggregated flows management.

The CN bearer service of the UMTS CN connects the UMTS CN Iu edge node (SGSN) with the CN gateway (GGSN) to the external network. The role of this service is to efficiently control and utilise the backbone network in order to provide the contracted UMTS bearer service. The UMTS packet CN shall support different backbone bearer services for variety of QoS.

Figure 3.1 Evolution of QoS architecture in GERAN based mobile networks.

The radio access bearer service is realised by a radio bearer service and an Iu bearer service. The role of the radio bearer service is to cover all the aspects of the radio interface transport.

The Iu bearer service together with the physical bearer service provides the transport between the radio access network and the CN. Iu bearer services for packet traffic shall provide different bearer services for variety of QoS.

Services supported in different releases

The evolution in the GERAN QoS architecture has allowed the introduction of new services along the different releases, due to the improvement of the QoS mechanisms.

Rel'97 is able to support non-real-time services (such as WWW, email, file transfer or wireless application protocol (WAP)) as well as some basic streaming services with low bit rate requirements (e.g. Internet radio) when the system load is small. In this release, it is not possible to provide guarantees in terms of throughput. Packet switching allows the resources to be shared by many users. For WAP browsing, this means that relatively high user densities can be sustained without having to make significant increases in network capacity. Furthermore, the rate of the bearer is not an issue since WAP is optimised to transfer data in a very compact format.

New real-time services such as video on demand (VoD) or multimedia streaming services are supported in Rel'99, since it is possible to provide a guaranteed bit rate based on the QoS profile requested by the application. Generally, multimedia streaming by definition is seen to include one or several media streamed or transported to the client over the network, such as audio streaming (offering, e.g. music playback at the terminal), streaming with an audio and a video component (e.g. news reviews, music videos) or audio streaming with simultaneous visual presentation comprising of still images and/or graphical animations, video clips presented at pre-defined order (e.g. surfing through an interactive map). Rel'4 will support real-time services such as voice over IP (VoIP) more efficiently due to features such as delayed temporary block flow (TBF) release which reduces the amount of TBF establishments, decreasing the experienced delay and network assisted cell change (NACC), which reduces the cell re-selection outage time. Conversational service as over the Gb interface are not supported in Rel'4.

Finally, Rel'5 allows the provision of session initiation protocol (SIP) based services (e.g. high-quality videoconferencing, presence service) in a non-transparent way, by adding an IP bearer service as well as new network entities (i.e. CSCF-PCF) [2]. SIP is a signalling protocol developed to set up, modify, and tear down multimedia sessions over the Internet [5]. It was designed to dovetail with other Internet protocols such as TCP, user datagram protocol (UDP), DNS and others. Besides, SIP incorporates elements of two widely used Internet protocols: hypertext transfer protocol (HTTP) used for web browsing and simple mail transfer protocol (SMTP) used for email. From HTTP, SIP borrows a client-server design and the use of uniform resource locators (URLs). From SMTP, SIP borrows a text-encoding scheme and header style in contrast to other signalling protocols more similar to signalling system number 7 (SS7), such as H.323.

3.2 Functional split in UMTS QoS

This section presents the functional split in UMTS networks as specified in 3GPP [1]. Control plane functions relate to the bearer level service management (e.g. bearer activation, resource reservation, deactivation), whereas user plane functions relate to the packet level service management (e.g. packet classifying, dropping, scheduling).

QoS management functions for UMTS bearer service in the control plane

The QoS management functions in the control plane support the establishment and modification of a UMTS bearer service by signalling/negotiation with the UMTS external services and by the establishment of all UMTS internal services with the required characteristics. The different QoS management functions for controlling the UMTS bearer service are shown in Figure 3.2.

The *service manager* and the *translation function* are introduced for the support of the IP bearer service. All the other functions are required for UMTS bearer service. An analogous UMTS bearer service is applicable to GERAN R'99 and Rel'4 over the Gb interface with the introduction of the Packet Flow Context."

The *service manager* (IP BS Manager) coordinates the functions of the control

Figure 3.2   QoS management functions in the control plane in GERAN Rel'5.

plane for establishing, modifying and maintaining the service it is responsible for. It provides all user plane QoS management functions with the relevant attributes. The service manager offers services to other components, it signals with peer service managers and uses services provided by other instances. The service manager may perform an attribute translation to request lower layer services. Furthermore, it may interrogate other control functions to receive permission for service provision.

The translation function converts between the internal service primitives for UMTS bearer service control and the various protocols for service control of interfacing external networks. The translation includes the converting between UMTS bearer service attributes and QoS parameters of the external networks service control protocol (e.g. between IETF TSPEC and UMTS service attributes). The service manager may include a translation function to convert between its service attributes and the attributes of a lower layer service it is using.

The admission/capability control maintains information about all available resources of a network entity and about all resources allocated to UMTS bearer services. It determines for each UMTS bearer service request or modification whether the required resources can be provided by this entity and it reserves these resources if allocated to the UMTS bearer service. The function checks also the capability of the network entity to provide the requested service, i.e. whether the specific service is implemented and not blocked for administrative reasons. The resource control performed by the admission control supports also the service retention.

The subscription control checks the administrative rights of the UMTS bearer service user to use the requested service with the specified QoS attributes.

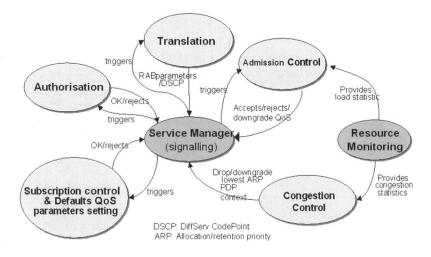

Figure 3.3 UMTS bearer service in the control plane.

Figure 3.3 (not provided by 3GPP standards) presents a slightly different view of the UMTS control plane bearer service components and their interaction. Note that the authorisation functionality is needed for SIP based services and shall be defined in 3GPP Rel'5 specifications.

Functions for UMTS bearer service in the user plane

User plane QoS management functions maintain the signalling and user data traffic within certain limits, defined by specific QoS attributes (Figure 3.4). UMTS bearer services with different QoS attribute values shall be supported by the QoS management functions. These functions ensure the provision of the QoS negotiated for a UMTS bearer service in Rel'5 using Iu interface. Analogous mechanisms exist in GERAN Rel'99 and Rel'4 using Gb network services. It shall be noted that the 3GPP user plane functional model is to some extent, similar to the IETF DiffServ management model.

The mapping function (mapper) provides each data unit with the specific marking required to receive the intended QoS at the transfer by a bearer service.

The classification function assigns data units to the established services of a MS according to the related QoS attributes if the MS has multiple UMTS bearer services established. The appropriate UMTS bearer service is derived from the data unit header or from traffic characteristics of the data.

The resource manager distributes the available resources between all services sharing the same resources. The resource manager distributes the resources according to the required QoS. Examples of resource management are scheduling, bandwidth management and power control for the radio bearer.

The traffic conditioner provides conformance between the negotiated QoS for a service and the data unit traffic. Traffic conditioning is performed by policing or

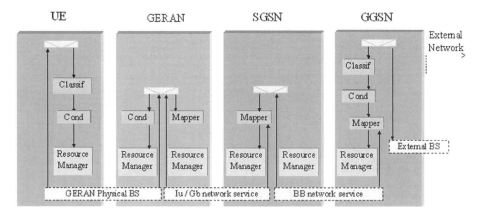

Figure 3.4   QoS management functions in the user plane in GERAN Rel'5 [1].

by traffic shaping. The policing function compares the data unit traffic with the related QoS attributes. Data units not matching the relevant attributes will be dropped or marked as not matching, for preferential dropping in case of congestion. The traffic shaper is used to bound or constrain the unpredictability of a certain traffic class.

UMTS QoS classes

The next important concept to be introduced in the QoS architecture is the traffic differentiation. Since Rel'99, 3GPP requires the traffic to be classified into four different classes. This section gives an introduction of the different traffic class characteristics.

When defining the UMTS QoS classes, the restrictions and limitations of the air interface have to be taken into account. It is not reasonable to define the same mechanisms used in fixed networks due to different error characteristics of the air interface. The QoS mechanisms provided in the cellular network have to be robust and capable of providing reasonable QoS resolution.

There are four different QoS classes (or traffic classes):

- conversational class
- streaming class
- interactive class
- background class

The main differentiating factor between these classes is how delay sensitive the traffic is: conversational class is meant for traffic which is very delay sensitive while background class is the most delay insensitive traffic class.

Conversational and streaming classes are mainly intended to be used to carry real-time traffic flows. Conversational real-time services, like video telephony, are the most delay sensitive applications and those data streams should be carried in conversational class.

Interactive and background classes are mainly meant to be used by traditional Internet applications like WWW, email, Telnet, FTP and News. Due to looser delay requirements, compared with conversational and streaming classes, both provide better error rate by means of retransmission schemes. The main difference between interactive and background classes is that interactive class is mainly used by interactive applications, e.g. interactive email or interactive web browsing, while background class is meant for background traffic, e.g. background email or file downloading. Responsiveness of the interactive applications is ensured by separating interactive and background applications. Traffic in the interactive class has higher priority in scheduling than background class traffic, so background applications use transmission resources only when interactive applications do not need them. This is very important in the wireless environment where the radio interface bandwidth is small compared with fixed networks.

Figure 3.5   Streaming service activation procedure.

Example of service activation procedure

An example of streaming service activation is described in this section. Figure 3.5 illustrates the different steps that form the streaming service activation within the UMTS network. An analogous procedure is applicable in Rel'99 and Rel'4 for a streaming service activation over the Gb interface, based on Packet Flow Control procedures.

Firstly, the user initiates the streaming client application, which requests a primary PDP context needed for real-time streaming protocol (RTSP) negotiation. The QoS requirements of the application signalling in the UE are mapped on QoS attributes. Since the primary PDP context is used for RTSP signalling, it requires high reliability. Thereby, a QoS profile with interactive traffic class, high priority and low error rate is appropriate for this kind of traffic. As the primary PDP context is not intended for real-time traffic, no resource reservations are needed in the CN.

After the SGSN has validated the service for that user by querying the home location register (HLR), local admission control is performed (e.g. based on the state of the buffers, the CPU load, etc.). Then, the SGSN maps the QoS attributes on to RAB QoS attributes and triggers a RAB assignment procedure in the RAN (note that the concept of RAB attributes are introduced for the Iu interface in GERAN Rel'5, though a similar procedure would be performed for the BSS PFC in the Gb interface for previous releases). In the RAN, admission control is mainly based on the availability of radio resources. Once accepted the new RAB request, RAB attributes are mapped on to proper radio bearer (RB) parameters used in the physical and link layers (e.g. retransmission requirements, etc.). A RB according to these parameters is established.

After RTSP session initiation signalling, the UE triggers two secondary PDP context activation procedures, one for unidirectional real-time protocol (RTP) traffic and the other one for bi-directional real-time control protocol (RTCP) traffic. The reason for the use of different secondary PDP contexts is that RTCP traffic must be separated from RTP if header compression is going to be applied for RTP/UDP/IP.

The UE converts user data application requirements for RTP and RTCP traffic into QoS profiles in order to activate the secondary PDP contexts. The SGSN validates the service request by checking if the service can be provided to this subscriber. In addition, the SGSN performs admission control based on the available bandwidth for the new PDP context. In case the flow is accepted, resource reservation is performed by decreasing the bandwidth quota by the guaranteed bit rate of the new PDP context. Allocation/retention priority parameter is given by the HLR to SGSN, which is used for admission precedence, i.e. to select which one is accepted among several users when streaming bandwidth is limited. The QoS profile is mapped on to RAB QoS attributes, which are sent to the GERAN. The RAB QoS attributes are exactly the same as UMTS QoS attributes but the values are not typically the same for the following parameters: residual BER, SDU error ratio and transfer delay, due to the packet lost and the delay inside the CN.

The GERAN performs an admission control and then does the mapping between RAB and RB parameters.

Both local admission control and resource reservation are performed in the GGSN in the same way as in the SGSN. The allocation/retention priority parameter is also used in a similar manner as for SGSN.

3.3 QoS control mechanisms in GERAN

This section focuses on possible implementation solutions of the different radio QoS mechanisms in GERAN. In addition, it is shown how to handle different services and users as well as the coordination between the different radio QoS functionalities in the different releases.

Effective QoS mechanisms are required in the radio access network in order to deliver the required QoS, providing service and user differentiation from the QoS attributes requested by the application and the subscription profile stored in the HLR, respectively. This is needed to be able to efficiently support a wide variety of services such as FTP, email, gaming or multimedia streaming.

These QoS mechanisms can be divided into control and user plane functions. During the RAB or PFC establishment procedure some control plane functions take care of the admission control of new connections, the selection of appropriate protocol transmission modes and error protection algorithms as well as the allocation of radio resources. Furthermore, the control plane preserves the negotiated QoS while the service is ongoing. The radio access network should be able to react against overloaded situations, as well as identifying potential downgrade in the QoS provisioning.

The goal in a QoS-aware environment is to allow predictable service delivery to certain types of traffic regardless of what other traffic is flowing through the network at any given time. It aims to create a multiservice radio access network where traditional bursty traffic may share the same resources with traffic with more rigorous latency, jitter, bit rate and/or packet loss requirements. The functions that take care of the QoS provisioning once the bearer is established belong to the user plane. An appropriate traffic conditioning and scheduling allows the differentiation between services though additional mechanisms are required in order to dynamically adapt the link against time variation radio conditions.

Figure 3.6 shows the major radio related functional blocks involved in the QoS provisioning. Radio configuration and performance indicator operating parameters are monitored and provisioned through the management interfaces. Monitored parameters include statistics regarding traffic carried at various QoS levels. These statistics may be important for accounting purposes and/or for tracking compliance to service level specifications (SLSs) negotiated with customers. A SLS is presumed to have been negotiated between the end user and the provider which specifies the handling of the end user's traffic, i.e. a set of parameters which together define the QoS offered to a traffic stream by the provider's network.

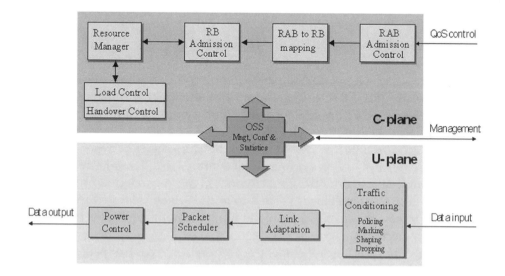

Figure 3.6   Radio QoS functional blocks coordination.

All the functionalities described in the following sections are supported in Rel'99, Rel'4 and Rel'5, while Rel'97 is out of the scope of the following sections. Note that RAB parameters are introduced in GERAN Rel'5 for the Iu interface, though the scope of the functionalities is also applicable to Rel'99 and Rel'4 by considering the PFC attributes instead of RAB attributes.

Control plane QoS mechanisms

Control plane QoS mechanisms are responsible for two main tasks: connection admission control and establishment, and QoS preserving mechanisms.

The first one is related to the RAB/PFC and RB establishment procedure once the SGSN has validated the service requested by the user equipment and has translated the UMTS QoS profile into RAB parameters.

The second task takes place once the bearers are established, additional mechanisms should try to preserve the negotiated QoS. For this purpose, load control function is responsible for controlling overload situations.

Admission control
The purpose of the admission control function is to accept or not new users in the radio access network. In this manner, the admission control tries to avoid overload situations basing its decisions on interference and resource availability measurements (Figure 3.7). It is performed at initial UE access (i.e. RAB assignment), handover and RAB reconfiguration (Rel'5). These cases may give different answers depending on user priority and load situation.

Admission control can take into account radio and transport resources, radio link control (RLC) buffers size and base station physical resources. When performing admission decision, this function requests resources from other entities and based on this information, different decisions can be taken, such as accept/reject the connection, queue the request or perform a directed retry, see section 6.3. The resource control performed by the admission control supports also the service retention through the *allocation/retention priority* attribute. This feature is service independent, providing users differen-

Figure 3.7   Admission control in GERAN.

tiation depending on their subscription profiles. High priority users are able to pre-empt lower priority users when admission control is performed and there is no available capacity. Furthermore, once they are admitted, this feature allows high priority users to retain their connections against lower priority users in case of overload.

In *circuit switched* the admission control decision is normally based on the availability of time slots in the target cell. However it is also possible that new calls are blocked even though there are time slots available to support it (this is called soft blocking). Soft blocking can be based on interference levels or quality measurements in surrounding cells. In case of Dynamic Frequency and Channel Allocation (see Chapter 8), the admission control based on interference levels can be done quite accurately.

In *packet switched* domain, admission control manages differently non-real-time and real-time connection requests. In case of non-real-time services such as interactive and background traffic classes, the experienced throughput decreases gradually when the number of users increase, up to the blocking situation. By limiting the non-real-time load it is possible to provide a better service to already admitted users.

Admission control evaluates whether there is enough capacity to provide the requested guaranteed bit rate for services that require guaranteed bit rate over shared channels (e.g. streaming traffic class). Since radio link conditions are unknown at connection set up phase and it can change when the user is moving, a statistical based admission control is required. The admission control can be performed with different levels of risk. This is done in terms of different blocking and user satisfaction probability, depending on the user subscription profile.

RAB/PFC to RB mapping

Different services must be conveyed over appropriate radio bearers that enable QoS requirements to be satisfied. To achieve it, this function takes care of selecting appropriate conditions for the transmission over a chosen logical channel (e.g. traffic channel (TCH), packet data traffic channel (PDTCH)) such as transmission modes (acknowledge, non-acknowledge), modulation and coding schemes (MCSs), error protection type (EEP, UEP), etc. For instance, the use of retransmissions causes additional delay, but at the same time provide much better reliability, which is needed for many data applications. The main input to select these conditions is the *traffic class* attribute though other attributes from the QoS profile such as bit rate, reliability and delay requirements should be taken into account in this procedure. Several RB parameter combinations are possible to configure the corresponding bearer for different traffic classes. A description of the different protocol of the GERAN architecture can be found in Chapter 2.

Conversational services are the most delay sensitive applications, hence retransmissions are not normally performed. Conversational packet switched calls are not possible over the Gb interface. Packet switched calls over the Iu-ps interface are likely to be fulfilled with non-transparent PDCP mode, which performs IP/UDP/RTP header adaptation. There is not an optimised voice bearer in physical layer (with optimised error protection schemes) for voice over IP (VoIP) in GERAN R'5, only generic conversation bearer is supported.

Streaming applications can be supported with acknowledged or non-acknowledged mode. High reliability requirements can impose the use of retransmissions though this implies an increase of the delay/jitter. Therefore, a possible solution to minimise the transmission delay can be to limit the number of retransmissions by selecting more robust MCS. By minimising the number of retransmissions and with proper scheduler algorithm design it is possible to fulfil delay requirements. However, limiting the number of retransmissions by selecting a more protective MCS has an impact on the throughput performance. It should be noted that incremental redundancy (IR) typically requires high retransmission rates and therefore limiting the number of retransmissions will impact on IR performance. In non-acknowledge mode, the link adaptation algorithm should select the highest MCS that matches the reliability QoS requirements (depending on the radio link conditions) and maximises the throughput. Though the use of shared MAC mode achieves a better resource utilisation, dedicated MAC may be used for certain services and/or users in order to provide a better quality in terms of delay, throughput and SDU error ratio.

Interactive and *background* services require the use of the acknowledged RLC mode in order to provide high reliable transport capability at the expense of introducing a variable throughput and delay. Retransmissions over the air provide much better reliability, needed for this kind of services. Since acknowledge mode is used, the link adaptation algorithm must select dynamically the MCS that achieves the maximum throughput per timeslot. These services can use dedicated or shared MAC mode though it is recommended to select shared MAC mode in order to minimise resources utilisation.

Figures 3.8 and 3.9 optimise the different network protocol transmission modes for each traffic class in GERAN Rel'99/Rel'4 (Gb interface) and Rel'5 (Iu-ps interface).

Resource manager
The optimisation of the radio resource management is a key issue in GERAN. An efficient resource management over the radio domain is one of the most important and effective QoS mechanisms. The resource manager has the main task of determining the needed resources to fulfil the QoS requirements of a new connection and verifying whether there are enough available resources in the

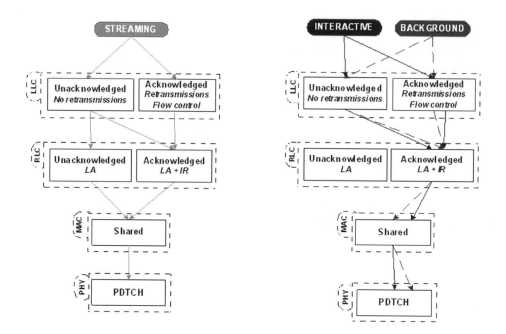

Figure 3.8   RB parameters mapping in GERAN Rel'99/Rel'4 (Gb interface).

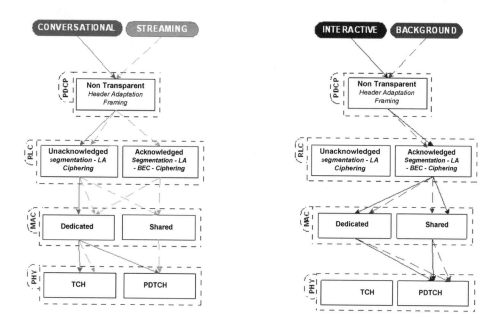

Figure 3.9   RB parameters mapping in GERAN Rel'5 (Iu-ps interface).

cell to support those requirements. Resource management is different for real-time and non-real-time services. Non-real-time services do not require any guaranteed bit rate so they can be allocated in any available channel. The main criteria used in channel allocation for these services is to allocate channels that give the maximum capacity for the TBF within the limits defined by the MS capability. However, real-time service allocation over dedicated channels consists of determining the availability of resources and reserving them based on guaranteed throughput requirements and mobile MS capability. In case of allocation over shared channels, additional information about the quality of the radio channels (throughput per timeslot) is needed in order to allocate necessary resources to guarantee QoS requirements.

The different options to allocate resources for real-time services are: to dedicate the channel for the real-time connection, to share the channel between a real-time connection and one or several non-real-time connections or to share the channel between more than one real-time connection, independently of the type of service. A better analysis of the different options to allocate resources for real-time services can be found at Chapter 8.

Load and handover control
Handovers can be triggered by different reasons. They are performed when the MS leaves the dominance area of the actual serving cell. In other cases, they are performed to when the current call is experiencing bad quality. A third kind of handover is referred to as the traffic handover when the current cell is congested whereas neighbour cells are not.

The aim of the load control is to control the load in overload situations, but it also has the ability to control it even before the overload situation happens, i.e. preventive load control. Whenever the network detects an overload situation in a cell, load control can trigger load reason handovers to adjacent cells and select which connections to move. Different types of services and user subscription profiles may be taken into account in this selection so that services and users with the lowest priorities are the most susceptible to be moved.

User plane QoS mechanisms

User plane QoS mechanisms take care of the QoS provisioning once the bearer is established. One of the tasks of the user plane in the radio domain is to provide service differentiation by means of a proper traffic conditioning and scheduling of the radio blocks similar to those performed in the IP transport domain following the DiffServ framework proposed in the IETF.

Furthermore, other tasks of the user plane are in charge of dynamically adapting the link to compensate fluctuations of the radio conditions as well as controlling output power.

Link adaptation

Wireless links are affected by radio link conditions. This implies that the performance in terms of throughput, delay and block error ratio is continuously varying. These variations in the QoS provisioning must be compensated as much as possible by selecting proper modulation and coding schemes. This mechanism is known as link adaptation (see Chapter 7 for more information).

More specifically, the link adaptation function is able to select dynamically the channel coding, channel mode (FR/HR), interleaving depth and modulation depending on the service requirements, system load, channel state and interference. The link adaptation algorithm may need cooperation with resource allocation to meet the negotiated radio bearer for real-time connections.

The throughput depends on the link quality in terms of carrier/interference (C/I) values. Thus, the more protective modulation and coding scheme, the less throughput for a given C/I condition. The performance of the ideal link adaptation corresponds to the envelope of all MCS curves, since its aim is to maximise the throughput.

The link adaptation algorithm in non-acknowledged mode should select the highest MCS that matches the reliability QoS requirements, depending on the radio link conditions. Since retransmissions are not performed in this mode, reliability requirements must be achieved using a proper coding scheme. However, in acknowledged mode the target is normally to achieve the maximum throughput/timeslot, that is, depending on the radio link conditions, the link adaptation algorithm should select the MCS that maximise the throughput/timeslot. Reliability QoS requirements in acknowledged mode can be achieved by means of retransmissions. Delay requirements impose limits to the throughput maximisation criteria.

Traffic conditioning

Any particular packet will experience different effects when it is sent via the radio interface. Traffic conditioner provides conformance between the negotiated QoS for a service and the data unit traffic. This function is performed by different mechanisms:

- policing function is the process of comparing the arrival of data packets against a QoS profile, and based on that, forwarding, marking, shaping or dropping them so as to make the output stream conformant to the profile;
- marker is responsible for tagging the packets exceeding the guaranteed bit rate in order to treat them accordingly;
- shaper is the procedure of delaying (queuing) packets within a traffic stream to cause it to conform to some defined QoS profile;
- dropper selectively discards packets that arrive at its input, based on a discarding algorithm.

Depending of the type of service, packet conditioner should work in a different way. For instance, since non-real-time services have no stringent delay requirements, packets exceeding the maximum bit rate can be shaped until they can be forwarded. However, real-time services are delay sensitive, hence these packets must be treated differently: packets between the guaranteed and the maximum bit rate can be marked so that the packet scheduler handles them accordingly as non-guaranteed traffic while packets above the maximum bit rate should be shaped.

Packet scheduler

Packet scheduler (see Chapter 7) is the core function in the QoS provisioning. Since the radio domain is most probably the bottleneck of the whole network, an efficient treatment of the radio blocks based on priorities is a key issue. This function takes care of scheduling radio resources for different RABs for both uplink and downlink directions. Based on the radio interface interference information, capacity requests from the MS (uplink) and received data in the buffers (downlink), packet scheduler will allocate radio resources for the RAB. This algorithm should take into account the QoS profile in order to prioritise between services and user subscription profiles (through the allocation/retention priority attribute).

Conversational services should be handled with dedicated radio resources due to their stringent delay requirements, although background traffic from the same user could be multiplexed on to the same radio channel. Therefore, an absolute priority packet scheduling policy could allow background traffic to be sent only during silent periods of conversational speech traffic.

Streaming services such as Video on Demand (VoD) requires QoS guarantees in terms of throughput. Some streaming traffic has a bursty nature and, because of this, shared channels are more adequate for this kind of traffic than dedicated channels. Hence, packet scheduler for shared channels must ensure certain bandwidth for streaming connections, providing a higher priority to the streaming traffic up to the guaranteed bit rate against other non-real-time connections. These priorities may depend on several parameters such as traffic class, QoS requirements, or some internal operator parameters. This separation between packets up to and above the guaranteed bit rate may be managed by a marking function, which differentiates precisely between both types of packets. Another option is the use of dedicated channels that have better link performance.

An example of packet scheduler for shared channels is depicted is the Figure 3.10. In this scheme, packet scheduling is ensuring certain bandwidth for streaming connections using a Deficit Round Robin (DRR) approach, which is responsible for assigning dynamically the allocated bandwidth for each flow. The deficit value is used for tracking the context lead/lack compared to the

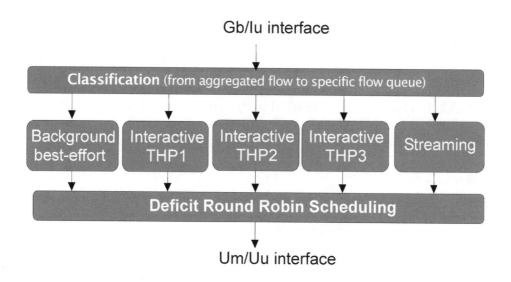

Figure 3.10   Example of packet scheduler's scheme.

guaranteed bitrate. As long as the deficit value is above zero, the scheduler shares the capacity to that context. For streaming services, a different priority is allocated when the deficit is above/below zero, corresponding to guaranteed/non-guaranteed bitrate, in a way that the streaming traffic up to the guaranteed bitrate is dealt with the highest priority, whereas streaming traffic exceeding the guaranteed bitrate up to maximum bitrate is managed as non-real time traffic.

The operator could control the packet scheduling not only based on specific parameters of the scheduling algorithm but also by means of the allocation/retention priority attribute (e.g. gold, silver and bronze). In that way, both service and user differentiation are performed.

Power control
The objective of the power control function is the reduction of interference levels while QoS of the connections is maintained, what causes an increase of capacity and an improvement of the spectral efficiency. Additionally, power control in uplink is used to reduce the power consumption of the MS.

Power control tries to keep the received level and the received quality in a range of good values (defined by an upper threshold and lower threshold) by changing the transmitter power 'on the other side' of the radio connection.

Power control could be used in order to distinguish between different type of users (e.g. gold, silver and bronze) and between different kind of services. Hence, with the use of different maximum base transceiver station (BTS) power values, it should be possible to prioritise between services, e.g. the use of the P0 parameter to decrease interference level to speech calls.

3.4 Quality of service policy management

Telecommunication mobile networks are suffering a big expansion over the world in terms of the growing diversity of devices and services. From a best-effort service model, mobile networks are intended to support different service levels for specific QoS requirements. These new technology requirements have caused a number of management-related problems in such areas as QoS. The main problem is the complexity for operators to manage their networks in a simple way, offering the possibility to regulate access to network resources based on categories such as user profiles or applications.

Policy management is a possible and optional way to solve the problem of managing complex networks. Basically, it offers a centralised QoS control point, which administrates the network in order to achieve a consistent service behaviour for the need of a specific QoS. That means that all QoS related functions distributed along the network could be managed from a central point of administration. Policy management is based on rules containing conditions and actions. For example, an operator could configure the network in order to provide the highest priority to those packets belonging to the 'gold' user profile.

In order to apply a proper policy management to the radio domain, different radio QoS functions must be modelled abstracting the implementation details from the parameters of interest for configuration and management. Once the manageable parameters are identified, the operator is able to build policy rules in the form 'if condition then action'. The semantics of a policy rule are such that if the set of conditions evaluates to TRUE then the set of actions is executed. The scope of these rules can vary: global policies have a global effect on the end result of applying the given policy, that is, the end result of such policy relies on it being executed globally across the entire path of network service; domain policies are meaningful within one domain of QoS mechanisms, e.g. radio domain; local policies are only meaningful within one network element. From the rule management viewpoint, it is very essential that the global and domain policies are administrated by a centre controlling point. Local policies can reside within the control of local element, so long as such local policies have no effect on the QoS service globally.

Summarising, rule based management is an optional tool for the network administrator to offer a variety of QoS differentiation to certain service/customer profiles.

References

[1] 3rd Generation Partnership Project; Technical Specification Group Services and Systems Aspects, 'QoS Concept and Architecture', TS 23.107 v3.6.0, 04-2001.

[2] 3rd Generation Partnership Project; Technical Specification Group Services and Systems Aspects, 'End-to-End QoS Concept and Architecture', TS 23.207 v5.1.0, 10-2001.

[3] Kilkki K., 'Differentiated Services for the Internet', Macmillan Technical Publishing, London, 1999.

[4] Braden, Zhang L., Berson S., Herzog S., Jamin S., 'Resource ReSerVation Protocol (RSVP) Version 1 Functional Specification', IETF RFC 2205, September 1997.

[5] Johnston Alan B., 'SIP. Understanding the Session Initiation Protocol', Artech House, Norwood, MA, 2000.

4

Mobile Station Location

Mikko Weckström, Maurizio Spirito and Ville Ruutu

The need to determine the mobile terminal position has been continuously growing during the last few years. Requirements for emergency services such as the North American E911 are one of the key drivers for the development of accurate position technologies such as the enhanced observed time difference (E-OTD). In Europe, the key drivers are the commercial services, even though, the European emergency services (E112) will utilise the same mobile location technologies.

Location system (LCS) functionality is going to enable a new set of services for cellular users in the coming years. With general packet radio system (GPRS), it is economically feasible for users to be permanently connected to Internet based services. With LCS, the available services can have new location related information providing new revenue generation opportunities for wireless operators.

For the network infrastructure, this means as well new challenges. In order to be able to locate a number of mobile terminals continuously, some new components and signalling protocols are required. Since the network has not been designed for accurate positioning, there has to be some new ways to combine the radio measurement data or carry out some new measurements. In the future, GSM EDGE radio access network (GERAN) and UMTS terrestrial radio access network (UTRAN) systems will offer the LCS services with a variety of technologies such as E-OTD, assisted GPS positioning system (A-GPS) and IPDL-OTDOA. The full capability will be available by the end of 2004. Current GSM networks will offer the basic capabilities mostly via cell identity (CI) based methods. During 2002 there will be some E-OTD based services other than the emergency positioning service in the US which is the main driver for E-OTD due to the Federal Communications Commission (FCC) regulations.

In the following sections the different technologies used for the mobile positioning will be presented, performing some analysis on the achieved accuracy of the different methods. More detailed information can be found from some of the sources listed in the References section [1,2].

4.1 Applications

There are a large number of applications that the location functionality enables, and the methods described in this section are just a subset of the potentially available ones. Applications are as well divided into three different service levels. These levels are as described in the Location Interoperability Forum (LIF): basic service level, enhanced service level and extended service level. Figure 4.1 illustrates these levels.

The service levels are related mainly to accuracy. The method for the basic service level is usually Cell Identity (CI) based but some other methods can be used as well. The other service levels use more accurate methods. The enhanced service level uses the E-OTD and the extended service level the A-GPS. But more important is that these service levels enable the support of different applications. Some of these applications can be found in Figure 4.2.

As accuracy increases, more applications are available, but with a higher cost of the LCS solution. These applications also have evolution paths from the network sub subsystem (NSS) based architecture towards base station subsystem (BSS) and mobile station (MS) based solutions, that will ensure larger location capacity in the network. All of these applications will connect to the system via standard interfaces, such as LIF application programming interface (API). The API allows different application providers to easily connect to the system and offer their applications.

Service level	Accuracy/Architecture	Application
Basic service level	400 - 1500 m / NSS	Friend finder, location sensitive billing
Enhanced service level	40 - 150 m/ NSS/BSS	Emergency, Tracking Finding
Extended service level	< 50 m/ BSS/MS	Accurate navigation, emergency, safety

Figure 4.1 LCS Service levels.

Figure 4.2   Segmentation examples of mobile location services.

4.2 Location architectures

This section will introduce the relevant LCS architectures. The first method implemented is usually the CI based method enabling the basic service level. In this case the architecture is usually NSS based and hence offers moderate capacity with low cost. The NSS architecture relies on the standard interfaces that enable the information to be collected which is required to calculate the position estimate. The NSS architecture can use information such as CI or timing advance (TA) and in some case measurements of the signal strength received by the handset (RXLEVs) as well. Basically, the more information the better the accuracy is. The basic description of the NSS architecture can be found in Figure 4.3 where the serving mobile location centre (SMLC) and the gateway mobile location centre (GMLC) are connected to the mobile services switching centre (MSC).

The location interfaces are the Ls for the SMLC and the Lg for the GMLC. These interfaces are based on the standards and hence should be the same for all cellular network suppliers. This should help in the case of multi-vendor networks where the cellular network consists of many different suppliers' equipment.

In Figure 4.3, the first element that is specific for positioning is the location measurement unit (LMU). The LMU is measuring the timings of the base transceiver stations (BTSs) and reporting these values to the SMLC to make it possible to calculate the position estimate together with the timing reports sent by the mobile. The SMLC is then sending the (x, y) coordinates to the GMLC that is sending the information to and from the application. The Ls and Lb interfaces are between SMLC and base station controller (BSC) or MSC, and the Lg inter-

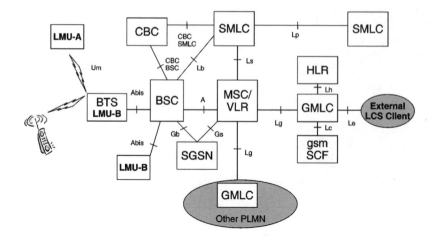

Figure 4.3 Generic LCS architecture.

face is between the GMLC and MSC. These are the LCS specific signalling system number 7 (SS7) interfaces. The Lb and Ls have in essence similar stack structure, but the information content is a bit different. In the case of the Lb, the NW database content, as the SMLC is serving only one BSC, is smaller and therefore it enables greater location calculation capacity in the SMLC. The next step in architecture evolution is the BSS based architecture that enables the system to use the radio interface information more efficiently and therefore does not have to use the signalling necessarily for the measurement information transport towards to the MSC. The BSS architecture utilises more efficiently the measurement data that are specific for enhanced methods like E-OTD. Figure 4.4 displays the basic BSS based architecture.

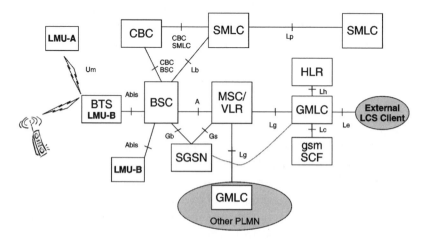

Figure 4.4 BSS LCS architecture.

As it can be seen the system utilises mostly standard interfaces. The interfaces to support the GPRS are however proprietary (dotted line from serving GPRS support node (SGSN) to GMLC). The interface between the GMLC and SGSN has not yet been standardised and work is ongoing in 3GPP.

The next architecture is MS based. It uses the assistance data that are generated in the cellular network as well as the MS capability of doing LCS measurements. In this architecture the network is sending the pre-calculated assistance information to the mobile. The SMLC calculates the assistance data that are sent to the mobile via standard messages. The calculation of the location is done in the mobile terminal. Mobile assisted E-OTD and A-GPS both belong to this category.

Figure 4.5 illustrates the basic signalling protocols for the NSS and BSS based method.

4.3 Location methods

There are many location techniques devised for application in mobile communications networks. Most of them estimate the MS coordinates by processing with multilateration techniques absolute distances and/or relative distances between the MS and multiple base stations. A general analysis and comparison of such multilateration techniques can be found in [3,4]. In this section the structure of LIF is followed going through first the basic service level and then enhanced service level. Finally the A-GPS method for extended service level will be presented. There are MS based and MS assisted methods, so when describing the appropriate method the assistance order is also described. Figure 4.6 illustrates the position accuracy related to these different methods.

Figure 4.5   Basic signalling structures of NSS and BSS architectures.

Figure 4.6   Accuracy of LCS methods.

Basic service level

Timely deployment of LCS for all MSs, including the legacy ones, requires that mobility management information, already available in cellular networks, should be used for implementing location technologies capable of offering basic service level accuracy. Such technologies are of crucial importance from a commercial point of view. In fact, they allow operators and service providers to start offering location based services to all customers with minimal additional costs, while waiting for location technologies capable of supporting enhanced and extended service levels to be available commercially. From a technical point of view, even when more accurate technologies will be fully available, basic service level location methods will still be needed; for instance when they allow to achieve the requested location accuracy or when more accurate techniques are not supported in the whole network. Beside, basic service level location technologies may serve as *back-up methods*, since the enhanced and extended service level methods can fail if, for instance, an insufficient number of base stations/satellites is available for measurements.

Measurements available for the implementation of basic service level location methods in a GSM network are CI, TA, and strength of signals received by the handset (RXLEVs). Along with the estimated coordinates, all methods provide a definition of a *confidence region*, which defines a geographical region where the actual location of the target terminal can be found with a given probability (confidence coefficient).

Cell identity (CI) location method
In cellular networks, base stations serve only a limited geographical area, thus the identity of the base station serving a MS provides the simplest location information. In GSM the serving base station identity is available with the parameter known as CI.

One way to determine a CI based location estimate is to assume that the mobile is at the serving cell's antenna coordinates. That is the most likely estimate when the serving cell is omni-directional. For a sector cell a more accurate CI based location can be estimated at the centre point of the serving sector. To allow this calculation, however, orientation and basic radiation properties of the serving base station antenna, and cell size are needed in addition to the serving cell's antenna coordinates. The basic principle of the CI location method is represented in Figure 4.7.

Cell identity and timing advance (CI + TA) location method
In addition to CI, also TA for the serving base station is available in GSM. The TA represents the round trip delay between the mobile and the serving base transceiver station (BTS) expressed as an integer multiple of the GSM bit period. The TA has been introduced in the GSM system to avoid overlapping of bursts transmitted by different users in the same cell at the serving base station site [5]. According to the GSM specifications the TA is an integer number between 0 and 63. A simple model for a TA measurement can be expressed as TA = round$\{d_{BTS}/(cT_b/2)\}$, where T_b = 3.69 μs is the GSM bit period, c the speed of radio waves, and d_{BTS} the distance between handset and serving base station estimated by the system. In the CI + TA location method an estimate of the distance between handset and serving base station, d_{TA}, is obtained from a TA made available by the network as follows:

$$d_{TA} = \begin{cases} \dfrac{1}{4}(cT_b/2) & \text{if TA} = 0 \\ \text{TA}(cT_b/2) & \text{if TA} > 0 \end{cases}.$$

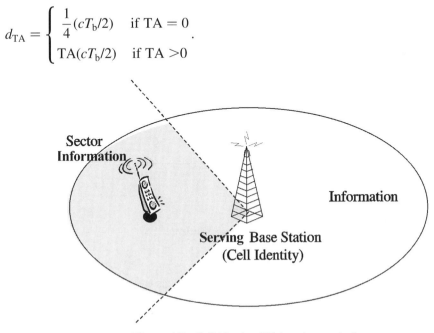

Figure 4.7   Cell identity (CI) location method.

The TA information as such cannot be used to improve the location accuracy when the serving cell is omni-directional; in fact in such circumstances and in the absence of any additional information, the most likely handset location is at the serving cell's antenna coordinates. On the other hand, if the cell is a sector cell the handset's location can be estimated on a circular arc having radius equal to the distance estimated from TA (d_{TA}), centre at the serving cell's antenna coordinates. The serving cell's antenna azimuth and radiation pattern determine the orientation and angular width of the arc, respectively. The radial thickness of the arc is determined by the TA measurement error margin, which is comparable to the TA resolution of half the bit period ($cT_b/2 \approx 550$ m). The performance of the CI + TA location method can be eventually improved by additionally taking into account the reliability of d_{TA} as an estimate of the distance between terminal and serving base station. The CI + TA location principle is represented in Figure 4.8.

Cell identity, timing advance and RXLEVs (CI + TA + RXLEV) location method

The CI + TA + RXLEV method estimates the handset's coordinates by combining CI, TA (used to estimate the distance between handset and serving base station) and RXLEV information. RXLEVs are measurements of the strength (i.e. power) of signals received by the mobile terminal from the serving cell and from up to six strongest neighbour cells, introduced in the GSM system to support mobility management procedures.

The level of a signal received by a mobile terminal, or more precisely the attenuation that signal has experienced, depends also on the reciprocal position of the terminal and the base station from which the signal was trans-

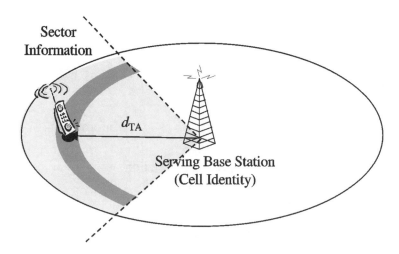

Figure 4.8   Cell identity and timing advance (CI + TA) location method.

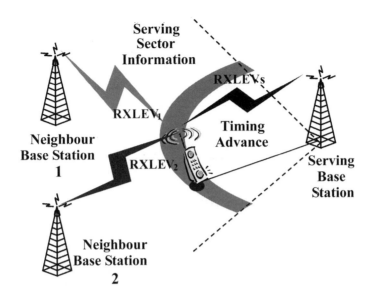

Figure 4.9 Cell identity, timing advance and RXLEVs (CI + TA + RXLEV) location method.

mitted. By using suitable propagation models, it is thus possible for instance to estimate the distance between handset and each transmitting base station. Each estimated distance defines a circle on which the handset may be located. When at least three distinct circles are available, the handset location can be estimated at their intersection. The CI + TA + RXLEV location principle is represented in Figure 4.9.

Enhanced service level

Enhanced service level location methods offer better accuracy compared with those belonging to the basic service level category. However, the cost of better performance is a more complex system requiring additional equipment.

Enhanced observed time difference
Like other enhanced service level location methods described in this chapter, E-OTD relies on timing measurements of cellular radio signals in the GSM network. In E-OTD the MS measures the time difference between the receptions of bursts transmitted from the reference base station and a neighbour base station, i.e. the observed time difference OTD $=t_{rxN} - t_{rxR}$, when the MS receives signals at the moments t_{rxR} and t_{rxN} from the reference base station and a neighbour base station, respectively (Figure 4.10). Normally the reference base station is the serving base station of the MS, since it is expected to have the best radio link. The OTD measurement is repeated between the reference base

Figure 4.10 Basic E-OTD measurement by a mobile station.

station and other neighbour base stations in order to collect a set of OTD measurements based on which the location can be estimated.

The reason why this method is called enhanced OTD is that GSM specifications have included the observed time difference measurement feature already before any location support. However, this is intended for pseudosynchronous handovers, and has a low measurement resolution that is inadequate for location purposes [6].

The basic equation for E-OTD location is OTD = RTD + GTD. OTD is composed of two parts. The first one is called the real time difference (RTD). It is the synchronisation difference between the reference and neighbour base stations, i.e. the relative difference in transmission times of their bursts. If the reference base station transmits its burst at the moment t_{txR}, and the neighbour base station at the moment t_{txN}, then RTD $= t_{txN} - t_{txR}$. If the base stations are synchronised and transmit at the same time, RTD is zero. The second part of OTD is the geometric time difference (GTD). It describes the contribution of different propagation times (different distances) between the MS and two base stations to OTD.

GTD is the important variable, since it includes actual information about location: GTD $= [d_{mN} - d_{mR}]/c$ (Figure 4.11). Here d_{mN} is the distance between the MS and the neighbour base station, d_{mR} is the distance between the MS and the reference base station, and c is the speed of radio waves.

With the E-OTD method the MS measures OTD, and assuming that RTD is known, it is possible to determine GTD. Geometrically GTD defines a set of points in the plane that have the property that the distance difference from such a point to the neighbour and reference base stations, respectively, is constant and equal to GTD $\times c$. But this is nothing but a definition of a hyperbola having its foci at the positions of the two base stations. Thus a GTD value limits the possible location for the MS observing a constant OTD value between the two base stations to be on a hyperbola.

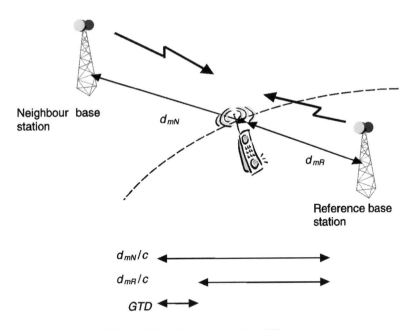

Figure 4.11 Geometrical time difference.

When at least two hyperbolas are obtained, the location estimate can be found at their intersection. In some cases two hyperbolas can have two intersections. Then a unique solution requires one additional hyperbola, or other additional information (e.g. coverage area of the reference cell) is needed to select one of the intersections. This is demonstrated in the Figure 4.12. If only hyperbolas 1 and 2 are available, there are two possible intersections, A and B. Adding hyperbola 3 to the picture it becomes evident that the intersection B is the correct answer. If hyperbola 3 is missing, then information about the coverage area of the reference base station (grey area) can be used to select intersection B.

The MS reports together with OTD measurements also the quality figures related to each measurement using the standardised radio resource LCS protocol (RRLP). These quality figures are the number of single OTD measurements for a certain base station pair, and the standard deviation of them. These can be used in location calculation to estimate the reliability of each hyperbola and to form a confidence area. Typically the confidence area for E-OTD is an ellipse that describes an area inside with a certain location probability (e.g. 90%).

E-OTD requires that RTD values are known. One possibility is that the whole network is synchronised so that RTD values are always zero. A more practical approach is to measure RTDs. According to GSM standards RTD values can be measured using LMUs. RTD values can be calculated from the equation: RTD = OTD − GTD, where OTD is measured by the LMU, and GTD can be calculated from the known coordinates of the LMU and base stations.

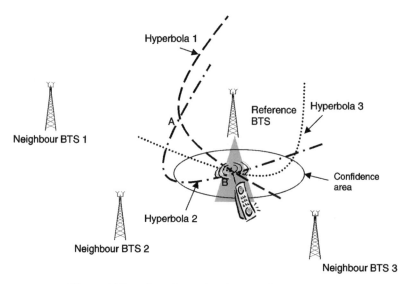

Figure 4.12.   Location estimation based on hyperbolas.

According to GSM specifications, a LMU can also attach to the measurement report the AT timestamp for the last measured burst from the reference base station. When this is done, the RTD values can be called absolute time difference (ATD) values. AT timestamp is given using the global positioning system (GPS) as a reference clock, i.e. a LMU needs to have a GPS clock available. At the moment only GPS is mentioned in the standards as a candidate for the reference clock. AT timestamp allows comparison of measurements from different LMUs directly if all LMUs have the GPS clock as a common reference clock. This allows, depending on implementation, easier calculation procedure and, e.g. for A-GPS location method delivery of GPS time to the MS.

The above-mentioned approach of defining hyperbolas for location estimation from OTD measurements is called hyperbolic E-OTD. Another possibility mentioned in GSM specifications is so-called circular E-OTD. The basic equation for each base station is $d_{bM} - d_{bL} = c^* (t_{rxM} - t_{rxL} + t_{offset})$. Here d_{bM} and d_{bL} are the distances from a base station to the MS and the LMU, respectively. t_{rxM} is the reception time of a signal from the base station at the MS, and t_{rxL} is the reception time of a signal from the base station at the LMU. t_{rxM} and t_{rxM} are measured against the internal clocks of the MS and the LMU, and there normally exists a time-offset t_{offset} between these clocks. As earlier, c denotes the speed of radio waves. The MS and the LMU are needed measure at least three base stations so that it is possible to solve the MS location and the unknown clock offset t_{offset}. This approach is called circular, since after solving a group of at least three equations, the distances between the MS and the base stations is known allowing circles around the base stations to be defined. Then the location estimate of the mobile can be found at the intersection of the circles. In practice circular E-OTD uses the same OTD measurement reported by the MSs and LMUs as the hyperbolic E-OTD, and it is possible to obtain from the group of circular equations the hyperbolic equations.

There are two variants of the E-OTD method defined in GSM specifications: MS assisted E-OTD and MS based E-OTD. In the case of MS assisted E-OTD the serving mobile location centre (SMLC) sends a request for OTD measurements to the MS. The request can include a list of base stations that the mobile should measure. The MS can also be told so-called expected OTD values, which are estimates of OTD values relative to the reference base station. The intention is to help the MS to perform measurements by giving the MS some idea when signals from neighbour base stations should arrive. Once the MS has performed OTD measurements it reports them to the SMLC, that calculates the location estimate.

In the MS based E-OTD the mobile performs the location estimation itself using base station coordinates and RTD values received from the network. The MS can get this information from the E-OTD request, by specifically asking for the information, or from broadcast sources.

E-OTD requires the MS to support the needed measurement functionality. In practice this can be done in modern GSM MSs with only software changes. This is why E-OTD is seen as a good candidate for mass location purposes since it is easy to include in every MS. Since each mobile performs the necessary measurements itself, location capacity is not limited by the measurements as in the case of methods relying on network based measurement equipment. E-OTD measurements can be performed relatively fast, e.g. 1- or 2-s measurement normally should be sufficient.

E-OTD is relatively accurate. Simulations performed in T1P1 (GSM 05.50) show that E-OTD accuracy varies much depending on environment, number of hyperbolas, etc. For example in an urban area 67% accuracy is approximately 200 m with two hyperbolas, and approximately 100 m in a suburban area. Some field tests indicate 67% accuracy up to 50–150 m, depending on environment. Some field test results are introduced later in Section entitled *E-OTD field test* in this chapter.

Location accuracy for E-OTD depends on several issues. The quality of measurements in the MS and LMUs is naturally one of the factors. The physical environment is another. Multipath propagation and especially non-line-of-sight conditions can decrease accuracy. The number of different base station sites that the MS can receive determines the number of hyperbolas available. More hyperbolas mean better accuracy. Poor geometry of base stations relative to the MS can degrade accuracy. For example if all base stations happen to be more or less aligned with the MS (e.g. on a straight highway), the accuracy transversal to the line formed by base stations and the MS is poor. Insufficient number of LMUs might mean that there is no full RTD information that can make the accuracy worse. It is also extremely important to have correct and updated network information available for good accuracy. Inaccuracies in base station coordinates directly cause inaccuracy, and incorrect base station identity and channel information can lead to wrong identification of base stations.

Time of arrival

In this method, time of arrival (TOA) of random access bursts in the random access channel from the MSs is measured by LMUs in different positions in the network (Figure 4.13). The mobile is made to perform an asynchronous handover in order to make it send usual random access bursts. Thus if the mobile is in idle state, a signalling channel is opened to it first. The difference in TOA values measured by LMUs 1 and 2 in two different positions determines a hyperbola: $c^*(\text{TOA}_1 - \text{TOA}_2) = d_{m1} - d_{m2}$, where c is the speed of radio waves, and d_{m1} and d_{m2} denote the distance from the MS to LMUs 1 and 2, respectively. Then location estimation proceeds the same way as described in the case of E-OTD.

The accuracy of TOA is comparable to that of E-OTD [7]. TOA can be implemented so that no changes in the mobile are required. However, TOA requires LMUs capable of TOA measurements and in addition also a common clock in LMUs so that the TOA measurements are comparable with each other. Coordination between base stations is needed in order to get the LMUs to measure the MS when it transmits random access bursts. This means that extra signalling load is generated. Also the capacity is limited since several LMUs are needed to locate one MS at a time.

At the moment there seems to be lack of interest in TOA solutions by operators, whereas, e.g. certain US operators have announced the adoption of E-OTD

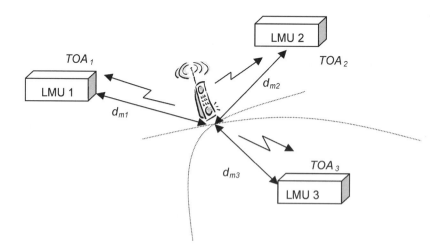

Figure 4.13   Time of arrival method.

for E911 purposes. TOA offers the possibility to locate even existing mobiles, which is beneficial for applications like E911. However, the limitations of TOA seem to restrict its attractiveness.

Extended service level

Methods belonging to the extended service level offer the best performance. This is achieved with both extra equipment in the network and additional hardware in the MS.

Assisted GPS

Global positioning system (GPS) is a satellite based, widely used location system. It relies on a fleet of medium earth orbit (MEO) satellites that transmit two L-band spread spectrum signals. A GPS receiver measures relative times of arrival of signals sent simultaneously from at least three (2D location) or four (3D location) satellites. Based on these measurements the distances between the satellites and the receiver can be estimated. Each distance defines a sphere around the position of a satellite, and by finding their intersection the location is estimated.

A straightforward way to use GPS for mobile location is to integrate a GPS receiver into a MS. However, this approach has its limitations, since GPS was not originally planned to be used indoors or in an urban environment. GPS signals are weak, and a normal stand-alone GPS receiver needs line of sight to satellites. Integration of a GPS receiver in a handset increases also the manufacturing price and power consumption. Antenna integration can also be a problem.

Wireless networks can deliver assistance information to the mobile and its GPS receiver in order to improve its performance and to reduce some above-mentioned problems. Assistance data can consist of reference position and time, differential corrections, ephemeris data, ionosphere models, almanac information, GPS time transfer data, UTC models, satellite acquisition and health data. Assistance data fastens acquisition, decreases power consumption, increases sensitivity, and expands coverage area (e.g. urban environment). As mentioned earlier, LMUs can be used to help to generate assistance data for GPS, namely reference time.

In GERAN and UTRAN standards there are basically three variants of GPS supported. The first one is simple stand-alone GPS, where the MS includes a GPS receiver that operates basically independent of the cellular network. The two other variants are two versions of A-GPS, where the network participates in GPS location (Figure 4.14). In MS based network assisted GPS, the MS receives assistance data from the network, but calculates the GPS location estimate itself. In MS assisted network based GPS the MS performs actual measurements, and reports them to the network that calculates the location estimate.

Figure 4.14   Assisted GPS.

GPS has good accuracy, up to 2 m with differential corrections. Also removal of the selective availability (SA) feature in GPS signals means that basic GPS can have accuracy up to 10–20 m.

4.4 LCS performance

Basic service level performance

Location system implementation

A simplified logical architecture of an implementation of a location system incorporating the basic service location methods described in Section 4.2 is represented in Figure 4.15. Beside the 'classical' GSM network elements (BTS, BSC, MSC, visiting location register (VLR) and home location register (HLR)) new elements introduced in the GSM LCS specifications Rel'98 and Rel'99 are shown, namely SMLC and GMLC. The SMLC is the network element, or corresponding functionality, where the actual location calculation is performed. The GMLC acts as interface between the GSM network and external LCS clients.

To perform a location calculation, CI, TA and RXLEVs must first be made available in the network side and then delivered to the SMLC. In GSM, the TA is available in the network only for a mobile terminal in *dedicated mode* (i.e. a mobile communicating with the serving base station using a dedicated channel). A GSM phone measures RXLEVs both in dedicated mode and in *idle mode* (i.e. the phone is turned on but no two-way communication takes place) however, it reports RXLEVs to the fixed part of the network only in dedicated mode. In order to obtain TA and RXLEVs from a handset in idle mode, and thus to be able to locate it, the MSC can establish a dedicated connection unnoticeable to the user. In this way, the TA can be determined and the mobile terminal is forced to report RXLEVs to the network, thus location calculation is enabled.

The forced measuring of TA and reporting of RXLEVs for GSM handsets in idle mode, as described above, is in line with subscriber privacy mechanisms defined in GSM LCS specifications release 1998. The target handset may be positioned only if allowed in the handset subscription profile, unless required by local regulatory requirements. The LCS client access to location information is restricted unless otherwise stated in the target terminal subscription profile, thus the subscriber has full control of delivering his/her location information. For emergency services, the target terminal making an emergency call can be positioned regardless of the privacy attribute value of the GSM subscriber associated with the target terminal making the call. The GMLC authenticates all applications and encrypts the traffic towards location based applications.

Experimental performance
This section presents experimental performance of CI, CI + TA and CI + TA + RXLEV location methods incorporated in the system shown in Figure 4.15. A more extended analysis of the data can be found in [8]. CI, TA and RXLEV measurements were collected from Network 11, in a city centre and a suburban area in early 2001. Measurements in the 900 MHz band were collected driving a certain number of test routes. Inside a vehicle a laptop, connected to a GPS receiver and a GSM phone, was periodically storing in log files GPS reference locations as well as *measurement reports* and time stamps. One *measurement report* includes the data a mobile terminal is forced to report to the network in the system implementation of Figure 4.16; namely, CI, TA, and one RXLEV for the serving cell and one RXLEV for each of the strongest neighbour cells (up to six). CI, TA and RXLEV data were gathered in 76 009 (102 269) locations in an urban (suburban) area from 44 (46) serving cells and 76 (83) neighbour cells.

Figure 4.15 Architecture of a location system implementation.

Figure 4.16 Location error statistics.

The analysis of the data collected during the test routes shows that in the suburban area half of the TA values were between 1 and 2 while in the urban area the TA values were mainly 0 and 1. In the urban area, TA values up to four were measured, but TA $\geq$ 3 were measured only at very few locations. In roughly 55% of all measurements the TA was zero. The number of neighbours in the urban area was in over 80% of the cases at least four; the mean RXLEV values were between $-$ 60 dBm for the first neighbour cell and $-$ 80 dBm for the sixth neighbour cell. In the suburban area, TA values up to four were measured, the number of neighbours was as high as in the urban area but, due to a less dense network, the mean RXLEV values were between 5 and 10 dBm lower than in the urban area.

The measurements collected have been processed off-line with the CI, CI + TA and CI + TA + RXLEV location methods described in Section 4.2. Each method determines a location estimate by processing only one measurement report. The *location error*, defined as the distance between a GPS reference location and the corresponding location estimate, is used as a measure of location accuracy. Location error statistics are shown in Figure 4.16.

It can be seen that the CI + TA + RXLEV method outperforms the other two methods in the urban area, while in the suburban area it performs essentially as well as CI + TA. Both CI + TA and CI + TA + RXLEV methods are significantly more accurate than the CI method in both environments for all percentiles. Therefore, CI + TA and CI + TA + RXLEV methods do not only reduce the average location error, but the maximum error as well. For example, when comparing CI and CI + TA + RXLEV methods, the 99% percentile reduces from 2 to 1.2 km in the suburban area, and from 770 to 620 m in the urban area. Controlling the maximum location error is crucial; in fact, larger than expected location errors may affect negatively the customer's perception of service reliability and can decrease the customer's interest for LCS.

The absolute values of the location errors depend strongly on the network as well as the environment where the measurements are performed. The results shown in Figure 4.16 can be used to estimate the accuracy of the methods in different conditions if the network dependence is somehow reduced. This can be done by properly scaling the location error. A good candidate for the scaling factor is the cell size, which is however not a clearly defined quantity. Since in the CI method the handset location estimate is always at the base station coordinates, a statistically meaningful definition for the average cell size (i.e. cell radius) in each environment is the 95% percentile of the CI location error. In the following, the accuracy performance is further analysed in terms of *relative location error*, defined as the location error scaled to the cell size defined above (e.g. 603 m in urban and 1345 m in suburban area).

It can be seen from the results in Figure 4.17 that for a given percentile level, the CI + TA relative accuracy is better in the suburban than in the urban area, while the CI + TA + RXLEV relative accuracy is almost the same in both environments. This clearly indicates that the location accuracy obtained by the CI + TA + RXLEV method scales according to the cell size. In other

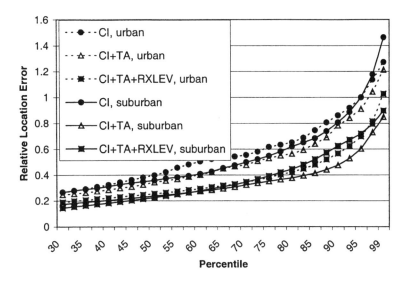

Figure 4.17   Relative location error statistics.

words, in networks with smaller cells the absolute CI + TA + RXLEV accuracy would most likely improve as compared to the accuracy shown in Figure 4.16. The scalability of the CI + TA + RXLEV accuracy with the cell size is very relevant from the commercial application viewpoint. This is because the location accuracy requirements in densely populated areas, where cells are small, are high due to the proximity of the points of interest (e.g. restaurants). In contrast, in less densely populated areas, where cells are large, the points of interest (e.g. gas stations) are more separated and the accuracy requirements can be relaxed.

The relative accuracy of CI, CI + TA and CI + TA + RXLEV methods for different values of measured TA is compared in Figure 4.18 (for illustration purposes, the 67% percentile is shown). The CI location error increases for increasing TA values. This is due to the fact that the TA gives an indication of the distance between handset and serving base station but the CI location estimate is always at the serving base station site, irrespective of the measured TA. Similarly, the CI + TA location accuracy degrades as TA increases (the only exception being for TA = 4, which is however not statistically meaningful). In fact the geographical region in which a mobile terminal can measure a certain TA value increases in size with the TA but the CI + TA location estimate is always placed along the serving cell's antenna azimuth, at a distance from the serving base station derived from the TA. The degradation of CI + TA + RXLEV accuracy for increasing TA values is much smaller as compared with the other two methods.

The accuracy results provided in this section show that irrespective of the environment, CI + TA and CI + TA + RXLEV methods significantly reduce

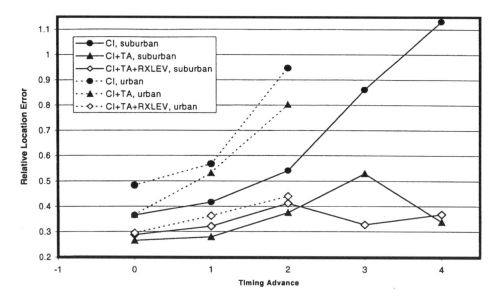

Figure 4.18   67% percentiles of the relative location error as a function of measured TA.

both average and maximum errors obtained with the CI method. Additionally, the CI + TA + RXLEV location error turns out to scale along with the cell size making this method very suitable for services such as finding/guidance and notification, as it delivers highly accurate location estimates in densely populated areas where the demand for accurate location information is strong. Finally, the CI + TA + RXLEV accuracy proves not to deteriorate as much as the CI and CI + TA accuracy, when the distance between the MS and the serving base station (or equivalently the TA) increases. A more extensive analysis of the experimental results [6] proves also that, especially in the urban environment, an increase in the number of measured neighbours leads to an improvement in the CI + TA + RXLEV location accuracy.

Enhanced service level performance

E-OTD field test

In this section, some results are described from an overlay E-OTD test system in Network 7 (Figure 4.19). The system consists of a demonstrator serving mobile location centre (SMLC-D), a modem gateway, a demonstrator mobile station (MS-D) and several demonstrator location measurement units (LMU-D).

Normal GSM data calls are used to deliver information between the SMLC-D and the LMU-Ds and MS-D via the modem gateway. LMU-D, MS-D and SMLC-D correspond functionally to the standardised solution, but signalling solution is proprietary. The MS-D includes a GPS receiver that is used to define the real position.

In the typical measurement procedure the SMLC-D sends measurement commands to LMU-Ds that start to send RTD measurements periodically

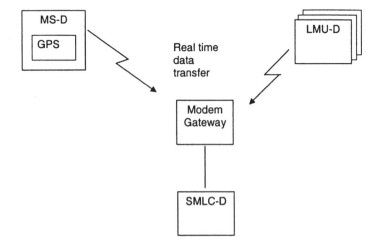

Figure 4.19   E-OTD field test system.

back. Therefore, SMLC-D is aware of the RTD situation all the time. Then, SMLC-D sends an E-OTD measurement command to the MS-D to be located. The mobile performs E-OTD measurements for one location that takes 1–2 s, and sends the measurement results back to the SMLC-D, that calculates the E-OTD position estimate.

The test area was a suburban commercial and residential environment. The typical distance between cells in the area is 200–1000 m. The area has apartment and office buildings with typically six to eight storeys, but also smaller residential houses with one to two floors.

To give an example, a test run was done with seven LMU-Ds. The MS-D was moving a 15 km route with velocity 0–50 km/h. The distance between the E-OTD location estimate and the real location, i.e. the location error, had the distribution as shown in Table 4.1.

Another example test run was done with six LMU-Ds. Now the MS-D was moving a shorter, 5 km, route. The location errors were as in Table 4.2. Better accuracy is probably due to the environment that was mainly with low residential houses (1–2 floors).

In the third example test run the MS-D was moving again a shorter 5 km route (Table 4.3). Now the area was densely built-up mainly with six floor apartment buildings. The location errors indicate less accurate results than in the previous runs. This is due to the less favourable environment.

Table 4.4 illustrates a stationary test in a single location (four LMU-Ds were on). In this case the location errors are very consistent.

Table 4.1   E-OTD field test example (moving MS)

Number of measurements	Median accuracy	67% accuracy	90% accuracy
267	58 m	90 m	232 m

Table 4.2   E-OTD field test example (moving MS)

Number of measurements	Median accuracy	67% accuracy	90% accuracy
150	35 m	42 m	85 m

Table 4.3   E-OTD field test example (moving MS)

Number of measurements	Median accuracy	67% accuracy	90% accuracy
200	87 m	120 m	240 m

Table 4.4   E-OTD field test results (stationary MS)

Number of measurements	Median accuracy	67% accuracy	90% accuracy
45	28 m	31 m	33 m

References

[1] M. Pent, M. A. Spirito, E. Turco, "Method For Positioning GSM Mobile Stations Using Absolute Time Delay Measurements,"IEE Electronics Letters, 20th Nov. 1997, Vol. 33, No. 24, pp. 2019-2020.

[2] M. A. Spirito, A. Guidarelli Mattioli, "Preliminary Experimental Results of a GSM Phones Positioning System Based on Timing Advance," in Proceedings of 1999 IEEE Vehicular Technology Conference VTC 1999-Fall, Sept. '99, Amsterdam, the Netherlands, pp. 2072-2076.

[3] Spirito M. A., 'Accuracy of Hyperbolic Mobile Station Location in Cellular Networks', IEE Electronics Letters, Vol. 37, No. 11, 24 May 2001, pp. 708–710.

[4] Spirito M. A., 'On the Accuracy of Cellular Mobile Station Location Estimation', IEEE Transactions on Vehicular Technology, Vol. 50, No. 3, May 2001, pp. 674–685.

[5] ETSI TC-SMG, European Digital Telecommunications System (Phase 2); Radio Subsystem Synchronization (GSM 05.10), ETSI – European Telecommunications Standards Institute.

[6] Spirito M. A., 'Further Results on GSM Mobile Stations Location', IEE Electronics Letters, Vol. 35, No. 11, 27 May 1999, pp. 867–869.

[7] ETSI TC-SMG, European Digital Telecommunications System (Phase 2), Background for Radio Frequency (RF) Requirements (05.50), ETSI – European Telecommunications Standards Institute.

[8] M.A. Spirito, S. Pöykkö, O. Knuuttila, 'Experimental Performance of Methods to Estimate the Location of Legacy Handsets in GSM', Proceedings of IEEE Vehicular Technology Conference – VTC 2001 (2001) 2001.

Part 2

GSM, GPRS and EDGE Performance

This second part of the book focuses on performance delivered by global system for mobile communications (GSM)/enhanced data rates for global evolution (EDGE) networks. Performance is measured in two different and complementary ways. First, field trial results help to get a realistic understanding of basic GSM performance capabilities. Second, simulation results envisage performance of features not available yet in real networks. In any case, the methodology used in the performance analysis is consistent throughout the book.

The basics of GSM/EDGE radio performance, the methodology followed to characterise network performance and measure spectral efficiency and the definition of the GSM basic set of functionality and its associated baseline performance are the topics covered in the first chapter of this part.

Chapter 6 presents GSM/adaptive multi-rate (AMR) voice performance. It starts from basic enhanced full-rate (EFR) speech codec performance and later on analyses additional capacity enhancement techniques already available in many of the existing GSM networks. GSM system performance with AMR codec is thoroughly presented in this chapter. Finally further evolution of the speech services in Rel'5 and beyond is addressed in its last section.

General packet radio system (GPRS) and enhanced GPRS (EGPRS) packet switched performance is studied in Chapter 7. The analysis covers link performance, radio resource management algorithms, system level performance and end-to-end performance considering of different applications over (E)GPRS. This chapter gives indications of how (E)GPRS services can be introduced in an existing GSM and how mixed (E)GPRS and speech traffic can be planned and dimensioned.

Dynamic frequency and channel allocation is an advanced radio resource management technique that pushes the limit of GSM network capacity. Chapter 8 describes this technique and provides accurate performance figures with EFR and AMR codecs.

Chapter 10 focuses on the GSM EDGE radio access network (GERAN) link budget calculation and describes the different techniques that can be applied to enhance both uplink (UL) and downlink (DL) performance.

Chapter 11 tackles control channel performance in GSM/EDGE from two points of view: signalling channels link level performance and capacity of the different channel configurations.

Chapter 12 covers GSM innovative automation functionality to be applied in the planning, parameter optimisation and troubleshooting tasks.

5

Basics of GSM Radio Communication and Spectral Efficiency

Juan Melero, Jeroen Wigard, Timo Halonen and Javier Romero

The purpose of this chapter is to introduce the principles of GSM/EDGE radio performance, defining the tools and methodology utilised throughout the book to study the performance enhancements provided by the multiple voice and data functionality presented. All the analysis presented in different chapters are based on commercial networks trials and simulation results. Appendix E describes the simulation tool and configurations used for the simulation based analysis and Appendix F presents the trial partners mentioned across the book as *Network #*.

Section 5.1 describes GSM radio interface basic link performance including a description of the transmission and receiving chain and the basic GSM standardised functionality, such as frequency hopping, power control and discontinuous transmission. In Section 5.2 the most relevant key performance indicators to monitor network performance are presented. Section 5.3 defines the spectral efficiency and the different existing ways to quantify it and Section 5.4 describes the effective frequency load (EFL) measurement methodology, utilised across the whole book to characterise networks performance and measure spectral efficiency from real networks trial data and simulation results. Finally, Section 5.5 defines the baseline performance of GSM systems analysing their achievable performance.

5.1 GSM radio system description

This section introduces very briefly the principles of GSM radio communication, its channel structure and basic functionality, such as frequency hopping, power control and discontinuous transmission. An introduction about their impact in link performance is included as well. It focuses on the basic GSM speech service since the equivalent (E)GPRS analysis is covered more extensively in Chapters 1 and 7. For a more detailed description of the subjects introduced in this section, the reader is recommended to consult [1,2].

Basic channel structure

The GSM standard is based on multi carrier, time division multiple access and frequency division duplex, MC/TDMA/FDD [1]. Table 5.1 shows the frequency bands which have been defined for GSM.

The carrier spacing is 200 kHz allowing, for example, for 124 and 374 radio frequency channels in the 900 and 1800 MHz bands, respectively, thus leaving a guardband of 200 kHz at each end of the sub-bands. Each radio frequency is time divided into TDMA frames of 4.615 ms. Each TDMA frame is subdivided into eight full slots. Each of these slots can be assigned to a full-rate traffic channel, two half-rate traffic channels or one of the control channels. A slot is equal to one timeslot on one frequency. The time and frequency structure is displayed in Figure 5.1.

The data transmitted in one slot is denoted a burst. There are five different types of bursts: the normal burst, the access burst, the frequency correction burst, the synchronisation burst and the dummy burst [3]. The format and information of the individual burst depends on the type of channel it belongs to.

Channels can be described at two levels: at the physical and the logical level. A physical channel corresponds to one timeslot on one carrier, while a logical channel refers to the specific type of information carried by the physical channel.

Table 5.1   GSM standardised frequency bands

GSM frequency band	Available frequencies	Where available
400 MHz	450.4–457.6 MHz paired with 460.4–467.6 MHz or 478.8–486 MHz paired with 488.8–496 MHz	Europe
800 MHz	824–849 MHz paired with 869–894 MHz	America
900 MHz[a]	880–915 MHz paired with 925–960 MHz	Europe, Asia Pacific, Africa
1800 MHz	1710–1785 MHz paired with 1805–1880 MHz	Europe, Asia Pacific, Africa
1900 MHz	1850–1910 MHz paired with 1930–1990 MHz	America

[a] Including the extended GSM (EGSM) band.

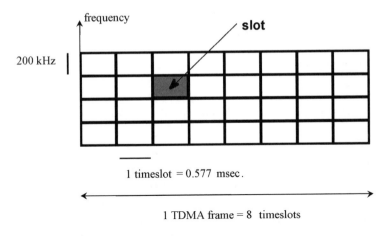

Figure 5.1 The multiple access scheme in GSM.

Different kind of information is carried by different logical channels, which are then mapped or multiplexed on physical channels.

Logical channels can be divided into two groups: control channels and traffic channels. Traffic channels are used to carry user data, which can be either speech or data. A logical traffic channel for speech and circuit switched data in GSM is called TCH, which can be either full rate (TCH/F) or half rate (TCH/H). A logical traffic channel for packet switched data is referred to as packet data traffic channel (PDTCH).

The full-rate traffic channel (TCH/F) is a 13 kbit/s coded speech or data channel with a raw data rate of 9.6, 4.8 or 2.4 kbit/s. The half rate supports 7, 4.8 and 2.4 kbit/s [3]. The TCH channels support bi-directional transmission in order to carry speech and data communication.

A PDTCH/F corresponds to the resource allocated to a single mobile station on one physical channel for user data transmission. Due to the dynamic multiplexing on to the same physical channel of different logical channels, a PDTCH/F using GMSK modulation carries information at an instantaneous bit rate ranging from 0 to 22.8 kbit/s. A PDTCH/F using 8-PSK modulation carries information (including stealing symbols) at an instantaneous bit rate ranging from 0 to 69.6 kbit/s. These bit rates will then be mapped into certain user payload depending on the channel coding scheme applied.

A PDTCH/H corresponds to the resource allocated to a single mobile station on half a physical channel for user data transmission. The maximum instantaneous bit rate for a PDTCH/H is half that for a PDTCH/F.

All packet data traffic channels are unidirectional, either uplink (PDTCH/U), for a mobile originated packet transfer or downlink (PDTCH/D) for a mobile terminated packet transfer.

T: TDMA frame for TCH
A: SACCH frame for TCH
- : idle frame

Figure 5.2   Organisation of TCH and the corresponding control channels [3].

Signalling and controlling information is carried on the other type of logical channels, the control channels. In GSM there are common as well as dedicated control channels. These channels are described in Chapter 11.

The GSM specifications describe which physical channels to use for each logical channel [3]. Several combinations of the different channels are possible. Figure 5.2 shows the organisation of the TCH/F speech channel and corresponding slow associated control channel (SACCH). This 26-frame multiframe only considers one timeslot per TDMA frame.

From the 26 frames, 24 are used for traffic. One frame is used for the SACCH channel, while the last one is an idle frame. During this idle frame time interval period a mobile can receive other control channels and measure the received signal level from neighbouring cells. The complete GSM frame, timeslot and burst structure is shown in Figure 5.3.

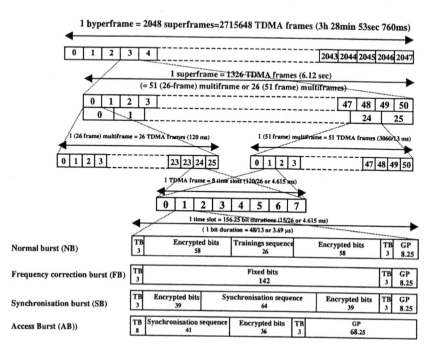

Figure 5.3   Timeframes, timeslots and bursts in GSM.

Transmitting and receiving chain

Figure 5.4 shows the transmitting and receiving chain of a GSM receiver. Several successive operations have to be performed to convert a speech signal into a radio signal and back.

The following operations take place on the transmitting side:

- *Source coding* converts the analogue speech signal into a digital equivalent.
- *Channel coding* adds extra bits to the data flow. This way redundancy is introduced into the data flow, increasing its rate by adding information calculated from the source data, in order to allow detection or even correction of bit errors that might be introduced during transmission. This is described in more detail below.
- *Interleaving* consists of mixing up the bits of the coded data blocks. The goal is to have adjacent bits in the modulated signal spread out over several data blocks. The error probability of successive bits in the modulated stream is typically highly correlated, and the channel coding performance is better when errors are decorrelated. Therefore, interleaving improves the coding performance by decorrelating errors and their position in the coded blocks.
- *Ciphering* modifies the contents of these blocks through a secret code known only by the mobile station and the base station.
- *Burst formatting* adds synchronisation and equalisation information to the ciphered data. Part of this is the addition of a training sequence.
- *Modulation* transforms the binary signal into an analogue signal at the right frequency. Thereby the signal can be transmitted as radio waves.

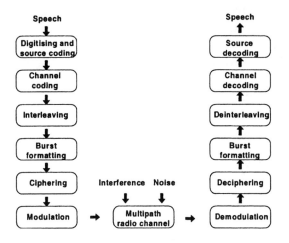

Figure 5.4   Flow diagram of the link operations according to [4].

The receiver side performs the reverse operations as follows:

- *Demodulation* transforms the radio signal received at the antenna into a binary signal. Today most demodulators also deliver an estimated probability of correctness for each bit. This extra information is referred to as soft decision or soft information.
- *Deciphering* modifies the bits by reversing the ciphering code.
- *Deinterleaving* puts the bits of the different bursts back in order to rebuild the original code words.
- *Channel decoding* tries to reconstruct the source information from the output of the demodulator, using the added coding bits to detect or correct possible errors, caused between the coding and the decoding.
- *Source decoding* converts the digitally decoded source information into an analogue signal to produce the speech.

In GSM the data bits are coded. The channel coding introduces redundancy into the data flow, by increasing the bit rate. For the TCH/FS mode, there are three bit-coding classes. The class 1a bits have a 3-bit cyclic redundancy check (CRC) and all class 1 bits are encoded by a convolution code. The class 2 bits remain unprotected. Note that the coding is different in case of a PDTCH, which is used to carry packet switched data. This is described in Chapter 1.

The reordering and interleaving process mixes the encoded data block of 456 bits, and groups the bits into eight sub-blocks (half bursts). The eight sub-blocks are transmitted on eight successive bursts (interleaving depth equals 8). The channel coding can be seen in Figure 5.5, while the reordering and interleaving is displayed in Figure 5.6.

Due to multipath propagation, the erroneous received bits tend to have a bursty nature. The convolutional code gives the best performance for random positioned bit errors, and therefore reordering and interleaving is introduced in the GSM signal transmission flow. However, the reordering/interleaving only improves the coding performance if the eight successive bursts carrying the data information of one speech block are exposed to uncorrelated fading. This can be ensured either by spatial movement (high user speed) or frequency hopping.

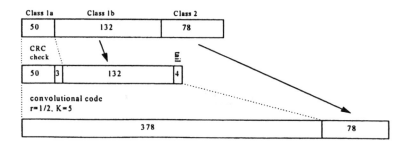

Figure 5.5   Channel coding of the TCH/FS.

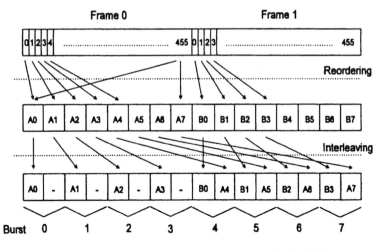

Figure 5.6 Reordering and interleaving of the TCH/FS.

Propagation effects

Radio propagation in mobile communications is complex. The transmitted signal from the base station may have been exposed to a combination of various propagation phenomena such as free space attenuation, diffraction, reflection and other propagation effects before the mobile station receives it. A statistical approach is therefore mostly used for characterisation of the mobile radio channel:

- Mean path loss based on some nth power law with range and an absolute power level at a given distance.
- Variations in the local mean path loss, normally modelled by a lognormal distribution.
- Fast fading superimposed on the slow variation of the local mean signal. Rayleigh or Rician distributions are often used to characterise the fast fading profile.

Path loss

A signal transmitted by a radio connection is attenuated along the propagation path. The power captured by a receiving antenna decreases with increasing distance to the transmitter. This is caused by the fact that the signal is transmitted not only in a straight line, but as a sphere, which radii is the separation between the transmitter and the receiver. There is less power on the same piece of surface of a larger sphere than a smaller sphere. In cities there are also losses in power due to reflection, diffraction around structures and refraction within them. The attenuation of received power with distance is called path loss.

Terrain features affect this propagation loss. Therefore, when the path loss has to be predicted, all the terrain features should be taken into account to get an accurate result.

Path loss is relatively easy to calculate when free space is assumed and when there is a direct path between the two antennas (a line-of-sight path). Then the path loss is given by [5]

$$L_{dB} = 32 + 20 \log f + 20 \log d$$

where f is the frequency in MHz and d is the distance in km.

The equation above gives the best-case result for the path loss. In reality the path loss always will be higher because of terrain features.

A simple model for the mean path loss as a function of the distance, including terrain features, is given by [6]

$$L_{dB} = \gamma \cdot 10 \log d + k$$

where d is the distance between base station and mobile station, γ is the propagation path slope, and k is a constant.[1] For free space propagation γ equals 2 (as in the equation above). For land mobile communications the path loss slope γ is typically within the range of 3–5.

Fading

A signal nearly never arrives by only one path. Due to scattering from the terrain obstacles, there is more than one path by which the signal arrives at the destination.

Radio signals therefore arrive at the mobile receiver from different directions, with different time delays. They combine at the receiver antenna to give a resultant signal, which depends crucially on the differences in path length that exists in the multipath field, since the various spectral components that comprise the transmitted signal can combine constructively or destructively. As the mobile station moves from one location to another the phase relationship between the spectral components within the various incoming waves changes, so the resultant signal changes rapidly. It is worth noting that whenever relative motion exists, there is also a Doppler shift in frequency components within the received signal.

Especially in built-up areas this can cause problems, because the antenna might be below the surrounding buildings, so that there is no line-of-sight transmission path. This situation can be seen in Figure 5.7.

Fading can be classified in two parts: shadow fading (slow fading) and multipath fading (fast fading).

---

[1] The constant k includes the dependency of mobile and base station height, frequency, land coverage, etc.

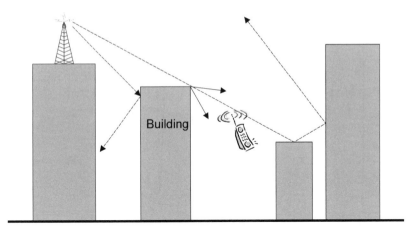

Figure 5.7 Radio propagation in urban areas.

Shadow fading is caused by shadowing from the terrain obstacles and generates a variation in the received signal level local mean. It typically exhibits a statistic lognormal distribution, i.e. a normal pdf (probability density function) of dB values [7]:

$$p(m) = \frac{1}{\sqrt{2\pi\sigma^2}} e^{-(m-\bar{m})^2/2\sigma^2}$$

where $\bar{m}$ is the mean value of the random variable m, and σ is the standard deviation – both expressed in dB. Typical measured standard deviation values for the lognormal fading are in the range 5–10 dB.

The multipath fading component is caused by the combination of the signal arriving via different paths. Measurements suggest that the received signal at each location is made up of a number of replicas of the signal with random amplitude and phase [8]. The amplitudes and phases are assumed to be statistically independent, with the phase of each replica distributed uniformly between 0 and 2π. The amplitudes are assumed to sum up to the mean value of the envelope. Figure 5.8 illustrates typical envelope (received signal strength) variations as a mobile station moves.

The pdf of fast fading in urban environments can mostly be described by a Rayleigh distribution

$$p(r) = \frac{r}{\sigma^2} e^{-r^2/2\sigma^2} \qquad r \geq 0$$

where σ^2 is the mean local signal level.

In line-of-sight (LOS) conditions a Rician distribution characterises better the fading profile.

Figure 5.8 Example of signal envelope variations caused by multipath propagation.

Basic TCH link performance with frequency hopping

This section describes the basic link level performance of TCH link performance GSM. The PDTCH one, which is used for packet switched traffic, is studied in detail in the first section of Chapter 7.

Specifications of the required performance of individual GSM mobile stations in terms of interference tolerance and sensitivity for certain *test conditions* are described in [10]. The required performance in the case of interference can be seen in Table 5.2 in the case of frequency hopping (FH) and non-hopping (NH). It is shown for the typical urban (TU) [4] channel profile with two different speeds (3 and 50 km/h). The channel profile and speed combinations are denoted TUx, where x is the speed. The factor α is a value that lies between 1 and 1.6, depending on the channel condition.

GSM specifications support a slow FH mode, which means that the frequency of the transmission is changed on a burst by burst basis. The uplink and down-link frequencies are always duplex frequencies. Two different kind of frequency hoppings are specified for GSM: random and sequential hopping.

Two parameters describe the hopping frequency used in each frame: the mobile allocation index offset (MAIO) and the hopping sequence number (HSN). The MAIO can take as many values as there are frequencies in the

Table 5.2 Reference interference performance at 9 dB C/I [GSM0505]

	TU3 NH	TU3 FH	TU50 NH	TU50 FH
Max. FER	$21\alpha\%$	$3\alpha\%$	$6\alpha\%$	$3\alpha\%$

hopping list and is used to indicate the offset within the hopping list used to identify the frequency used. The hopping list is given by the mobile allocation (MA) list. The HSN can take 64 different values defining the hopping sequence used. If the HSN is set to zero, the hopping mode is sequential, while the frequencies vary pseudo-randomly when it differs from zero.

The FH algorithm is described in [3]. For a given set of parameters, the index to an absolute radio frequency channel number (ARFCN) within the mobile allocation is obtained with the following algorithm, where FN is the frame number and N is the number of frequencies:

if HSN = 0 (cyclic hopping) **then**:
MAI, integer (0..N − 1) : MAI = (FN + MAIO) modulo N
else:
MAI, integer (0..N − 1) : MAI = (S + MAIO) modulo N

where S is calculated from the HSN, time parameters, the number of frequencies and a predefined table, which can be found in [3].

Two sequences with the same HSN, but different MAIOs will never overlap each other, i.e. the hopping sequences are orthogonal. Two channels having different HSN, but the same frequency list and the same timeslot, will interfere each other only in $1/m$ of the bursts, where m is the number of frequencies in the hopping sequence. This way the interference can be averaged out over the network [2].

Some control channels have some restrictions with regards to FH. For example the BCCH control channel, which takes up one timeslot on the BCCH transceiver (TRX) cannot use FH. The reason for this is that the BCCH frequency has to be fixed in order for the mobiles to be able to find it.

The activation of FH in GSM improves link performance due to two main effects, frequency diversity and interference diversity.

Frequency diversity
Frequency hopping increases the fading decorrelation among the bursts that compose each frame. This is due to the different fast fading profiles of different frequencies. As the decorrelation increases the efficiency of the channel coding will increase.

Figure 5.9 shows the gain from sequential and random hopping under *test conditions* [9]. The TU channel profile with 3 and 50 km/h has been used. The maximum hopping frequency diversity gain of 7 dB for slow moving mobiles is achieved with sequential hopping and eight frequencies (equal to the interleaving depth). This is also found in [4]. The gain of random hopping for TU3 would slowly continue increasing with larger number of frequencies getting closer to the maximum value of 7 dB. The largest relative gain increase is already achieved with a limited number of frequencies, i.e. four frequencies achieve

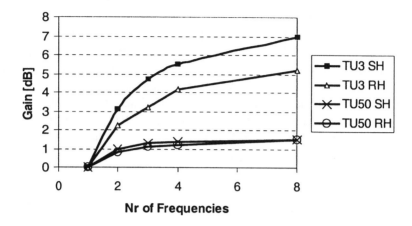

Figure 5.9 Gain at 2% FER from random and sequential hopping (RH and SH) for a TU3 and TU50 in *test conditions*.

over 75% of the gain of eight frequencies. It also can be seen that for the TU50 *test conditions* the gain from FH is marginal. This is caused by the fact that the higher speed of the mobiles already gives frequency diversity without FH. The bit error rate (BER) is the same for all hopping configurations, since the BER is calculated before the decoding with no gain from FH.

Random hopping gives approx. 1–2 dB lower gain than sequential hopping for low number of hopping frequencies [10]. However, random hopping still provides the best overall network performance once the improvement from interference diversity is taken into account [11] and therefore it is the most utilised hopping scheme in practical deployments.

The previous figure assumed the different frequencies used had a fully uncorrelated fading profile. In practice, the hopping frequencies will have a certain correlation factor depending on the frequency separation between them and the channel profile in use [12]. Figure 5.10 represents the frequency diversity gain for TU3 and random FH when the frequencies in the hopping list are consecutive, i.e. the effective hopping bandwidth is limited to 200 kHz * (*number_of_- hopping_frequencies*). For configurations between four and eight frequencies there is roughly 1 dB degradation.

Interference diversity
By using random FH or non-grouped frequency planning sequential hopping, FH also gives an interference diversity gain. This gain consists of two components:

- An interference averaging effect. This is illustrated in Figure 5.11, where the interference level of three mobiles is displayed with and without FH. Without hopping one mobile suffers from heavy interference. By using FH the heavy interference is spread out over the three mobiles. The result is equal

Figure 5.10 TU3 random hopping frequency diversity gain with correlated frequencies.

interference for the three mobiles. This interference spreading effect has again a favourable impact on the channel coding performance.

- Fractional loading gain. Fractional loading gain is obtained when not all frequencies are fully loaded. Figure 5.12 shows the fractional loading gain. The FER of a TU3 link with FH, power control and no synchronisation can be seen as a function of the mean carrier/interference (C/I). It can be concluded from this figure that the gain from fractional loading is around 3 dB, when going from a typical 15% load to a 100% load (without DTX). The results in the case of a DTX factor of 0.5 can easily be read from the figure by dividing the load by a factor of 2.

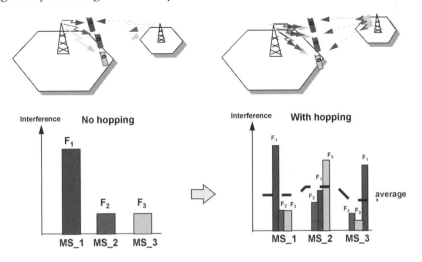

Figure 5.11 Interference averaging by random frequency hopping.

Figure 5.12 The FER in a TU3 full hopping link with six interferers and one background interferer for different loads in the case of power control and no synchronisation [11].

Discontinuous transmission (DTX)

Voice activity detection (VAD) is a feature in GSM, which can be used to omit transmission when a silent period is detected during a speech connection. This is called discontinuous transmission (DTX). The network operator controls the activation of DTX.

When a mobile station is not in DTX mode it sends and receives 100 TDMA bursts per SACCH multiframe (0.48 s), which can be used for estimation of the RXQUAL [13]. When a mobile station is in DTX mode, i.e. there is a silent period, only 12 TDMA bursts are sent in 480 ms. The burst transmission in DTX mode is shown in Figure 5.13.

Speech detection is carried out at the transmitting end by the before mentioned VAD, which distinguishes between speech superimposed on environmental noise and noise without speech being present. The output of the VAD is used to control the transmitter switch. If the VAD fails to detect every speech event, then the transmitted speech will be degraded due to clipping. On the other hand, if the VAD identifies noise as speech often, then the effectiveness of DTX is diminished. Both of these result in degraded performance [14].

Figure 5.13 Schematic time–power diagram of a SACCH multiframe in DTX mode [7].

At the receiving end, the background acoustic noise abruptly disappears whenever the radio transmitter is switched off. Since switching can take place rapidly, it has been found that this noise switching can have a very negative effect on the end user perception. In bad scenarios the noise modulation greatly reduces the intelligibility of the speech. This problem can be overcome by generating at the receiver a synthetic signal known as 'comfort noise' inserted at the receiver whenever the transmitter is switched off. If the comfort noise characterisation is well matched to that of the transmitted noise, the gaps between talk spurts can be filled in such a way that the listener does not notice the switching during the conversation. Since the noise constantly changes, the comfort noise generator should be updated constantly [15].

From the 12 bursts, which are sent in DTX mode, 4 are the SACCH frame, which is being used for signalling. The eight others contain the silence descriptor frame (SID frame) refreshing the comfort noise characteristics.

DTX has an influence on the estimation of RXQUAL. RXQUAL is measured by the mobile station for the downlink and by the base station for the uplink. The estimated RXQUAL values can be averaged before they are used in the power control and handover algorithm.

The estimation is done by evaluating the BER before decoding over a SACCH multiframe (0.48 s) and then maps the value over to an RXQUAL value. The way the BER is estimated is not specified, however its accuracy has to fulfil the specifications requirements [9].

Figure 5.14 shows one way of estimating the BER. *A* is the frame, which has been transmitted through a radio channel and which has deinterleaved, but not yet decoded. If this frame is compared with the original encoded frame (*B*), we get the actual BER before decoding, called *real BER*. The estimated BER is found

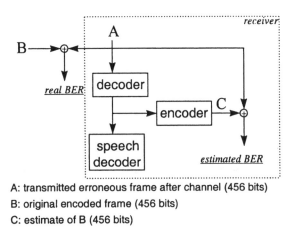

A: transmitted erroneous frame after channel (456 bits)
B: original encoded frame (456 bits)
C: estimate of B (456 bits)

Figure 5.14   BER estimation before decoding method.[a]

[a] There are other methods to estimate the BER and RXQUAL.

by encoding the received frame after it has been decoded and then comparing it with A, i.e. comparing A and C. C is the estimate of B. If the decoder can correct all the errors in A, C will be the same as B, so the estimated BER is the same as the real BER. However, if there are still errors after the decoding, the estimated and real BER might differ.

Since an RXQUAL value is calculated for every SACCH multiframe, the BER is estimated over 100 TDMA bursts in the case of non-DTX mode and 12 TDMA bursts if in DTX mode. The first is called RXQUAL-FULL and the second RXQUAL-SUB.

Figure 5.15 shows the difference between the real BER and the estimated BER as a function of the real BER respectively in the case of being not in DTX mode and in DTX mode. A GSM link simulator has been used to generate the results. Each of the operations in the GSM transmission path, including a fading radio channel and thermal noise (white Gaussian), is included. A typical urban channel (TU) [9] with speed of 50 km/h has been used with random hopping over eight uncorrelated frequencies. Every point in the DTX mode case is averaged over 12 bursts, while 100 bursts are used in the non-DTX mode case.

It can be seen that the estimation error is highest for high *real BER* and for the DTX mode case compared with the non-DTX mode case, i.e. the variance in DTX mode is higher. The reason for this is the lower number of bursts, which are used for the estimation.

The estimation error is not symmetric. The average estimation error is positive. In the non-DTX mode there are no negative points at all, while also in DTX mode the error is clearly more positive than negative. This means that the estimated BER is lower than the real BER.

An explanation can be found in Figure 5.14. As long as all errors in A are corrected in the decoder, no estimation error will be made. However, if there are still errors after the decoder, this will introduce errors in C. These errors are not

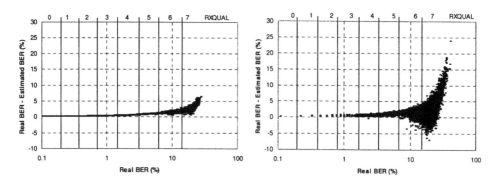

Figure 5.15 Error of estimated BER as a function of the real BER in non-DTX (left) and DTX (right) modes.

just errors on random places, since the error patterns in *B* and *C* are quite correlated. If the first 10 bits of *B* are corrupted and the decoder can only correct 5 of them, then it is very likely that there are some errors in the first 10 bits in *C*. The estimated BER will be lower, since some errors of *B* are also present in *C*, so they are not detected.

Sometimes however the error will be negative, especially for a high real BER. This is due to the fact that the error patterns become more random. No negative error can be seen in the case of non-DTX mode, since there the averaging over 100 bursts removes this, while when averaging over 12 bursts, some negative values still exist. However, if the real RXQUAL is compared with the estimated RXQUAL (based on the estimated BER), then it is found that only a small percentage of the values differ. The difference is never more than 1 RXQUAL value. When compared with the specifications [9], this means that the requirements are 100% fulfilled.

Power control

Power control is a feature, that tries to reduce the interference. In the uplink it has the additional advantage to save the battery power of the mobile station.

The values of the link quality (RXQUAL) and the link signal power (RXLEV) are used as input to the power control algorithm. Up and downlink power control act independently. The RXQUAL and RXLEV values can be weighted and averaged before being used in the power control algorithm. These averaged values are called RXLEV-AV and RXQUAL-AV. The power is regulated according to the following parameters:

- *U_RXQUAL_P*: if Px1 out Nx1 values (voting) of the averaged RXQUAL, (RXQUAL-AV) are lower (or better quality) than the *Upperthreshold RXQUAL-AV*, then the output power is decreased.
- *L_RXQUAL_P*: if Px2 out Nx2 values of the averaged RXQUAL, (RXQUAL-AV) are higher (or worse quality) than the *Lowerthreshold RXQUAL-AV*, then the output power is increased.
- *U_RXLEV_P*: if Px3 out Nx3 values of the averaged RXLEV, (RXLEV-AV) are greater than the *Upperthreshold RXLEV-AV*, then the output power is decreased.
- *L_RXLEV_P*: if Px4 out Nx4 values of the averaged RXLEV, (RXLEV-AV) are lower than the *Lowerthreshold RXLEV-AV*, then the output power is increased.

If one of the above determines that the power should be increased, the power is indeed increased, even though one of the parameters may determine that the power should be decreased. This way good quality and signal level is always guaranteed.

A power increase step and a power decrease step can be specified, but if the level or the quality of a connection drops rapidly, for example if a mobile station moves from outdoor to indoor, then it is possible to increase the power with one step to the desired level. The power control is only used once in a certain time interval, which can be specified, although the maximum performance is achieved at its maximum rate, once each 480 ms period.

5.2 Cellular network key performance indicators (KPIs)

Speech KPIs

The performance of cellular radio networks and, in particular, of speech services can be measured using multiple different key performance indicators (KPIs). Furthermore, many of these KPIs can be collected either by drive tests or from network management system (NMS) statistics. Since NMS statistics represent the overall network performance, they are more representative than drive test results. Drive test based KPIs should only play a relevant role to monitor the overall network performance when such KPIs are not available from NMS statistics. Additionally, statistics can represent busy hour or daily average performance. This book uses busy hour values, in order to analyse the performance in high load conditions.

Traditionally, cellular operators have measured and benchmarked their network performance using mostly BER and dropped call rate (DCR) to quantify the speech quality and the rate of lost connections, respectively. Additionally call success rate (CSR) and handover success rate (HSR) have been used to measure the performance of the signalling channels associated with call originations and handovers. There are however a number of additional KPIs that can be used to measure the network performance and in particular the speech quality.

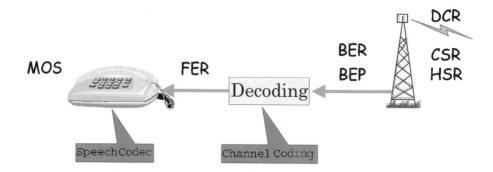

Figure 5.16 Main speech KPIs.

Figure 5.16 illustrates the different KPIs and the location in the receiving chain where they are measured. These indicators are not only utilised to measure the network performance but can be used as well by the network for multiple radio resource management functions, such as handovers, power control, adaptive multi-rate codec (AMR) link and channel mode adaptation, GPRS link adaptation, etc.

Mean opinion score (MOS)
Speech quality can be quantified using mean opinion score (MOS). MOS values, however, can only be measured in a test laboratory environment. MOS values range from 1 (bad) to 5 (excellent). They are generated from the subjective voice quality perceived and reported by the group of listeners participating in the test. Additionally, listening tests performed with different conditions (like speech material, language used and listening conditions) could lead to different MOS results. Therefore, absolute MOS values between the different tests can't be directly compared. A thorough description of this KPI and the methodology used to measure it can be found in [16].

There are however multiple tools in the market that try to measure and quantify the subjective speech quality of the link from data collected through drive tests using different algorithms.

Bit error rate (BER)
BER is a measurement of the raw bit error rate in reception before the decoding process takes place. Any factor that impacts the decoding performance, such as the use of FH or the robustness of the channel coding associated with a certain speech codec, will impact the correlation between BER and FER, or the perceived end user voice quality. Therefore, in order to properly measure the link quality using this KPI, this correlation will have to be considered for all the existing codecs and channel profiles. In GSM, BER is mapped into RxQual, which values range from 0 to 7 as displayed in Table 5.3.

Table 5.3   RXQUAL values and corresponding BER

RXQUAL	BER	Assumed value
0	< 0.2%	0.14%
1	0.2–0.4%	0.28%
2	0.4–0.8%	0.57%
3	0.8–1.6%	1.13%
4	1.6–3.2%	2.26%
5	3.2–6.4%	4.53%
6	6.4–12.8%	9.05%
7	> 12.8%	18.1%

Frame erasure rate (FER)

Frame erasure rate is a very powerful performance indicator since it is highly correlated with the final voice quality the end user perceives. Different speech codecs will have however, a slightly different FER to MOS correlation since the smaller the speech codec bit rate is, the more sensitive it becomes to frame erasures. According to the studies presented in [17], the FER values in which different speech codecs will start to experience MOS degradation and the rate of such degradation is quite uniform. Therefore, FER can be efficiently used as a speech quality performance indicator.

The traditional problem associated with the usage of FER as a KPI in GSM systems has been the lack of FER based NMS statistics. Therefore only drive test data have been available to measure the FER of the system. FER is however available to the network in the uplink (UL), and the downlink (DL) FER could be estimated from the available information in the network. Rel'99 specification includes the enhanced measurement report, which together with other enhancements brings the DL FER reporting. With this enhancement terminals will report DL FER, thus permitting the network to generate DL FER statistics based on these reports. The FER performance data presented in this section have been collected through drive tests.

Finally, it is important to define how FER data should be analysed, both in terms of the averaging to apply to the raw data and the threshold values that define the occurrence of bad quality samples. The impact of frame erasures in the speech quality degradation will be not only related with the average FER value but as well with the frame erasure distribution. The FER associated with real connections has a bursty nature. Once the channel coding is not able to restore the original information, the FER degradation is easily very high. Figure 5.17 displays a typical FER profile with 2 s (four 480 ms multiframe periods) averaging samples. When a single FER degradation event occurs, the FER values are likely very high, and it can have a significant impact on the average FER per call.

Figure 5.18 displays the same data processed with different FER averaging windows, in terms of percentage of samples with FER over a threshold. Surprisingly, the higher the averaging window, the higher the percentage of samples over the selected threshold. This is due to the previously displayed bursty nature of FER. The use of high averaging windows may lead to erroneous conclusions about the associated speech quality. Therefore, in order to better be able to determine the speech quality from the FER data, small averaging windows should be selected.

FER statistics should have a threshold defined that determines the level of non-acceptable speech degradation. For the recommended small average windows, this threshold will be in the range of 2–8%. The percentile of FER samples above this threshold will represent the proportion of bad speech quality

3500 segments of 2sec in 58 calls

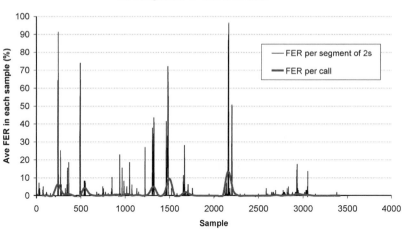

Figure 5.17 FER instantaneous distribution (Network 1).

samples. Figure 5.19 displays the outage of 2 s samples with FER over different thresholds.

Bit error probability (BEP)

Rel'99 includes an additional signal quality indicator, the bit error probability. The terminals will report the mean bit error probability and its coefficient of variation (standard deviation/mean value). According to [13] the received signal quality for each channel shall be measured on a burst-by-burst basis by the

Figure 5.18 FER performance with different averaging windows (Network 1).

Figure 5.19 Outage of FER samples for different thresholds (Network 1).

mobile station and base station subsystem (BSS) in a manner that can be related
to the BEP for each burst before channel decoding using, for example, soft
output from the receiver. The scale of the reported mean value will range
from 1 to 32 and that of the coefficient of variation will range from 1 to 8.

BEP has a higher accuracy than RXQUAL and its coefficient of variation can
be used to characterise the channel profile.

Dropped call rate (DCR)
DCR measures the percentage of connections lost and it has always been a very
important KPI since it is generally considered that a dropped call has a very
negative impact on the end user perceived quality of service (QoS). In GSM, the
mechanism of one of the main triggering events of a speech dropped call relies
on a counter (radio link timeout) that counts down or up depending on the
successful reception of the SACCH frames. This mechanism was designed to
drop connections which link had degraded below a certain threshold. Therefore,
although DCR is correlated with the end user perceived voice quality, it is rather
an indication of the performance of the SACCH signalling link.

There are different ways to calculate DCR. It can be calculated as dropped
calls per Erlang, dropped calls over originated calls and dropped calls over all
calls handled by each cell, including incoming handovers. In this book we will
use dropped calls over originated calls only, since it is the most common way
operators calculate DCR.

Figures 5.20 and 5.21 shows the DCR performance associated with multiple
mature live networks measured from NMS statistics and drive tests.

Figure 5.20   Real networks DCR from NMS statistics.

Although both the drive test and the NMS statistics belong basically to the same set of networks, the average DCR across all the measured networks is 1.8% in the case of NMS statistics and 2.8% in the case of drive tests. This difference is mostly related to the drive test route selection, which is generally designed to cover all the potential problematic areas, and has a high handover per call ratio. Additionally, there is a high variance in the drive test results due to the small amount of events (number of calls) collected to achieve a statistically significant result. Figure 5.22 shows the DCR values collected from NMS and drive tests for the same network during multiple days. The average DCR from drive tests was in this case 2.5 times higher than the one collected from NMS

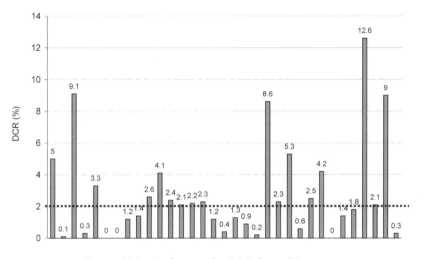

Figure 5.21   Real networks DCR from drive tests.

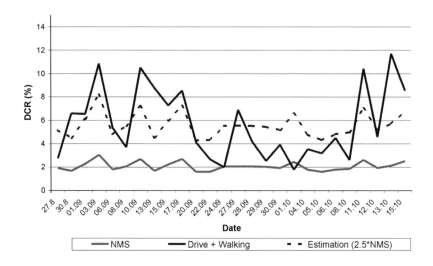

Figure 5.22   Drive tests and NMS statistics – DCR from a live network (Network 1).

statistics and there is an obvious high variance in the drive test data which may lead to erroneous conclusions.

As a conclusion, DCR should always be measured from NMS statistics, and well performing networks should have a DCR ranging from 1 to 2% depending on the baseline performance (see Section 5.5), features in use and the network load.

Call success rate (CSR) and handover success rate (HSR)
Finally, there are other KPIs, such as call success rate and handover success rate, which measure the success rates of the signalling process related with originations and handover, respectively. Figure 5.23 displays the values collected for different days and configurations from live networks for CSR and HSR. Well performing networks should have both CSR and HSR above 95%.

KPIs correlation
There is no bad performance indicator, since all of them point out different aspects of the network performance. The important factor is to be able to understand what each one represents and how is the correlation between different relevant indicators. Additionally, such correlation may be dependent on different factors. For instance, BER can be used to measure speech quality but in order to analyse the voice quality from the BER data, the correlation between BER and speech quality is needed. This correlation is not constant, since it is dependent on the decoding gain, which is dependent on the channel coding and channel profile in use. Figure 5.24 displays the significantly different correlation between BER

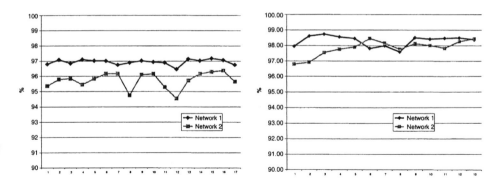

Figure 5.23 Live networks busy hour CSR (left) and HSR (right).

and FER for hopping and non-hopping channels, measured in a live network.

Finally, Figure 5.25 displays the correlation of the most common KPIs for the same network and different load levels. The BER and FER data have been collected from drive tests. The statistically significant NMS DCR statistics have been used. As presented in Figure 5.22, the DCR from the drive tests routes is higher than the overall network DCR, so the NMS DCR values need to be scaled accordingly in order to obtain the estimated DCR performance associated with the drive tests routes and perform the KPI correlation.

The percentage of 2 s samples with FER equal or higher than 4.2% is highly correlated to the percentage of dropped calls. For the purpose of this book, 2% DCR and 2% outage of 2 s FER samples with FER equal or higher than 4.2% (2% bad quality samples outage) have been selected in order to benchmark the

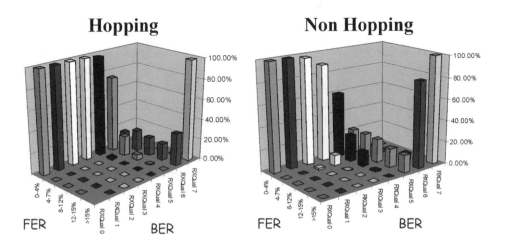

Figure 5.24 BER to FER correlation for different channel profiles (hopping/non-hopping) (Network 1).

Figure 5.25   Main KPIs correlation from a live network (Network 1).

network performance. From multiple analysis performed, BER and FER have shown a very linear degradation as the interference (load) increased. On the other hand, DCR tends to have a linear degradation up to a certain 'saturation point', when the degradation starts to become exponential. This is due to the nature of the radio link timeout and the other DCR triggering mechanisms.

Data KPIs

GSM/EDGE data services can be offered in circuit switched (CS) mode with high speed circuit switched data (HSCSD)/enhanced circuit switched data (ECSD), and in packet switched (PS) mode with (E)GPRS. This book will mostly focus on PS services delivered through the (E)GPRS PS core network. Measuring (E)GPRS performance is somehow different than measuring voice performance. The end user perception will be determined in most data services by the application interfaces. Such interfaces will filter events such as dropped connections, which in the case of speech services are critical in the end user quality perception, so end users only perceive delays and effective throughputs. There are many different data services with different performance requirements. The performance requirements are agreed between the network and the mobile station at the service establishment phase. The radio resource management should then do as much as possible to maintain the negotiated quality. Data KPIs can be measured at different levels and in different places. The radio access network has control of the performance below logical link control (LLC) level (or packet data convergence protocol (PDCP) layer in GERAN Rel'5), but it cannot control what is happening on the core network side. The most important (E)GPRS radio link performance parameters in the radio access network are reliability, throughput and delay.

Reliability
It describes the maximum probability of erroneous radio link control (RLC) delivery to the LLC layer. Reliability depends mainly on the operation mode of the RLC layer. In most of the cases, errors are not tolerated by data applications and retransmissions are needed. In order to avoid further delays, retransmissions are done preferably in the RLC layer by operating in acknowledge (ACK) mode. In RLC ACK there is a small probability of undetected block errors due to the limited CRC field length (16 and 12 for GPRS and EGPRS, respectively) included in each radio block. The residual BER in RLC ACK mode is 10^{-5} and 2×10^{-4} for GPRS and EGPRS, respectively. There are some delay sensitive data services that operate in UNACK mode and can handle a certain number of erroneous RLC blocks. The mode of operation is negotiated at the service establishment phase.

Throughput
Throughput is measured as the amount of data delivered to the LLC layer per unit of time, i.e. RLC payload. Throughput is only relevant when the amount of data that is being transmitted is large enough and can be easily measured by the network over the duration of a temporary block flow (TBF) as displayed in Figure 5.26. Each TBF has associated a measured throughput sample. Each network will have a different throughput probability distribution depending on the load and network configuration. Instead of using the whole probability distributions it is more practical to compute the average and percentile throughput values. The typical variation with load of the average TBF throughput can be seen in Figure 5.27. Throughput percentile values characterise the tails of the throughput distribution. Tenth percentile (minimum throughput achieved by 90% of the TBFs) is considered in Chapter 7 as one of the main quality criteria. Real-time services with guaranteed throughputs require instantaneous throughput measurement at RLC level that can be done with a running average filter.

User throughput can be also measured at application layer, for example, by measuring the downloading time of a large file using FTP protocol. This end-to-

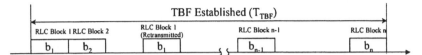

Figure 5.26   Temporary block flow throughput.

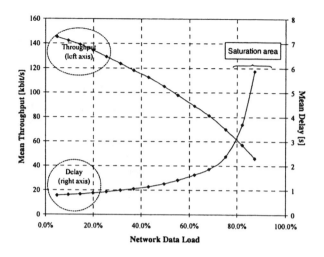

Figure 5.27 Typical TBF throughput and LLC variation with load (EGPRS simulations).

end measurement should be done in one of the extremes of the communication (either at the server or at the mobile station (MS)). This measurement can be done with a simple application that monitors application layer throughputs. For real-time services, like streaming, instantaneous throughput can be measured with a running average filter at application level.

Delay

Delay is the time it takes for a LLC protocol data unit (PDU) to be transferred from the serving GPRS support node (SGSN) to the mobile station or from the mobile station to the SGSN. LLC level delay includes the required time to re-transmit erroneous RLC blocks. LLC delay probability distribution can be measured at the PCU. Delay is relevant for short packets transmission and for real-time data services. In RLC ACK mode, the delay is very dependent on the radio access round trip time (RA RTT) (see Section 7.6), and it may have a considerable impact on application performance. RA RTT defines the total time since a RLC packet is transmitted until ACK/NACK information is received. The polling strategy (see Chapter 7) plays an important role in RA RTT. Its typical value is of the order of 250 ms including TBF establishment delay. With adequate network architecture optimisation it is possible to minimise RA RTT. As the data load increases, the LLC delay increases smoothly until the network is close to saturation point, where LLC delays are so large that it is not possible to offer any data service efficiently. The evolution of the delay with load is displayed in Figure 5.27. For real-time data services working in RLC NACK mode, the delay is minimised by allocating the maximum priority in the scheduler.

Delay can be measured at application layer as well. A common way of measuring (E)GPRS network response time is by sending PING commands to a server. A heavy loaded network may have too long response times that may disrupt application and transport layer protocols.

There are some data performance indicators that are defined at network level and do not refer to the performance or individual users. Some of the most important are: (E)GPRS load, throughput per cell, timeslot (TSL) capacity, percentage of satisfied users, TBF blocking and throughput reduction factor.

(E)GPRS load

A useful way to measure data load is to measure the average amount of TSL utilised by (E)GPRS services, which can be expressed in terms of 'Data Erlangs' or 'TSL utilisation'. This measurement includes user data, retransmission and control information sent over PDTCH channels. This definition is very useful to understand how many TSL are, on average, fully utilised by (E)GPRS services. Both *Data Erlangs* and *TSL utilisation* can be easily calculated in real (E)GPRS network from the available counters. *Data Erlang* can be defined as:

Data Erlang= (Total number of transmitted radio blocks

$\times$ 0.02 s)/Duration of busy hour (s)

Note that 1 Erlang corresponds to a continuous transmission of 1 radio block every 20 ms.

A Data Erlang is made of initial RLC block transmissions, successive retransmissions of previously sent RLC blocks, and control RLC blocks. Retransmissions are done in RLC acknowledgement mode. Either the network or the mobile station retransmit RLC blocks that have not been acknowledged. Retransmissions can be divided in two classes: regular retransmissions and pre-emptive retransmissions. Regular retransmissions are done when RLC blocks have been negatively acknowledged. Pre-emptive retransmissions are done for a radio block that has not been positively or negatively acknowledged (no information has been received since the radio block was sent). Those blocks are called pending-ACK blocks. Pre-emptive retransmissions of pending-ACK blocks are done when there is no other RLC block to transmit under the following two conditions. The first one is when the RLC protocol is stalled and the transmitting window can't advance. The second one is when there is no more user data to transmit but acknowledgement information has not been received yet. Pre-emptive retransmissions are optional for both network and MS side. Although pre-emptive retransmissions are good from the RLC protocol point of view, they actually increase the Data Erlang in the network even though only a small fraction of the pre-emptive retransmissions are useful.

Figure 5.28 Kbps/cell and TSL capacity as a function of Data Erlangs (EGPRS simulations).

Throughput per cell/sector

Throughput per cell is a measure of the amount of bits that can be transmitted in a cell. It is normally measured in kbits/s/cell units. (E)GPRS networks can easily calculate this indicator by measuring how much information has been transmitted during a certain period. Throughput per cell doesn't take into account retransmissions nor control information below LLC layer. Figure 5.28 shows an example of variation of throughput per cell as a function of *Data Erlangs*.

TSL utilisation

A measurement of how much hardware (HW) resources are being utilised by (E)GPRS service is given by the *TSL utilisation* concept. TSL utilisation is defined as

TSL utilisation (%) = *Data Erlang*/Available TSL for (E)GPRS.

  TSL utilisation takes into account the average number of TSLs that are available for (E)GPRS. It is a measure of how much the network is loaded with data services. Networks with *TSL utilisation* close to 100% are close to saturation and end user performance is likely to be very poor. In Figure 5.27, the network data load is measured as a percentage of *TSL utilisation*.

TSL capacity

TSL capacity is a measure of how much data the network is able to transmit with 1 Data Erlang. It is a measure of how efficiently HW resources are being utilised to transmit data. TSL capacity depends mainly on network interference levels. It also may depend on RLC protocol efficiency. It can be calculated as

TSL capacity = Throughput per cell/*Data Erlangs*.

TSL capacity depends on the RLC algorithm design (e.g. the use of pre-emptive transmission or not, polling strategy, etc.) and data traffic burstiness. An RLC efficiency factor can be introduced as:

TSL capacity = RLC efficiency factor

 × Throughput per cell/Optimum Data Erlangs

where Optimum Data Erlangs is the Data Erlangs calculated excluding non-useful retransmission of pending-ACK blocks. Therefore, it assumes optimum RLC protocol behaviour. RLC efficiency factor tends to be lower with GPRS (mainly due to RLC window stalling, see Section entitled *Polling and acknowledgement strategy* in 7.2) and in the case of bursty traffic (see Section entitled *Effect of traffic burstiness* in 7.4). EGPRS with non-bursty traffic has an RLC efficiency factor close to 1 as it doesn't suffer that much from stalling.

A typical TSL capacity can be seen in Figure 5.28. The more loaded the network the higher the interference levels and the lower the TSL capacity.

Satisfied users

Even in best-effort services it is necessary to introduce a network level indicator of end-to-end performance. The user satisfaction criteria depend on the type of services. For best-effort data, a combination of minimum TBF throughput and maximum LLC delay can be considered. The network can produce statistics of TBF throughput and, possibly, LLC delays. The degree of user satisfaction is measured as a percentage of samples that fulfil the user satisfaction criteria. Operators' experience and customer feedback will help to identify the percentage and the satisfaction criteria to be associated with a good performing network, in the same way that nowadays 1% DCR is considered to be good performance for speech services. In Chapter 7, different satisfaction criteria will be considered with 90% of good quality samples.

Satisfied users and throughput per cell are closely related in best effort networks. The more loaded the network the higher the throughput per cell but, on the other hand, a lower percentage of satisfied users can be achieved.

TBF blocking

TBF blocking probability shows how close cells are to saturation. When the blocking limit is reached, packets have to be dropped in the core network. Blocking changes exponentially with network load, so once a network is experiencing blocking, small increases in the offered load will be translated into much higher blocking probabilities. With high TBF blocking the performance collapses, and LLC delay increases exponentially, as displayed in Figure 5.29. Although there are seven TSLs available for (E)GPRS services in the network in this example, the maximum Data Erlangs will never reach 7. The EGPRS

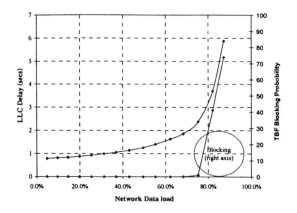

Figure 5.29 LLC delay degradation with TBF blocking (EGPRS simulations).

network should be dimensioned to ensure a low TBF blocking probability. This is a similar concept to speech blocking dimensioning, where the number of TSLs are dimensioned to achieve a maximum blocking probability of 2% (typically).

Throughput reduction factor
Throughput reduction factor (shortly, reduction factor) is a measure of how the TBF throughput is reduced when the same TSLs are shared by more than one user. By knowing TSL capacity and reduction factor, it is possible to compute average user throughput as

TBF throughput = TSL capacity × Number of allocated TSL × Reduction factor.

Low loaded cells have a reduction factor close to 1, meaning that very few TSLs are shared between the users. Ideally, reduction factor wouldn't depend on the number of allocated TSLs. In practice with limited available TSLs and real channel allocation algorithms there are some differences. In any case, reduction factor concept is a simple and useful tool for dimensioning (E)GPRS networks.

Figure 5.30 shows reduction factors computed in an EGPRS network with eight TSLs available for EGPRS. Best-effort traffic is assumed. When the number of available TSLs increases, the available resources are used more efficiently and the same reduction factors can be achieved with higher *TSL utilisation*. This can be used to measure trunking efficiency gains. An in-depth (E)GPRS trunking performance analysis is performed in Chapter 7.

5.3 Spectral efficiency

This section introduces the concept of spectral efficiency and the different ways used to quantify it. Cellular operators invest money to get frequency spectrum licences for a certain number of years. It is therefore very important to ensure

Figure 5.30   Typical reduction factor vs. *TSL utilisation* curves (EGPRS simulations).

maximum return on such investment. Spectral efficiency quantifies the amount of traffic a network can carry for a given spectrum. Furthermore, it is a measure of the radio performance efficiency, thus higher spectral efficiency will imply higher quality of service provided to the end users for a given traffic load.

Figure 5.31 shows the performance characterisation, in terms of load and quality, of two different networks, or a single network with a different set of functionality. Network B is more spectral efficient than Network A because, for a certain benchmarked quality, Network B can carry more traffic, and for a certain traffic load, Network B has a better performance. If certain functionality that increases the spectral efficiency is introduced in Network A, the quality will immediately improve and considering the previous quality as the benchmarking quality level, there will be additional traffic load that can be carried by the network. The relative value of this additional traffic load is the capacity increase provided by the introduced functionality.

Figure 5.31   Spectral efficiency concept.

Even though the most traditional way to express spectral efficiency has been Erlangs per MHz per square kilometre (Erl/MHz/km^2), there have been many other ways to quantify the same concept in the wireless industry. Some of the ways to measure it are:

- effective reuse
- fractional load
- frequency load
- effective frequency load (EFL)

It is important at this point to highlight the terminology used across this book, since some commonly used terms have a different meaning in different regions worldwide:

- *Cell/sector.* Both terms represent the same concept. They refer to the logical entities that have a single BCCH associated with each one of them. Several cells/sectors can be grouped under the same site if they are fully synchronised.
- *Site.* A site refers to a single or multiple cells/sectors sharing the same physical location and being fully synchronised. The radio performance benefits from synchronised cells/sectors are better handover performance of synchronised handovers and the possibility of using MAIO management across the synchronised cells/sectors.

Effective reuse

One of the common ways used to try to quantify the spectral efficiency is the effective reuse concept. Effective reuse describes how often the same frequency can be reused in the network. Basically, the capacity of the network can be estimated, if the potential effective reuse of a particular solution is known. It is calculated as

$$R_{\text{eff}} = \frac{N_{\text{freqsTOT}}}{N_{\text{TRXave}}}$$

where R_{eff} = effective reuse; N_{freqsTOT} = total number of used frequencies; N_{TRXave} = average number of TRXs per sector.

Effective reuse is a measure of the rate of the available spectrum by the number of TRXs per sector. However, this measure has some limitations:

- The deployed number of TRXs does not imply the TRXs are fully loaded.
- The relationship between the number of TRXs and the maximum traffic carried will be conditioned by the blocking dimensioning criteria, which may vary from network to network.

- The number of Erlangs per TRX for the same blocking rate will increase as the TRX configuration increases.

Therefore, effective reuse is not an accurate indicator to measure the spectral efficiency since the correlation between number of TRXs and carried traffic is not fixed.

Fractional load

Another way to measure the efficiency of hopping networks, related to the effective reuse, is the fractional load. Fractional loading means that the cell has been allocated more frequencies than the number of deployed TRXs. This is only possible for RF hopping TRXs. The fractional load reflects the ability to load the hopping network with certain TRX configuration. It however has the same effective reuse associated limitations. It can be calculated as

$$L_{frac} = \frac{N_{TRX}}{N_{freqs/cell}}$$

where

L_{frac} = fractional load
N_{TRX} = number of TRXs in a cell
$N_{freqs/cell}$ = number of frequencies allocated to a cell (frequency hopping list length).

Frequency allocation reuse

Frequency allocation reuse indicates how closely the frequencies are actually reused in a network. It is calculated as

$$FAR = \frac{N_{freqsTOT}}{N_{freqs/MA}}$$

where

FAR = frequency allocation reuse
$N_{freqsTOT}$ = total number of used frequencies
$N_{freqs/MA}$ = average number of frequencies in MA lists.

This term measures the ability of hopping networks to implement very tight reuses. If the network doesn't utilise fractional loading, the frequency allocation reuse is the same as effective reuse. FAR also gives good indication about frequency allocation, but does not provide a good measure for spectral efficiency since it doesn't reflect how loaded the network is.

Frequency load

The frequency load can be written as

$$L_{freq} = L_{HW} \cdot L_{frac}$$

where

L_{freq} = frequency load
L_{HW} = the busy hour TSL occupation
L_{frac} = fractional load

This indicator reflects how loaded a frequency of a sector is, i.e. how much traffic is carried by the available spectrum. Therefore, it is a good indication of the network spectral efficiency. However, it is dependent on the deployed reuse. In order to remove this limitation, the definition of a spectral efficiency indicator that can be used in any network configuration, the effective frequency load (EFL) measure, is introduced.

Effective frequency load

The effective frequency load (EFL) quantifies how loaded each frequency available in the system is [18]. It is independent of the frequency reuse and of the TRX configuration deployed. It can be calculated as

$$EFL(\%) = \frac{\text{Frequency load}}{\text{FAR}}$$

or

$$EFL(\%) = \frac{L_{HW}}{R_{eff}}$$

which can be translated to

$$EFL(\%) = \frac{Erl_{BH}}{Tot\# \text{ freq}} \frac{1}{Ave\#\left(\dfrac{TSL}{TRX}\right)}$$

where

Erl_{BH} = busy hour carried traffic (Erlangs)
Tot# freq = total number of available frequencies in the system
Ave#(TSL/TRX) = average number of TSLs per TRX available for traffic usage

EFL can be directly translated into other spectral efficiency indicators. For example, it can be translated into Erl/MHz/sector, using the following formula:

Spectral efficiency (Erl/MHz/sector) $\cong 40 \times$ EFL (%)

where Ave#(TSL/TRX) has been considered to be 8.

EFL can be easily calculated from the network and it provides a direct measure of how loaded the frequency spectrum is. If one TRX was deployed in every sector for each available frequency and all the TRXs were fully loaded (8 Erlangs/TRX) the EFL would be 100% and the associated spectral efficiency (Erl/MHz sector) would be 40. In this book EFL has been selected as a measure of traffic load and spectral efficiency. As presented in the next section, an EFL based methodology has been developed to characterise real networks performance and quantify their spectral efficiency for any benchmarking quality.

FR and HR mixed traffic scenario

HR EFL = HR Erlangs/(Tot# freq $\times 2 \times$ aver#(TSL/TRX)).

Total EFL can be defined as

Total EFL = FR EFL + HR EFL.

The overall performance depends on the share of HR and FR users. Relative performance analysis with different features or different networks can be done only with the same traffic share.

EFL for data services
EFL describes how much the frequencies are loaded with (E)GPRS traffic. It can be calculated using the *Data Erlangs* KPI defined in the previous section. In the FH layer, the TSL capacity depends only on the EFL value, independently of the number of hopping frequencies or the amount of *Data Erlangs*. This principle is illustrated in Figure 5.32. The figure shows the evolution of TSL capacity as EFL increases. It can be noticed that networks with different number of hopping frequencies have the same performance.

EFL for mixed voice and data services

When voice and data services are supported together in the hopping layer, it is possible to define a total Erlang and a total EFL:

Total Erlangs = Voice Erlangs + Data Erlangs

Total EFL = Voice EFL + Data EFL.

EFL analysis can be done considering Voice and/or Data KPIs vs Total EFL. In mixed scenarios the service performance depends on the share of Data and Voice Erlangs.

Figure 5.32   TSL capacity evolution with EFL in an EGPRS network.

5.4 EFL trial methodology

The spectral efficiency of deployed cellular networks is very hard to quantify. This is due to the fact that cellular networks have a pseudo-static operating point in terms of traffic load. The performance measurements collected will identify the performance of the network associated with the traffic load operating point. However, this information is not sufficient to quantify the spectral efficiency of the measured network, i.e. it is not possible to quantify the maximum capacity the network can support with a certain benchmark quality. In order to do so the network should be operating in maximum load conditions.

Network performance characterisation

The methodology presented in this section allows the performance characterisation of deployed cellular networks. Once this is done, it is possible to estimate the load associated with any benchmark quality. It is therefore possible to calculate the maximum network capacity for a given benchmark quality or how much the network quality will degrade as load increases. As described in the previous section, the EFL is directly proportional to the traffic and inversely proportional to the available bandwidth. In order to characterise the performance of a network the quality associated with different EFL values needs to be identified. In real networks the traffic load is fixed but the number of frequencies deployed can be modified in order to increase the EFL. During trials the network performance can be characterised by modifying the number of frequencies used.

Figure 5.33 displays the characterisation of Network 1 carried out during extensive trials. Both the BCCH and hopping layer have been characterised since each one of these layers displays a very differentiated performance. There is hardly any degradation on BCCH performance as load increases since the BCCH TRX is always transmitting full power. On the other hand the hopping layer shows a clear degradation as load increases. The overall network performance will be the combination of the performance of both layers weighted by the traffic distribution. In the hopping layer, each point represents different hopping configurations, modifying the number of frequencies from 23 to 14. In the BCCH layer the different traffic load operating points are achieved by means of different channel allocation priority strategies, which define the traffic distribution across the BCCH and hopping layers.

As shown in the figure, even though Network 1 had a particular traffic load operating point, once the network performance is characterised, the performance associated with any traffic load and the maximum traffic load for certain quality requirements (spectral efficiency) can be calculated. This has obvious benefits in order to deploy the most adequate technology strategy according to the required quality and the forecasted traffic growth.

Trial area definition

The trial area has to be representative of the network to be characterised. In this book, results from multiple trials will be presented. In order to be representative

Figure 5.33   Characterisation of Network 1 performance.

of the networks to be characterised, most of them where performed in the high traffic level areas of the network and had an adequate size (30–40 sites).

The trial area was always divided into the main trial area and a boundary area. All the results presented and analysed correspond to the main trial area (15–20 sites). The boundary area (15–20 sites) implements the same functionality and configuration as the main trial area but with a slightly modified planning ensuring the main trial area is isolated from the external network and serving as a transition buffer to the different planning of the external network. Figure 5.34 illustrates the trial area definition for Network 1.

Each network configuration trialled, using a different set of functionality, was tested for as long as required in order to collect all the information required to characterise the performance associated with such configuration. All available performance indicators were collected, processed and analysed, but the results presented in this book are usually provided in terms of dropped call rate and frame erasure rate.

The information presented from the trials, in terms of traffic load and quality, represents the average value of the network during busy hour period. This average value however will have a distribution across the network that should be understood in order to properly analyse the results. Figures 5.35 and 5.36

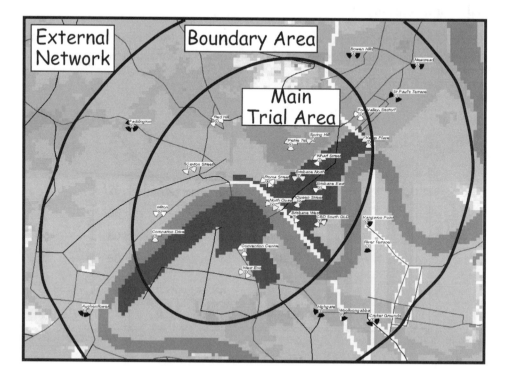

Figure 5.34 Network 1 trial area definition.

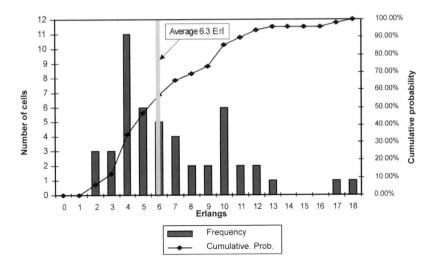

Figure 5.35 Traffic load distribution in Network 1 main trial area.

represent the distribution of traffic load and dropped calls across the sectors of the Network 1 trial area.

Methodology validation

The presented methodology is based on the assumption that the performance of a network is fixed for a given EFL value. The high EFL trialled values were obtained decreasing the utilised spectrum. This means that doubling the load or halving the number of utilised frequencies should have the same impact on

Figure 5.36 Dropped call rate distribution in Network 1 trial area.

Figure 5.37 EFL network characterisation with different bandwidths based on simulations.

performance since they generate equivalent EFL values. In other words the performance characterisation of the network should be independent of the available bandwidth.

Figure 5.37 presents the results of a simulation study which utilised different bandwidths for the hopping layer. It is clearly seen that the network performance characterisation is equivalent for all bandwidths with the exception of the 2 MHz case. This is due to lower hopping gains associated with this case. However, as explained in Section 5.1, the additional gains from FH gets marginal over a certain number of frequencies in the hopping list. Therefore, the methodology is accurate as long as the number of frequencies of the configurations utilised to characterise the network does not impact the performance of the basic functionalities deployed in the network. The deviation suffered from this effect will worsen the calculated performance as compared with the real performance.

5.5 Baseline network performance

The baseline performance of a network is defined in this book as its performance with the basic set of standard GSM functionality. Such functionality has been considered to be RF hopping with advanced reuse schemes, MAIO management, DL and UL power control and DL and UL discontinuous transmission (DTX). Chapter 6 will analyse the performance enhancements associated with this set of functionality. All the performance analysis carried out throughout this book is going to show the achievable performance with different features and functionality based on relative gains obtained from simulations and trials on top of the GSM achievable baseline performance, which is based on real networks performance and presented in this section.

The baseline performance of any cellular network, using any technology, is going to be heavily conditioned by the initial network planning, the traffic distribution and the complexity of the propagation environment. In terms of network planning, the most important factors are the antenna heights and locations. In order to ensure a satisfactory baseline performance the antenna locations should be carefully selected. High antennas may provide low-cost coverage for initial deployment but will have a negative impact on interference limited environments, limiting the achievable spectral efficiency. Antennas below roof levels, which signal propagation is contained by the surrounding buildings have proven to provide best performance for interference limited environments and hence highest spectral efficiency. Following this principle, microcellular networks, due to their characteristic antenna locations will be able to reach a higher spectral efficiency through the implementation of tighter effective reuses. Different networks will have very different performance depending on these factors.

The network performance characterisation is generally quite similar across different networks. As presented in the previous section, the generation of this characterisation requires to measure the performance associated with multiple effective frequency load (EFL) values. Figure 5.38 presents the hopping layer characterisation for two networks with exactly the same set of functionality deployed (RF hopping 1/1, MAIO management, UL PC and UL DTX). There is a significant baseline performance offset but both networks display similar quality degradation slope as traffic load increases.

In order to estimate the achievable GSM baseline network performance, the measured characterisation of a live network with the defined basic set of GSM

Figure 5.38 Live networks hopping layer characterisation.

standard functionality will be used. Figure 5.39 displays the hopping layer performance associated with mature macrocellular networks using such defined basic set of GSM standard functionality. Network 1 performance has been fully characterised. From Networks 3, 4 and 5 only the performance information from the traffic load operating point is available. The dotted line represents the estimated characterisation for Network 5 according to the previous analysis. Considering 2% DCR as the benchmark quality point, the displayed estimation indicates that 12% EFL in the hopping layer is achievable for Networks 5. Networks 3 and 4 measured performances are located in the area marked as *higher performance*, which indicates that these networks should achieve as well the previously mentioned achievable baseline performance. This is the value used throughout the book as the achievable GSM baseline performance for 5 MHz bandwidth deployments.

In terms of BCCH achievable performance, the assumption throughout the book is that a BCCH reuse of 12 can be achieved guaranteeing adequate control channels performance. With this reuse the speech traffic is likely to experience worse performance than the benchmarking quality (as displayed in Figure 5.33), but since this targeted quality is related to the overall performance, with the right traffic load and traffic distribution between BCCH and hopping TRXs, the targeted requirement will be fulfilled. Further functionality, such as the transmit diversity techniques presented in Chapter 10, AMR or the 'BCCH underlay' presented in Chapter 9 may enhance the BCCH performance, either improving the quality or allowing for even tighter reuses to be deployed.

The previous analysis was performed for the case of 5 MHz deployment, however, the available deployment bandwidth will condition the performance

Figure 5.39   GSM baseline network performance for 5 MHz bandwidth deployments.

Figure 5.40 Performance gains relative to the 5 MHz case.

of the hopping layer. As presented in Section 5.1, the gains from FH, frequency and interference diversity, are going to be dependent on the number of hopping frequencies in use. Figure 5.40 illustrates the hopping performance differences, relative to the 5 MHz case, for the 3.6 and 10 MHz bandwidth cases.

References

[1] Mouly M., Pautet M. B., 'The GSM System for Mobile Communications', 1992.
[2] Nielsen Thomas T., Wigard Jeroen, 'Performance Enhancements in a Frequency Hopping GSM Network', Kluwer Academic Publishers, Dordrecht, 2000.
[3] ETSI TC-SMG, GSM 05.02: European Digital Cellular Telecommunications System (Phase 2) – Multiplexing and Multiple Access on the Radio Path, September 1994.
[4] COST231, Digital Mobile Radio towards Future Generation Systems, COST 231 Final Report, 1999.
[5] Mehrotra A., 'Cellular Radio Performance Engineering', Artech House, London, 1994.
[6] European Radiocommunications Office (ERO), Final Document for the ERO Workshop on Traffic Loading and Efficient Spectrum Use, 1998.
[7] Mogensen P., Jensen C., Andersen J. B., '1800 MHz Mobile Netplanning based on 900 MHz Measurements', COST 231 TD(91)-008, Firenze, January 1991.
[8] Madsen B., Data Receiver for the Pan European Mobile Telephone System (GSM), Dancall Radio A/S, September 1990.
[9] ETSI TC-SMG, GSM 05.05: European Digital Cellular Telecommunication System (Phase 2) – Radio Transmission and Reception, March 1995.

[10] Mogensen P., Wigard J., Frederiksen F., Performance of Slow Frequency Hopping in GSM (Link-level). In COST 231, September 1995.

[11] Wigard J., Mogensen P., Johansen J., Vejlgaard B., 'Capacity of a GSM Network with Fractional Loading and Random Frequency Hopping', IEEE Proc. PIMRC, Taipei, 1996, pp. 723–727.

[12] Klingenbrunn Thomas, Mogensen P., 'Modelling Frequency Correlation of Fast Fading in Frequency Hopping GSM Link Simulations', IEEE Veh. Technol. Conf., 1999, pp. 2398–2402.

[13] ETSI TC-SMG, GSM 05.08: European Digital Cellular Telecommunication System (Phase 2) – Radio Sub-system Link Control, March 1996.

[14] ETSI Section 6.12, Comfort Noise Aspect for Full Rate Speech Traffic Channels, January 1993.

[15] Mehrotra A., 'GSM System Engineering', Artech House, Boston, 1997.

[16] ITU-T P.800, TELECOMMUNICATION STANDARDIZATION SECTOR OF ITU (08/96), SERIES P: TELEPHONE TRANSMISSION QUALITY, Methods for Objective and Subjective Assessment of Quality.

[17] ETSI TC-SMG, GSM 06.75: Performance Characterization of the GSM Adaptive Multi-rate (AMR) Speech Codec.

[18] Wigard Jeroen, Mogensen Preben, Michaelsen Per Henrik, Melero Juan, Halonen Timo, 'Comparison of Networks with Different Frequency Reuses, DTX and Power Control using the Effective Frequency Load Approach', VTC2000, Tokyo, Japan.

6

GSM/AMR Voice Performance

Juan Melero, Timo Halonen, Jari Hulkkonen, Jeroen Wigard,
Angel Rivada and Martti Moisio

Despite the expected growth of data services, conventional speech will be the dominant revenue generation service for many years. As voice traffic continues growing, many networks are facing capacity challenges, which may be addressed with the right technology deployment. This chapter presents the global system for mobile communications (GSM) voice performance, highlighting the spectral efficiency enhancements achievable with different functionality. Section 6.1 introduces the performance improvements associated with basic baseline GSM speech performance functionality (as defined in Chapter 5), which includes frequency hopping (FH), power control (PC) and discontinuous transmission (DTX). In Section 6.2 frequency reuse partitioning methods are presented. Section 6.3 presents trunking gain GSM functionality such as traffic reason handover or directed retry. Half rate is introduced in Section 6.4. Finally, Rel'99 adaptive multi-rate (AMR) is thoroughly analysed in Section 6.5 and Rel'5 enhanced data rates for global evolution (EDGE) AMR enhancements are introduced in Section 6.6. The presented performance results in this chapter include comprehensive simulations and field trials with real networks around the world.

6.1 Basic GSM performance

This section will describe the network level performance enhancements associated with the basic set of standard GSM functionality defined in the previous chapter: FH, PC and DTX. Section 5.5 introduced the GSM baseline network performance based on the deployment of this basic set.

Frequency hopping

The principles of operation of frequency hopping and the associated link level gains were introduced in Chapter 5. This section will focus on hopping deployment strategies and overall network performance. Frequency hopping can be deployed using two differentiated strategies: Baseband (BB) hopping and radio frequency (RF) hopping. Both kinds of hopping are illustrated in Figure 6.1.

In BB hopping, the number of hopping frequencies is limited to the number of deployed transceivers (TRXs), since the hopping is executed by the call being switched across the TRXs, which frequencies are fixed. The broadcast control channel (BCCH) can be included in the hopping sequence.[1]

When RF hopping is used, the call stays on one TRX, but the frequency of the TRX changes every frame. The number of hopping frequencies is only limited by the number of available frequencies, thus providing greater deployment flexibility.[2] In practice, the BCCH TRXs have fixed frequencies since the BCCH has to be transmitted all the time. These frequencies are generally not included in the hopping sequence since the overall performance improves when the hopping and non-hopping bands are not mixed. This means that when RF hopping is deployed, there are usually two effective layers per sector, the BCCH TRX (non-hopping) and the non-BCCH TRX (hopping).

Figure 6.1   Baseband and RF hopping.

---

[1] Some terminals experience problems with BB hopping when BCCH is included in the hopping list and DL power control is used due to the wide range of signal fluctuation.

[2] GSM specifications limit the maximum number of hopping frequencies per hopping list to 64.

When the TRX configuration is very large across the whole network (over 64 TRXs per sector) BB hopping will perform similarly to RF hopping. Additionally, the benefits from BB hopping are applicable to the BCCH TRX as well, and for large configurations the combining loss is potentially lower since BB hopping can use remote tune combiners (RTCs) while RF hopping requires wideband combining. However, in most practical deployments, RF hopping will outperform BB hopping due to its flexibility and robustness since its performance is not conditioned by the number of TRXs deployed or the accuracy of the frequency planning. The additional frequency diversity gains from RF hopping and the use of cross polarisation antennas and air combining will compensate for the potential additional combining losses. All the studies throughout this book have considered RF hopping deployments although the performance of BB hopping would be equivalent under the previously stated conditions.

Figure 6.2 displays the performance of the RF and BB hopping layers for Network 1 in terms of dropped call rate and frame erasure rate. Since the average number of TRXs per sector was 2.5, BB hopping gains were in this case not comparable to those of RF hopping.

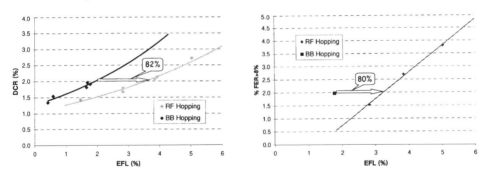

Figure 6.2   Network 1 RF hopping vs. BB hopping DCR and FER trialled performance.

With a hopping list of n frequencies, $64 \times n$ different hopping sequences can be built. These are mainly defined by two parameters, as was explained in Chapter 5. Firstly, the mobile allocation index offset (MAIO), which may take as many values as the number of frequencies in the hopping list. Secondly, the hopping sequence number (HSN), which may take up to 64 different values. Two channels bearing the same HSN but different MAIOs never use the same frequency on the same burst, while two channels using different HSNs only interfere with each other $(1/n)$th of the time.

One of the main hopping deployment criteria is the frequency reuse selection. The different frequency reuses are characterised with a reuse factor. The reuse factor indicates the sector cluster size in which each frequency is used only once. The reuse factor is typically denoted as x/y, where x is the site reuse factor and y the sector reuse factor. This means that a hopping reuse factor of 3/9, corre-

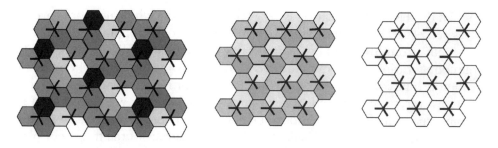

Figure 6.3   Hopping standard reuses: 3/9, 1/3 and 1/1.

sponds to a cellular network with nine different hopping lists which are used every three sites. Frequency hopping started to be used in the mid-1990s. Initially conservative reuses, such as 3/9, were used. However, 3/9 minimum effective reuse was limited to nine. Later on 1/3 proved to be feasible and finally 1/1 was considered by many experts as the optimum RF hopping frequency reuse. Figure 6.3 illustrates the hopping *standard* reuses of 3/9, 1/3 and 1/1.

As the frequency reuse gets tighter, the interference distribution worsens, but on the other hand the associated frequency hopping gains increase. Figures 6.4 and 6.5 display the performance of hopping reuses 1/1 and 3/9. The results are based on simulations using Network 1 propagation environment and traffic distribution. Reuse 3/9 shows better carrier/interference (C/I) distribution and therefore bit error rate (BER) performance. However, reuse 1/1 has a better overall frame error rate (FER) performance due to the higher frequency and interference diversity gains. As a conclusion, the most efficient hopping reuse will be dependent on the generated C/I distribution and the associated link level gains.

Reuse 1/3 has proven to have equivalent or even slightly higher performance in deployment scenarios with regular site distribution and antenna orientation. However, most practical networks are quite irregular both in terms of site distribution and antenna orientation. Reuse 1/1 has the advantage of better

Figure 6.4   C/I and BER Network 1 simulation performance.

Figure 6.5   FER Network 1 simulation performance.

adaptability to any network topology and provides higher hopping gains, especially in the case of limited number of available hopping frequencies. Finally, the MAIO management functionality can be used in the 1/1 deployment to ensure there is no interference between sectors of the same site.

The MAIO management functionality provides the ability to minimise the interference by controlling the MAIOs utilised by different TRXs within a synchronised site containing multiple sectors. There are as many possible MAIOs as number of frequencies in the hopping list. Therefore, the number of TRXs that can be controlled within a site without repetition of any MAIO value is limited by the number of hopping frequencies. As the effective reuses decrease, adjacent and co-channel interference will start to be unavoidable within the sites. When this intra-site inter-sector interference is present, the

Figure 6.6   Network 2 RF hopping 1/3 and 1/1 trialled performance.

use of different antenna beamwidth will impact the final performance. Appendix A analyses the MAIO management limitations, the best planning strategies and the impact of different antenna beamwidths.

Figure 6.6 displays the trial performance of RF hopping 1/3 and 1/1 when deployed in Network 2. In this typical network topology both reuses had equivalent performance.

The logic behind 1/1 being the best performing hopping reuse relies on the higher hopping gains for the higher number of frequencies. As presented in Chapter 5, most of the hopping gains are achieved with a certain number of hopping frequencies and further gains associated with additional frequencies are not so high. On the other hand reuse 1/1 has the worst possible interference distribution. In order to get most hopping gains, improve the interference distribution and keep the high flexibility and adaptability associated to 1/1 reuse, the innovative *advanced* reuse schemes (*n/n*) were devised. The idea is to introduce a site reuse in the system but keep the 1/1 scheme within each site. The end result is an improved C/I distribution while maintaining most of the link level gains. Reuse 1/1 would belong to this category with a site reuse of 1. Figure 6.7 illustrates some examples of *advanced* reuse schemes.

The optimum frequency hopping *advanced* reuse scheme will be dependent on the number of available frequencies. For example, if 18 hopping frequencies are available, and assuming that 6 frequencies provide most of the hopping gains, the optimum scheme would be 3/3. On the other hand, the MAIO management limitations presented in Appendix A are determined by the number of hopping frequencies within the site and as *n* (reuse factor) increases the efficiency of MAIO management will decrease. Therefore, the optimum *advanced* reuse scheme will be determined by both the total number of available hopping frequencies and the existing TRX configuration.

Figure 6.8 shows the performance achieved by 1/1 and 2/2 reuses both in terms of dropped call rate (DCR) and FER in Network 1 trials. The measured capacity increase ranged between 14 and 18%. In this scenario, with 23 hopping frequencies available, reuse 3/3 is expected to outperform both 1/1 and 2/2.

Figure 6.7 *Advanced* reuse schemes. 1/1, 2/2 and 3/3.

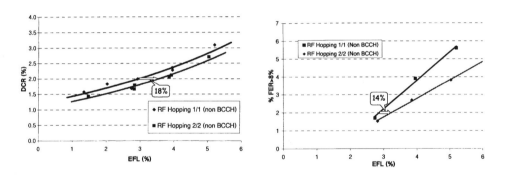

Figure 6.8 Network 1 RF hopping 1/1 vs. 2/2 trialled DCR and FER performance.

As a conclusion, the best performing hopping schemes are the RF hopping *advanced* reuse schemes. Additionally, these schemes don't require frequency planning or optimisation in order to deliver their maximum associated performance. Other hopping deployment strategies can yield equivalent results, but have intensive planning and optimisation requirements in order to achieve their expected performance and are not so flexible in order to be upgraded or re-planned. Examples of these other strategies could be BB hopping with large TRX configurations or heuristic planning,[3] which require complex planning tools and are very hard to optimise. These deployment strategies can make use of the automated planning functionality described in Chapter 12 in order to achieve high performance without intensive planning and optimisation effort.

Power control

Power control is an optional feature in GSM that reduces the interference in the system by reducing both the base station and mobile station transmitted power. Additionally, it contributes to extend the terminal's battery life. The power control algorithm is described in more detail in Chapter 5. Power control can be efficiently combined with frequency hopping.

The control rate of power control is 0.48 s (once per slow associated control channel (SACCH) multiframe). For relatively slow moving mobiles, the interference reduction gain from power control can be substantial, whereas for fast moving mobiles it may not be possible to efficiently track the signal variation. Hence, conservative power settings must be used, which limits the interference reduction gain. The average power control gain in carrier interference (CIR), measured by network simulations including frequency hopping and for a mobile speed of 50 km/h, has been found to be approximately 1.5–2.5 dB at a 10% outage level [1].

---

[3] Irregular hopping lists in order to optimise the C/I distribution.

Figure 6.9   Network 1 DL PC trialled received signal level offset.

Figure 6.9 displays the impact of downlink power control (DL PC) in the distribution of received signal level. The 10 dB offset shows the efficiency of DL PC. Much higher dynamic ranges (range of power reduction) do not provide additional gains and, in some cases, can cause some problems since the DL PC speed may not be fast enough in some cases to follow the signal fading variations.

The power control settings have an important effect on the DL PC performance. From multiple analysis carried out, the DL PC performance reaches its optimum value when the following criteria are fulfilled:

- fast averaging and voting parameters are set in order to define a fast algorithm;
- the power step should be smaller (around 2 dB) when decreasing the power and higher (around 4 dB) when increasing it;
- the dynamic range should be large enough (16–20 dB) but the largest value is not recommended.

The network level performance enhancements achieved with the introduction of DL PC has been measured in real networks. Figure 6.10 displays the measured gain from DL PC both in terms of DCR and FER in Network 1. A 50% capacity increase over the hopping network can be generally considered when DL PC is activated in the network.

Discontinuous transmission

An effective way to decrease the interference and thus increase the network performance is to switch the transmitter off when speech is not present. This is known as discontinuous transmission (DTX). In a normal conversation, each person talks, on average, less than 50% of the time. DTX functionality is described in detail in Chapter 5.

Figure 6.10   Network 1 DL PC DCR and FER performance gains.

Results from simulations have shown a linear relationship between the DTX factor and the C/I distribution improvement when combined with random FH [1]. This implies that a DTX factor of 0.5 gives an interference reduction of approximately 3 dB [2]. When the SACCH and noise description frame are considered, the reduction is down to 2.5 dB. The gain from DTX is not dependent on the network load and location of the mobile stations, as is the case for power control. Uplink DTX has, as UL PC, the second aim to extend battery power in the mobile station.

When network operators turn on DTX in their network, they should carefully analyse the BER (RXQUAL) statistics since the BER estimation accuracy is affected by the introduction of DTX, as presented in Chapter 5. The BER is estimated over 100 time division multiple access (TDMA) bursts in the non-DTX mode and over 12 TDMA bursts in the DTX mode. The first is called RXQUAL-FULL and the second RXQUAL-SUB.

Network simulations showed that the higher variance causes the RXQUAL-SUB values to be more spread when compared with the RXQUAL-FULL values. While the network actually gets better with DTX, this cannot directly be extracted from the RXQUAL network statistics since they are worse with DTX [3]. Figure 6.11 shows the distribution of RXQUAL-FULL and RXQUAL-SUB values from simulations (1/3 and a 3/9 reuse high loaded hopping network) and real network measurements. The amount of RXQUAL values equal to 6 and 7 is higher when using the SUB values. This is caused by the high variance of the SUB values. In [4] simulation results can be found for different frequency reuses with and without power control and DTX. This effect will impact not only the generated statistics but also the overall network performance as well since multiple radio resource management (RRM) algorithms, such as handover and power control, rely on accurate BER information.

The impact on network performance when DTX is introduced has been analysed both with simulations and trials. Due to the different issues mentioned, such as definition of realistic voice activity factor, inaccuracies in the RXQUAL

Figure 6.11   Simulations (left) and Network 8 (right) RXQUAL-FULL and RXQUAL-SUB distribution.

reporting and additional interference from SACCH and noise description frame, the realistic gain from DTX is hard to quantify via simulations and will be limited in practical deployments. Figure 6.12 displays the Network 1 measured performance enhancements, both in terms of DCR and FER, with the introduction of DL DTX in the hopping network. A 30% capacity increase can be considered when DL DTX is activated in a hopping network using power control.

The conclusion of this section is that the use of RF hopping, deployed with the *advanced* reuse schemes and making use of the MAIO management functionality yields the best hopping performance without demanding planning requirements. Additionally, DL PC and discontinuous transmission work efficiently in hopping networks and combined together double the previous hopping network capacity. The results from this section are considered in this book to define the GSM baseline configuration for speech services providing the baseline performance. Further sections of this chapter will analyse the additional enhancements other functionality will provide on top of the defined GSM baseline performance and, unless otherwise stated, this baseline configuration will always be in use in the simulations and trial results presented.

Figure 6.12   Network 1 DL DTX DCR and FER performance gains.

6.2 Reuse partitioning

The basic concept of reuse partitioning is to decrease the effective reuse utilising channel allocation schemes able to assign connections to different layers using different reuses. Therefore, the available spectrum of a radio network is divided into different bands, which constitute different layers with different reuses. In this book, the layer with looser reuse that ensures full coverage is called *overlay* and the different layers with tighter reuses, which provide higher effective capacity, are called *underlay* layers. Mobile stations close to the base station can use the underlay network, while mobile stations on the edge of the cell usually utilise the overlay layer. The dynamic allocation of mobile stations to the underlay layers provides an effective capacity increase, since these layers are deployed with tighter reuses.

There are many different practical implementations for reuse partitioning, such as concentric cells,[4] multiple reuse pattern (MRP),[5] intelligent underlay overlay (IUO)[6] and cell tiering.[7] All these are based on two or more layers deployed with different reuses, and channel allocation strategies that allocate the calls into the appropriate layer according to the propagation conditions. However, each approach has its own special characteristics. For instance, IUO channel allocation is based on the evaluation of the C/I experienced by the mobile station, while concentric cells utilise inner and outer zones where the evaluation of the inner zone is based on measured power and timing advance. More insight into some of the different reuse partitioning methods, and especially about the combination of frequency hopping and frequency partitioning, is given in [2].

Many different studies from frequency reuse partitioning have been carried out with an estimated capacity gain typically of 20–50% depending on system parameters such as available bandwidth, configuration used, etc. This section does not focus on basic reuse partitioning, but rather analyses its combination with frequency hopping since the gains of both functionality can be combined together.

Basic operation

Frequency reuse partitioning techniques started to be deployed in the mid-1990s. At that time some terminals did not properly support frequency hopping, especially when it was combined with power control and discontinuous transmission. When frequency hopping terminal support started to be largely

---

[4] Motorola proprietary reuse partitioning functionality.
[5] Ericsson proprietary reuse partitioning functionality.
[6] Nokia proprietary reuse partitioning functionality.
[7] Nortel proprietary reuse partitioning functionality.

required by operators, the terminal manufacturers made sure these problems were fixed and frequency hopping started to be extensively deployed. It was therefore important to combine both frequency hopping and frequency reuse partitioning in order to achieve the best possible performance.

Basic reuse partitioning implements a multilayer network structure, where one layer provides seamless coverage and the others high capacity through the implementation of aggressive frequency reuses.

Figure 6.13 displays the basic principle of frequency reuse partitioning operation. The overlay layer provides continuous coverage area using conventional frequency reuses that ensures an adequate C/I distribution. Overlay frequencies are intended to serve mobile stations mainly at cell boundary areas and other locations where the link quality (measured in terms of C/I ratio or BER) is potentially worse. The underlay layer uses tight reuses in order to provide the extended capacity. Its effective coverage area is therefore limited to locations close to the base transceiver station (BTS), where the interference level is acceptable.

The channel allocation, distributing the traffic across the layers, can be performed at the call set-up phase or later on during the call by means of handover procedures. Different implementations base the channel allocation on different quality indicators. The best possible quality indicator of the connection is C/I. Therefore IUO, which makes use of such an indicator, is considered to be the best reuse partitioning implementation and will be the one further analysed across this section.

In IUO implementation, the C/I ratio is calculated by comparing the downlink signal level of the serving cell and the strongest neighbouring cells, which use the same underlay frequencies. These are reported back from the terminals to the network, and the C/I ratio is computed in the network according to

$$\frac{C}{I} = \frac{P_{\text{own\_cell}}}{\displaystyle\sum_{n=1}^{6} P_{i\_\text{BCCH}}}$$

Figure 6.13 Basic principle of reuse partitioning.

where $P_{\text{own\_cell}}$ is the serving cell measured power, corrected with the power control information, and P_{i_BCCH} is the power measured on the BCCH of interfering cell i.

Reuse partitioning and frequency hopping

Both frequency hopping and reuse partitioning reduce the effective frequency reuse pattern and thereby enable capacity gains in cellular mobile systems. The gains of hopping are associated with a better link level performance due to the frequency and interference diversity. On the other hand, the gains from reuse partitioning are related to an efficient channel allocation that ensures calls are handled by the most efficient layer/reuse. These gains can be combined since they have a different nature. In the IUO implementation, the combination of frequency hopping and reuse partitioning is defined as intelligent frequency hopping (IFH).[8]

Figure 6.14 displays the IFH concept and how the frequency band is divided into different layers/sub-bands. IFH supports BB and RF hopping. As with conventional frequency hopping, in BB hopping mode, the BCCH frequency is included in the hopping sequence of the overlay layer, but not in RF hopping mode.

Basic performance
The performance of IFH is going to be presented based on extensive trial analysis. Figure 6.15 illustrates Network 2 IFH performance. Network 2 traffic

Figure 6.14   IFH layers with RF and BB hopping.

---

[8] Nokia proprietary functionality.

Figure 6.15 Network 2 IFH performance increase.

configuration was quite uniform with four TRXs per cell across the whole trial area. There were two TRXs in both the overlay and the underlay layers. DL PC and DTX were not in use in this case.

These results suggest that IFH provides an additional capacity increase of almost 60%. However, the gains from IFH are tightly related with the gains of power control since both functionalities tend to narrow down the C/I distribution, effectively increasing the system capacity. Therefore it is important to study the performance of IFH together with DL PC in order to deduce its realistic additional gains on top of the previously defined GSM baseline performance.

Figure 6.16 Network 5 IFH and DL PC performance.

IFH with power control performance
Simulation analysis showed the gains from DL PC and IFH were not cumulative, so a careful analysis is required to find out the additional gain of IFH over the defined GSM baseline performance. The simulation analysis suggests this gain is roughly 25%. Figure 6.16 shows the results collected in Network 5 trials. Its configuration is equivalent to Network 2. The measured gain was between 18 and 20%. However, in order to achieve this gain, the power control and IFH parameters have to be carefully planned for the algorithms not to get in conflict.

As a conclusion, additional gains of the order of 20% on top of the GSM baseline performance are expected with the introduction of IFH. However, this introduction requires careful planning and parameter optimisation and the gains are dependent on network traffic load and distribution. Therefore, unlike frequency hopping, power control and discontinuous transmission, the use of reuse partitioning is only recommended when the capacity requirements demand additional spectral efficiency from the network.

6.3 Trunking gain functionality

This section describes various solutions that increase the radio network trunking efficiency through a more efficient use of the available hardware resources. This is possible with an effective share of the resources between different logical cells. Sharing the load of the network between different layers allows a better interference distribution to be achieved as well. This study focuses on the trunking gains obtained with directed retry and traffic reason handover functionality.

Directed retry (DR)

Directed retry (DR) is designed to improve the trunking efficiency of the system thus reducing the blocking experienced by the end users. As its name points out, this functionality re-directs the traffic to neighbouring cells in case of congestion in call set-up (mobile originated or mobile terminated). The cell selection is based on downlink signal level. The performance of this functionality is very dependent on the existing overlapping between cells since it is required that at least one neighbouring cell has sufficient signal level for the terminal to be redirected. The higher the overlapping, the higher the trunking efficiency gain. If the overlapping between two cells was 100%, the trunking efficiency of both cells would be ideally equivalent to that of a single logical cell with the resources of both cells. Figure 6.17 displays the gain of DR with different overlapping levels based on static simulations. The gain ranges from 2 to 4%. In order to achieve substantial gains, the required overlapping level is higher than the typical one of single layer macrocellular networks.

Figure 6.17   Performance of directed retry measured from static simulations. Four TRXs per cell and different overlapping level configurations.

Traffic reason handover (TRHO)

Traffic reason handover (TRHO), sometimes referred to as load sharing [5], is a solution that optimises the usage of hardware resources both in single layer and multilayer networks. The main difference with DR is that in the case of TRHO, when a certain load is reached, all the terminals connected to the serving cell can be subject to reallocation to neighbouring cells. This makes this functionality much more efficient than DR since there is a much higher probability of finding a suitable connection to be reallocated.

There are several different implementations that may impact the end performance. Some basic implementations are based on releasing the load of the serving cell by forcing a certain number of connections to 'blindly' be handed over to neighbouring cells. Some others change the power budget handover margins directing the mobile stations in the cell border to less loaded neighbouring cells. Finally, some advanced solutions can monitor the load in the possible target cells to ensure successful handovers and system stability.

The performance of TRHO functionality has been verified extensively with simulations and real network trials. Figure 6.18 shows static simulation results using advanced implementation of the functionality. The network configuration was based on 36 cells with 4 TRXs per cell and 15% overlapping with neighbouring cells, which is considered here as a typical macrocellular scenario. For a 2% blocking rate, TRHO increases 15% the capacity of each cell through an overall trunking efficiency gain. Additionally, DR does not provide substantial additional gain for large TRX configurations, although its use is recommended since both TRHO and DR are complementary functionalities.

Figure 6.18   TRHO and DR performance from static simulations.

6.4 Performance of GSM HR speech channels

As described in Chapter 5, a logical traffic channel for speech is called TCH, which can be either in full-rate (TCH/F) or half-rate (TCH/H) channel mode. When TCH/H is in use, one timeslot may be shared by two connections thus doubling the number of connections that can potentially be handled by a TRX and, at the same time, the interference generated in the system would be halved for the same number of connections. TCH/H channels are interleaved in 4 bursts instead of 8 as the TCH/F, but the distance between consecutive bursts is 16 burst periods instead of 8.

In the first GSM specifications, two speech codecs were defined, the full-rate (FR) codec, which was used by the full-rate channel mode, and the half-rate (HR) codec, which was used by the half-rate channel mode. Later on, the enhanced full-rate (EFR) codec, used as well by the full-rate channel mode, was introduced to enhance the speech quality. Finally, AMR was specified introducing a new family of speech codecs, most of which could be used dynamically by either channel mode type depending on the channel conditions.

In terms of link performance, the channel coding used with the original HR codec provides equivalent FER robustness to the one of the EFR codecs. This is displayed on the left-hand side of Figure 6.19. However, when the quality of the voice (mean opinion score (MOS)) is taken into account, there is a large difference between EFR and HR. This is displayed on the right-hand side of Figure 6.19. Considering the error-free HR MOS performance as the benchmark MOS value, the EFR codec is roughly 3 dB more robust, i.e. for the same MOS value,

Figure 6.19 Link level performance in terms of FER and MOS of FR and HR codecs in TU3 for ideal frequency hopping [6].

the EFR codec can support double the level of interference. On the other hand, as mentioned before, HR connections generate half the amount of interference.

Figure 6.20 displays network level simulations comparing the performance of EFR and HR. The quality benchmark criterion, bad quality samples outage, has been selected to be the outage of FER samples which degrade the performance below the error-free HR MOS. In the case of EFR, the considered FER threshold is 4.2%, whereas in the case of HR, this threshold is down to 1%. In terms of spectral efficiency, both HR and EFR have similar performance. For 2% bad quality samples outage, EFR outperforms HR, but when the quality requirements are not so tight, HR outperforms EFR. These results do not consider speech quality or hardware (HW) utilisation and cost. Obviously, the use of HR has an associated degradation in speech quality since the HR speech codec has a worse error-free MOS, i.e. for the remaining 98% of the samples the

Figure 6.20 HR vs. EFR network simulations (hopping layer).

speech quality of EFR would be higher than that of HR. On the other hand, the HW utilisation improves so the number of TRXs required for the same traffic is lower. Appendix B analyses the HR associated HW utilisation gains, providing guidelines for HR planning and dimensioning.

The main reason why the original HR codec has not been widely deployed has been the market perception of its speech quality degradation. However, some operators have successfully deployed HR as a blocking relief solution only utilised during peak traffic periods. The introduction of AMR will undoubtedly boost the use of half-rate channel mode since the speech codecs used in this mode are the same as the ones used in the full-rate channel mode, so with an efficient mode adaptation algorithm there should not be substantial MOS degradation when the half-rate channel mode is used.

6.5 Adaptive multi-rate (AMR)

Introduction

The principles of adaptive multi-rate (AMR) were described in Chapter 1. AMR is an efficient quality and capacity improvement functionality, which basically consists of the introduction of an extremely efficient and adaptable speech codec, the AMR codec. The AMR codec consists of a set of codec modes with different speech and channel coding. The aim of the AMR codec is to adapt to the local radio channel and traffic conditions by selecting the most appropriate channel and codec modes. Codec mode adaptation for AMR is based on received channel quality estimation [6]. The link level performance of each AMR codec, measured in terms of traffic channel FER, is presented in Section entitled *GSM AMR link level performance* in 6.5.

Increased robustness to channel errors can be utilised to increase coverage and system capacity. System performance is examined by using dynamic GSM system simulations. Mixed AMR/EFR traffic scenarios with different penetration levels are studied. Finally, the performance of AMR half-rate channel mode utilisation and its HW utilisation gains are carefully analysed.

GSM AMR link level performance

With AMR, there are two channel modes, full-rate and half-rate, and each of these use a number of codec modes which use a particular speech codec. Figure 6.21 includes the link level performance of all AMR full-rate codec modes: 12.2, 10.2, 7.95, 7.4, 6.7, 5.9, 5.15 and 4.75 kbit/s. Channel conditions in these simulations are typical urban 3 km/h (TU3) and ideal frequency hopping.

As displayed in Figure 6.21, robust AMR FR codec modes are able to maintain low TCH FER with very low C/I value. With AMR FR 4.75 codec mode

Figure 6.21   AMR full-rate link level TCH FER results (TU3, iFH).

TCH FER remains under 1% down to 3 dB C/I, whereas 8.5 dB is needed for the same performance with EFR codec. Therefore, the gain of AMR codec compared with EFR is around 5.5 dB at 1% FER. However, the gain decreases 1 dB when frequency hopping is not in use as presented in Figure 6.22 where the channel conditions were TU3 without FH.

Figure 6.23 shows the link performance of all AMR half-rate codec modes: 7.95, 7.4, 6.7, 5.9, 5.15 and 4.75 kbit/s. In the case of AMR half-rate codec

Figure 6.22   AMR full-rate link level TCH FER results (TU3 without FH).

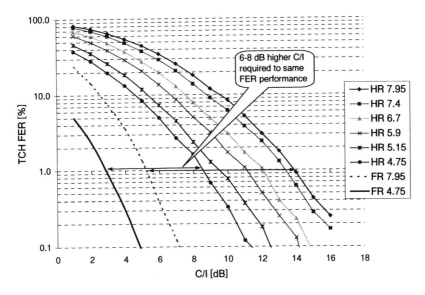

Figure 6.23   AMR half-rate link level TCH FER results (TU3, iFH).

modes, although the same speech codecs as in full-rate mode are in use, higher C/I values are required for the same performance. For example, there is around 8 dB difference between full-rate 7.95 kbit/s and half-rate 7.95 kbit/s. This is due to the different channel coding used. The gross bit rate of full-rate channel mode codecs is 22.8 kbit/s, whereas it is only 11.4 kbit/s for the half-rate channel mode codecs [6]. Therefore, for the same speech coding, half-rate channel mode has considerably less bits for channel coding and therefore it has higher C/I requirements. Finally, the half-rate 4.75 codec mode has a similar link performance to the full-rate 12.2 codec mode. All these points need to be taken into account when the channel mode adaptation strategy between full- and half-rate channel modes is defined.

Speech quality with AMR codec
With AMR there are multiple speech codecs dynamically used. Some of them tend to have higher speech quality (stronger speech coding), others tend to be more robust (stronger channel coding). Figure 6.24 shows the MOS performance of the AMR full-rate codec modes. These curves illustrate how each codec mode achieves the best overall speech quality in certain C/I regions. The codec mode adaptation algorithm should ensure the MOS performance of the overall AMR codec is equivalent to the envelope displayed in the figure. This envelope displays the performance of the AMR codec in the case of ideal codec mode adaptation, i.e. the best performing codec mode is selected at each C/I point. In the high C/I area,

Figure 6.24   EFR and AMR full-rate MOS results from [6].

EFR and AMR codecs obtain about the same MOS. However, with less than 13 dB C/I, the EFR codec requires approximately 5 dB higher C/I to obtain the same MOS than the AMR codec. Figure 6.25 shows equivalent MOS results for AMR half-rate channel mode. For the same MOS values, half-rate codecs have higher C/I requirements.

Figure 6.25   AMR half-rate, EFR, FR and HR MOS results [6].

GSM AMR system level performance

This section introduces different AMR implementation strategies and the associated performance gains. The performance evaluations are based on dynamic simulation results and in most cases refer to the performance of RF hopping configurations. The BCCH performance is evaluated separately.

In the mixed traffic case (EFR and AMR capable terminals), two deployment strategies are studied. The first one deploys AMR in a single hopping layer and achieves its gains with the use of more aggressive power control settings for AMR mobiles thus decreasing the interference level in the network. Due to better error correction capability against the channel errors, lower C/I targets can be set for AMR mobiles, hence tighter PC thresholds can be used. Another deployment strategy would be to make use of the reuse partitioning techniques to distribute the traffic across the layers according to the terminals' capabilities.

The use of half-rate channel mode is potentially an efficient way to increase capacity and HW utilisation in the network. However, when using half-rate codec modes, the connections require higher C/I in order to achieve the same quality as AMR full-rate, as displayed in Figure 6.23. Therefore, in order to determine the performance and resource utilisation capabilities of half-rate channel mode, a careful analysis is needed.

Codec mode adaptation

The codec mode adaptation for AMR is based on received channel quality estimation. The GSM specifications include an adaptation method having C/I based link quality estimation. In this method, the link quality is estimated by taking burst level C/I values. Then, FIR filtering is used with non-adaptive filters of order 100 for full-rate and 50 for half-rate channels [7]. The described C/I based quality evaluation method was used in the codec mode adaptation simulations in this study. The target of these simulations was to demonstrate the effect of codec mode adaptation in the system level quality and to find out optimum thresholds for the further performance simulations. Table 6.1 defines C/I thresholds used in the codec mode adaptation simulations.

The GSM specifications [7] define that a set of up to four AMR codec modes, which is selected at call set-up, is used during the duration of the call (the codec

Table 6.1 Simulated C/I threshold sets to demonstrate the effect of codec mode adaptation in FR channel

	set1	set2	set3	set4	set5
12.2–7.4 threshold	9 dB	11 dB	13 dB	15 dB	17 dB
7.4–5.9 threshold	5 dB	7 dB	9 dB	11 dB	13 dB
5.9–4.75 threshold	2 dB	4 dB	6 dB	8 dB	10 dB

Table 6.2 Defined mode sets for the performance simulations

AMR full-rate set	AMR half-rate set
12.2 kbps	
7.4 kbps	7.4 kbps
5.9 kbps	5.9 kbps
4.75 kbps	4.75 kbps

set can be updated at handover or during the call) [7]. The set of codec modes presented in Table 6.2 was included in the following performance evaluation simulations. In order to utilise the total dynamic range that AMR codec offers, the lowest 4.75 kbit/s and the highest 12.2 kbit/s modes were chosen in this set.

Figure 6.26 displays the TCH FER as a function of the C/I threshold sets included in Table 6.1. The relative usage of each codec mode is shown as well. The tighter the thresholds, more robust codec modes are used and lower FER can be achieved. On the other hand, tighter thresholds and robust codec modes may impact negatively the MOS performance. The 12.2 codec mode is clearly the most used mode. However, its usage drops rapidly when thresholds are tightened. With *set4* thresholds, for example, mode 12.2 is used less than 50% of the time and the usage of the mode 4.75 has increased over 10%. In order to find out the threshold settings with best speech quality performance a MOS evaluation is needed.

Figure 6.27 displays a MOS performance analysis for the different threshold settings. The best performing setting is *set3*, even though its average FER is higher than *set4* and *set5*. As explained in the Section entitled *Speech quality*

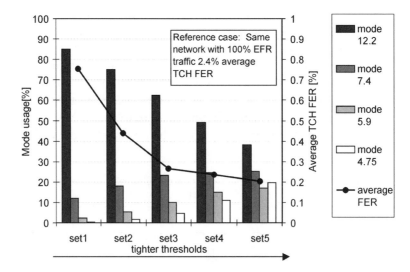

Figure 6.26 Average TCH FER and mode usage for different settings.

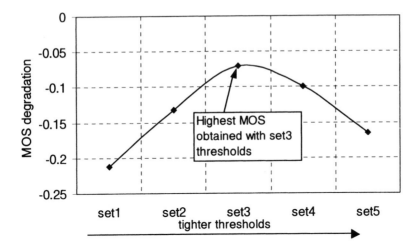

Figure 6.27 MOS performance as a function of threshold tightening.

with AMR codec in 6.5, the selection of the optimum thresholds will ensure that the AMR codec MOS performance follows the envelope of the ideal codec mode adaptation, i.e. best performing codec always used. In the following simulations the codec mode usage distributions are not presented, but this example gives the typical proportions likely to be seen in different AMR configurations.

Single layered scenario deployment
The previous section showed the speech quality increase accomplished with AMR codec. There are several possibilities to turn this increased quality to additional capacity in the network. Naturally, if there would be only AMR mobiles in the network, the network capacity could be easily increased by using tighter frequency reuses. However, most practical cases will have a mixture of AMR and non-AMR capable terminals in the network. In the mixed traffic scenario, one way to obtain the AMR gains would be through the adjustment of the power control settings so that AMR connections generate less interference. This can be effectively done defining different power control thresholds for AMR and non-AMR mobiles. The use of tighter power control thresholds for AMR mobiles will effectively lower the transmission powers used and therefore less interference will be generated. The following network simulations present the performance results using this technique.

Figure 6.28 displays the simulation results for a single layered hopping network with adjusted power control thresholds for AMR mobiles and with AMR penetration levels. Performance was measured in terms of FER. The number of bad FER samples using a 2 s averaging window was calculated. Samples with FER higher than 4.2% were considered bad quality samples.

Figure 6.28   Single layer AMR performance.

The use of AMR has a major impact on the system quality for the same traffic load. On the other hand, the use of AMR increases substantially the network capacity for the same quality criteria. Figure 6.29 illustrates the relative capacity gains for different AMR penetrations compared with the EFR traffic case. There is a 48% capacity gain with 50% AMR penetration, and up to 150% capacity increase in the case of full AMR penetration.

Since different speech codecs have different MOS performance, it is important to analyse the MOS based performance of AMR. In the system simulator, the MOS of the ongoing connections can be estimated by mapping the measured

Figure 6.29   TCH FER based AMR capacity gain.

Figure 6.30 MOS based AMR capacity gain.

TCH FER values to corresponding MOS values. MOS values presented in [6] were used in this mapping. Figure 6.30 shows that the AMR MOS based gain is still 150%. This could be expected from the link level results presented in the previous section where similar gains were pointed out for FER and MOS based link performance. At this point it can be concluded that the gain of AMR with full terminal penetration on top of the defined GSM baseline performance is 150%, both in terms of FER and MOS.

Multiple layered scenario deployment
The use of the AMR functionality, together with reuse partitioning techniques, allows the network performance in the mixed traffic scenario to be further increased. AMR mobiles can be handed over to the underlay layer, which deploy tighter reuses, while the non-AMR mobiles would stay in the overlay layer where the looser reuses guarantee adequate performance. In Figure 6.31, the performance of the simulated multiple layer network is compared with the single layer network. In the single layer case, if the quality of the EFR connections is to be guaranteed, the potential AMR gains can not be realised. In the multiple layer scenario, the AMR gains are realised and at the same time the quality of EFR connections guaranteed. This is translated to the slopes of the capacity increases displayed in the figure.

Half-rate channel mode
The MOS performance evaluation presented in Section entitled *Speech quality with AMR codec* in 6.5 demonstrated that AMR half-rate channel mode can be

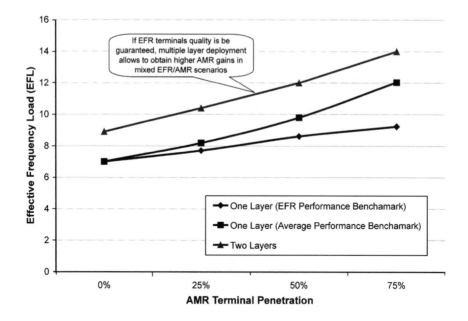

Figure 6.31   Single vs. multiple layer AMR performance. Capacity increase relative to the EFR single layer case.

used without any noticeable speech quality degradation in high C/I conditions, since the half-rate channel mode uses the same set of speech codecs as the full-rate mode. In order to efficiently deploy half-rate, it is essential to have a quality based channel mode adaptation algorithm that selects the adequate connections when adaptation from full- to half-rate channel is performed. In the following simulations, the channel mode adaptation was based on the measured RXQUAL values. All incoming calls were first allocated to full-rate channel, and after a quality measurement period, it was decided whether the connection continues in full-rate mode or a channel mode adaptation is performed.

The effect of AMR half-rate utilisation on the FER based performance is illustrated in Figure 6.32. Both AMR full-rate only and AMR full-rate and half-rate have equivalent performance. In other words, the use of HR neither improves nor degrades the overall network performance. The gain from AMR half-rate channel mode comes from the higher hardware utilisation, which leads to a lower number of TRXs required to support the same traffic load. When the required number of timeslots for different traffic loads is evaluated, this number is reduced by between 28 and 39% when half-rate channel mode is in use. This analysis is illustrated in Figure 6.33. The channel reduction rate is dependent on the percentage of half-rate channel mode capable connections, i.e. is dependent on the C/I distribution. In the lowest traffic case the usage of the half-rate channel mode was about 80% decreasing to 60% in the highest traffic point due to the worse C/I

Figure 6.32   AMR half-rate channel mode performance.

distribution of the higher load case. Appendix B presents in detail half-rate channel mode resource utilisation increase and dimensioning guidelines.

Since the 12.2 and 10.2 codec modes are not available when HR channel mode is in use, there will be some impact on speech quality when half-rate channel mode is introduced. Figure 6.34 presents the quality distribution of the connections as a function of MOS degradation, where the reference is EFR speech quality in error-free conditions. There is only an additional 0.2

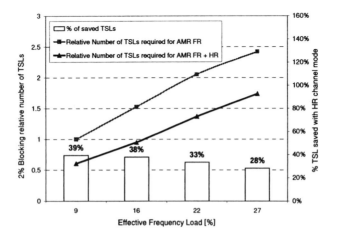

Figure 6.33   Half-rate gain in terms of timeslot requirements.

Figure 6.34 HR channel mode impact on MOS performance.

MOS degradation when AMR HR channel mode is used which is due to the unavailability of the high AMR speech codecs (12.2 and 10.2) in the HR channel mode.

Finally, the use of AMR HR channel mode requires fewer TRXs for the same traffic load so the effective deployed reuse is looser. Looser reuses have some positive performance side effects such as higher MAIO management control, and higher performance gains in case dynamic frequency and channel allocation (DFCA) (Chapter 8) is introduced. Therefore, the introduction of HR channel mode reduces the TRX configuration and indirectly increases the performance.

Figure 6.35 BCCH layer performance with AMR.

AMR codec performance in BCCH frequencies

Section entitled *GSM AMR link level performance* in 6.5 showed that AMR link level gain is slightly lower without frequency hopping. Figure 6.35 presents network simulation results for EFR and AMR codecs and different BCCH reuses. In these results, AMR BCCH reuses 9 and 12 outperform the EFR BCCH reuses 12 and 18, respectively. This means that BCCH reuses can be tightened in case of full or very high AMR penetration. The tightening of the BCCH reuse is of special importance in the narrowband deployment scenarios, as explained in Chapter 9. With the introduction of AMR, the signalling channels may become the limiting factor in terms of network performance. This subject is studied in Chapter 11.

6.6 EDGE AMR enhancements

Introduction

This section presents the performance analysis of the AMR enhancements included in Rel'5 and presented in Chapter 2.

GERAN Rel'5 specifies a new octagonal phase shift keying (8-PSK) modulated half-rate channel mode (O-TCH/AHS) for AMR narrowband speech. This new physical channel allows the use of all the available narrowband AMR speech codecs. Code rates range from 0.14 (AMR4.75) to 0.36 (AMR12.2).

Additionally, Rel'5 brings one additional method to increase the speech capacity of the network. A new *enhanced power control* (EPC) method can be used for dedicated channels with both Gaussian minimum shift keying (GMSK) and 8-PSK modulation.[9] EPC steals bits from the SACCH L1 header to be used for measurement reports (uplink) and power control commands (downlink). This way the PC signalling interval can be reduced from SACCH block level (480 ms) down to SACCH burst level (120 ms). The cost is decreased robustness of the SACCH channel (logical channel SACCH/H is replaced by new SACCH/TPH when using EPC). Link level loss of SACCH/TPH is 0.5–1.0 dB (TU3 ideal FH at 10% block error rate (BLER)) compared with SACCH/H. The corresponding new dedicated control channel EPCCH (enhanced power control channel) has no interleaving and is block coded with code rate of 1/4.

This section also describes a major speech quality enhancement feature of Rel'5, the wideband (WB) AMR speech channels. Link level results for WB-AMR in both GMSK full-rate and 8-PSK half-rate channels are presented.

---

[9] Note that this is not the same as fast power control (FPC), which is specified for ECSD in Rel'99. ECSD FPC uses a 20 ms signalling period.

[10] iFH assumes ideal hopping operation with all bursts having a fading profile fully decorrelated.

Figure 6.36 EDGE AMR half-rate link level performance.

Finally, some possible further speech enhancements beyond GERAN Rel'5 are discussed.

EDGE NB-AMR link level performance

Link level performance of 8-PSK AMR half-rate codec modes (O-TCH/AHS) is

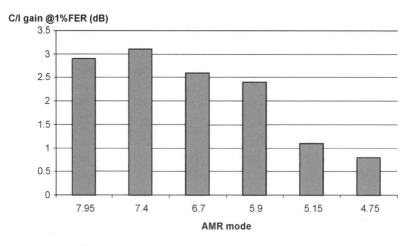

Figure 6.37 O-TCH/AHS vs. TCH/AHS link level gains at 1% FER.

0.5 dB (4.75 codec mode) to 3 dB (7.95 codec mode) better than the standard AMR half-rate codec modes (TCH/AHS) for TU3 ideal frequency hopping (iFH[10]). C/I requirements for 1% FER range from 8 to 13 dB. Figure 6.36 illustrates the 8-PSK AMR half-rate channel mode link performance.

Figure 6.37 shows the link level gains of 8-PSK AMR half-rate codec modes compared with the corresponding GSMK modulation modes. These link level gains are smaller when using non-hopping channels.

The support of the 10.2 and 12.2 codec modes and the gains experienced by the rest of the codec modes will increase the usage of half-rate and more importantly will contribute to a better half-rate channel mode speech quality, comparable with that of the full-rate channel mode.

Listening tests (33 listeners, clean speech) were conducted in order to verify the subjective performance of EDGE AMR half-rate compared with standard AMR half-rate. Results are collected in Figure 6.38. At each operating point (C/I), the AMR mode with highest MOS value is selected from GSM and EDGE channels. The results show a clear benefit for EDGE AMR half-rate over GSM AMR half-rate. Due to the higher AMR modes in EDGE, the subjective performance gains would be even higher in typical practical conditions, where C/I is high and background noise is present.

From all these results it can be concluded that EDGE AMR half-rate channels should be prioritised over GSM AMR half-rate channels, i.e. when EDGE AMR half-rate is available, there is no need to do channel mode adaptation to GSM AMR half-rate.

Figure 6.38 MOS test results between GSM AMR (TCH/AHS) and EDGE AMR half-rate (O-TCH/AHS). Best codec is selected for each operating point.

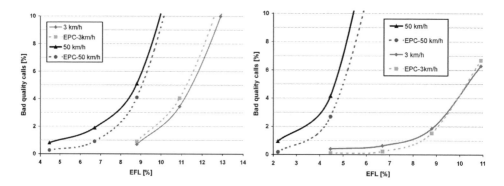

Figure 6.39   TCH-AHS EPC performance for slow fading standard deviation of 6 (left) and 8 dB (right).

EPC network performance

The 4-fold signalling rate of EPC can be transformed into system capacity gain. This gain depends heavily on the mobile station speed and shadow fading component of the signal. An extensive set of network level simulations were performed in order to study the effect of EPC on system level. The simulation setup was similar to the one used in the basic GSM Half Rate simulations presented earlier. Only non-BCCH layer was studied with 1/1 RF hopping and 4.8 MHz of total spectrum. DTX was not used and bad quality call criteria was 1% FER averaged over the whole call. In the simulations the EPC performance was compared to the standard PC performance, using the same RxQual/RxLev based power control algorithm. Figure 6.39 shows the performance results at different speeds and shadow fading profiles.

These results show that the gain from EPC is substantial for fast moving mobiles (around 20%) but quite small in the case of slow moving mobiles (the relative gain of EPC increases slightly also when the slow fading variation is higher). This points out the sufficient speed of the GSM standard power control mechanism to follow shadow fading variations for slow moving mobiles.

EDGE wideband AMR codecs

GERAN Rel'5 specifies new AMR wideband (WB) speech codecs, which present a major improvement in speech quality. With the introduction of a new 8-PSK half-rate physical channel (O-TCH/WHS), WB codec bit rates range from 6.6 to 23.85 kbit/s. The highest modes offer performance exceeding standard wireline quality. The usage of GMSK half-rate with WB codecs has not been standardised since only the two lowest modes would fit into the channel. Due to improved speech coding, WB codecs offer better quality than NB codecs with similar bit rates.

Figure 6.40   TCH/WFS (left) and O-TCH/WHS (right) link level performance (TU3 iFH).

Figure 6.40 shows, on the left-hand side, the link level performance of wide-band AMR codecs on the GMSK full-rate channel mode (TCH/WFS). The two highest wideband AMR codec modes (23.05 and 23.85) do not fit into the TCH/WFS channel. However, all the modes fit into the 8-PSK modulated half-rate channel mode (O-TCH/WHS). Link level performance of this channel mode is shown on the right-hand side of Figure 6.40. The C/I requirements for 1% FER range from 9 to 19 dB.

It is interesting to compare at this point the C/I requirements of TCH/WFS and O-TCH/WHS channels. The comparison at 1% FER limit is presented in Figure 6.41. GMSK full-rate codec modes are 2–5 dB more robust than 8-PSK half-rate modes. Note that 19.85 mode is the only TCH/WFS codec mode that contains non-protected Class 2 bits (see [8]). This is the reason why it performs *better* than 18.25 in terms of FER calculated for the most important Class 1 bits (naturally, the cost is larger residual BER for Class 2 bits).

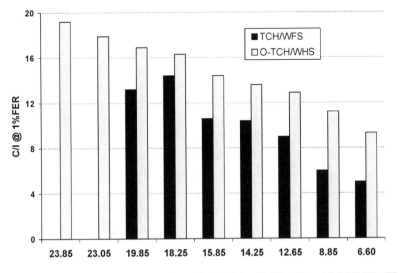

Figure 6.41   C/I requirement for 1% FER for WB-AMR in GMSK FR and 8-PSK HR (TU3 iFH).

Figure 6.42 Left: comparison of TCH/WFS and O-TCH/WFS. Right: link level performance of O-TCH/WFS (TU3 iFH).

The use of full-rate 8-PSK channels for wideband AMR codecs is an interesting option. Thanks to the increased number of bits, the coding rate can be considerably decreased for O-TCH/WFS compared with TCH/WFS, which means higher robustness, especially for the higher AMR modes. The left-hand side of Figure 6.42 shows the CIR gain of O-TCH/WFS over TCH/WFS at 1% FER in the TU3 ideal FH channel. On the right, the link level performance of the O-TCH/WFS channel is displayed.

Possible Rel'6 further extensions

An interesting further extension for GERAN speech is *quarter rate* channels [8]. The three times higher gross bit rate of 8-PSK modulation makes it possible to divide the physical resources further, allowing four users to share a single time-slot. The resulting 17.1 kbit/s gross bit rate allows all the NB-AMR codecs to be used and more than half of the WB codecs. Physical resources are doubled compared with half-rate and the actual gain is even higher due to additional trunking gain. However, an efficient channel mode adaptation algorithm is needed to switch between different modes. Mode adaptation is realised by triggering an intra-cell handover, which cause frame losses due to establishment of FACCH channels. Simulations suggest that switching to QR should be made only if the measured channel quality is good enough and TRX load high enough (60–75%). Channel packing (packing, e.g. four QR users in the same channel) should be used to maximise the number of free timeslots.

References

[1] Johansen J., Vejlgaard B., 'Capacity Analysis of a Frequency Hopping GSM System', M.Sc.E.E. Thesis Report, Aalborg University, June 1995.

[2] Nielsen Thomas T., Wigard Jeroen, 'Performance Enhancements in a Frequency Hopping GSM Network', Kluwer Academic Publishers, Dordrecht, 2000.

[3] Wigard Jeroen, Nielsen Thomas Toftegaard, Skjærris Søren, Mogensen Preben, 'On the Influence of Discontinuous Transmission in a PCS1900/GSM/DCS1800 Type of Network', VTC'99, Houston.

[4] Wigard Jeroen, Mogensen Preben, Michaelsen Per Henrik, Melero Juan, Halonen Timo, 'Comparison of Networks with different Frequency Reuses, DTX and Power Control using the Effective Frequency Load Approach', VTC2000, Tokyo, Japan.

[5] Watanabe F., Buot T., Iwama T., Mizuno M., 'Load Sharing Sector Cell in Cellular Systems', IEEE Proc. PIMRC '95, pp. 547–551.

[6] GSM 06.75 (ETSI TR 101 714): 'Performance Characterization of the GSM Adaptive Multi-rate (AMR) Speech Codec'.

[7] ETSI TC-SMG, GSM 05.09.

[8] Bellier T., Moisio M., Nikula E., Pirhonen R., Rantala H., 'Quarter Rate AMR Channels – A Speech Capacity Booster in GSM/EDGE', WPMC'00, Bangkok, Thailand, 12–15 November 2000.

7

GPRS and EGPRS Performance

Javier Romero, Julia Martinez, Sami Nikkarinen and Martti Moisio

The present chapter covers performance analysis of general packet radio service (GPRS) and enhanced GPRS (EGPRS) as they are specified in the Rel'97 and Rel'99 specifications. The analysis is carried out considering of potential GSM EDGE radio access network (GERAN) deployment scenarios and covers most of the practical issues related to packet switched data dimensioning in a GERAN system. Rel'99 introduces the enhanced data rates for global evolution (EDGE) radio interface, which is the gateway to enhanced spectrum efficiency (SE) and data rates. Packet switched (PS) capabilities of later releases (Rel'4 and Rel'5) rely on the EDGE radio interface. The transition to EDGE will be smooth for most operators, which have an already deployed global system for mobile communications (GSM)/GPRS network. Most of them are introducing GPRS gradually, keeping the network changes as small as possible. In many networks, a limited GPRS implementation has been deployed in order to provide initial service. The performance of these cases has been carefully studied in this chapter.

There are many cases and many deployment scenarios to consider. Current operational GSM networks utilise different deployment and frequency planning strategies. Baseband hopping has been used by many operators, while fractional reuse networks (with radio frequency (RF) hopping) is now being widely adopted, especially in new network deployments. Most of the analysis of this chapter can be applied to any network.

Section 7.1 introduces the basic link performance, which is key to understand system level performance. Achievable throughput depending on radio-link conditions is studied under interference and noise limited scenarios. The benefits of Link Adaptation (LA) and incremental redundancy (IR) are studied, both with and without frequency hopping.

Section 7.2 describes the main radio resource management (RRM) functionality required in (E)GPRS. Algorithm design and its impact on performance are discussed. Polling, acknowledgement strategy, LA, IR, channel allocation, scheduler and power control functionality are covered.

GPRS and EGPRS system level capacity and dimensioning guidelines are covered in Sections 7.3 and 7.4. Most of the analysis is performed by dynamic simulations with a sophisticated simulation tool described in Appendix E. Hopping and non-hopping scenarios are studied. Spectral efficiency with different quality criteria is discussed as well. The performance analysis covers from pure soft-capacity deployments to hardware (HW) limited scenarios. Two different sets of coding schemes (CS) are considered for GPRS: CS-1–2 and CS-1–4. The impact of the different mobile station timeslot (TSL) capabilities is studied. Finally, throughput distribution maps and the effect of different PS traffic patterns are included as well.

Section 7.5 studies system performance in mixed packet switched and circuit switched traffic cases. The interaction between CS and PS traffic is described. A procedure to compute the maximum data capacity while maintaining speech quality is introduced as well. Section 7.6 presents (E)GPRS performance estimation based on collected field trial results from a major European city. Section 7.7 discusses application performance over (E)GPRS describing transport control protocol (TCP)/Internet protocol (IP) behaviour in (E)GPRS networks and possible TCP enhancements. Finally, Section 7.8 presents some GPRS trial results from field trials.

7.1 (E)GPRS link performance

In this section, the performance of one single radio link is studied under interference and noise limited conditions. The performance is studied in terms of block error rate (BLER) and throughput per TSL for the different (modulation and) coding schemes, (M)CS. It should be noted that the core of EGPRS data capacity improvement is purely based on link level enhancements, and therefore, it is of outmost importance to understand the link level performance before studying system level capacity issues.

Introduction

Link level simulations are used to evaluate the performance of the physical layer. These link level results can be utilised as an input to further studies including network level simulations, see Appendix E. The link is a one-way connection between one transmitter and one receiver. A general simulation chain is shown in Figure 7.1.

The simulation chain assumes a certain modulation method (GMSK or octagonal phase shift keying (8-PSK)), channel coding, channel estimation and receiver algorithm. The input parameters are channel type (e.g. typical urban, rural area, ...), mobile speed, frequency band, the use of frequency hopping, average signal level relative to average interference level (CIR) or average energy per bit relative to noise density (E_b/N_0). Output parameters are the channel bit error rate before decoding (raw BER), information bit error rate after decoding (BER), block error rate after decoding (BLER) and throughput.

In GSM specifications the minimum BTS and MS performance for (E)GPRS is specified. The receiver sensitivity and minimum received CIR for a reference BLER value and different reference channel models are specified for the mobile and the base station in [1]. In the next subsections simulated link level results of (E)GPRS will be presented. Note that exact results may change depending on the considered receiver and transmitter implementation.

This section presents (E)GPRS link level results for a typical urban (TU) channel model, described in GSM specifications in [1]. The simulated band is 900 MHz, but the same results could be applied for the 800 MHz band. For 1800 and 1900 MHz bands, the same results could be applied dividing the MS speed by 2 (as the fast fading process is two times faster due to the halved wavelength). This means that the results for the TU3 (typical urban at 3 km/h) channel in the 900 MHz band are the same as the TU3 channel in the 800 MHz band; and also the same as the TU with a MS speed of 1.5 km/h for 1800 and 1900 MHz bands.

Simulation results are shown for interference and noise limited scenarios. Interference limited simulations correspond to MS testing set-up in the interference limited scenarios described in GSM specifications [2].

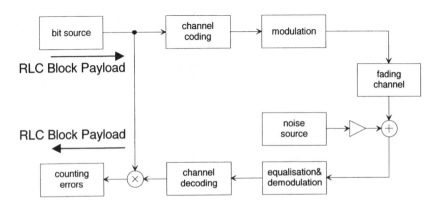

Figure 7.1 General structure of a link level simulation chain.

(E)GPRS peak throughputs

(E)GPRS transmits segments of logical link control (LLC) blocks with a maximum bit rate that depends on the (M)CS utilised. The amount of LLC bits transmitted in a radio block is given by the amount of channel coding applied on each (M)CS. Table 7.1 summarises the amount of LLC bits in a radio block of 20 ms and the peak throughput offered to the LLC layer for the different (M)CS.

Interference limited performance

Radio link is interference limited when the interference levels are well above the receiver sensitivity level. Most of the networks are interference limited as the same frequencies are reused in different cells, in order to efficiently utilise the available spectrum. An interference limited network is usually considered to be capacity limited, meaning that the interference level is setting the limits of the network spectral efficiently. On the other hand, noise limited networks are considered to be coverage limited, meaning that the network has cell range rather than spectrum efficiency limitations.

Table 7.1   (E)GPRS peak throughput values

	MAC header	RLC header	RLC data			
			LLC segment			
	Coding scheme	Modulation	RLC blocks/ radio blocks	FEC code rate	User bits/20 ms	Bit rate (bps)
GPRS	CS-1		1	0.45	160	8000
	CS-2		1	0.65	240	12 000
	CS-3	GMSK	1	0.75	288	14 400
	CS-4		1	n/a	400	20 000
EGPRS	MCS-1		1	0.53	176	8800
	MCS-2		1	0.66	224	11 200
	MCS-3		1	0.85	296	14 800
	MCS-4		1	1.00	352	17 600
	MCS-5	8-PSK	1	0.38	448	22 400
	MCS-6		1	0.49	592	29 600
	MCS-7		2	0.76	448 + 448	44 800
	MCS-8		2	0.92	544 + 544	54 400
	MCS-9		2	1.00	592 + 592	59 200

BLER *performance*

Some of the received radio link control (RLC) blocks can be erroneous. Link level robustness depends on the coding rate and the modulation format. Different MCS achieve different BLER as it can be seen in Figure 7.2. In general, the higher the (M)CS the higher the BLER probability for the same CIR. The change from MCS4 to MCS5 may break this rule.

Minimum base transceiver station (BTS) and MS performance in interference limited scenarios have been included in GSM specifications. Minimum performance is specified as the minimum C/I required to achieve 10% BLER for different channel conditions, see Table 7.2. Minimum performance can be exceeded by both MS and BTS depending on the receiver implementation.

Figure 7.2 (E)GPRS link performance. Interference limited. No hopping. TU3 channel. 900 MHz band.

Table 7.2 Minimum C/I for BLER < 10% in interference limited scenarios. 900 MHz band. From GERAN specifications [1][a]

Type of channel	Propagation conditions				
	TU3 (no FH)	TU3 (ideal FH)	TU50 (no FH)	TU50 (ideal FH)	RA250 (no FH)
PDTCH/CS-1	13 dB	9 dB	10 dB	9 dB	9 dB
PDTCH/CS-2	15 dB	13 dB	14 dB	13 dB	13 dB
PDTCH/CS-3	16 dB	15 dB	16dB	15 dB	16 dB
PDTCH/CS-4	19 dB	23 dB	23 dB	23 dB	[b]
PDTCH/MCS-1	13	9	9	9	9
PDTCH/MCS-2	15	13	13	13	13
PDTCH/MCS-3	16	15	16	16	16
PDTCH/MCS-4	21	23	27	27	[b]
PDTCH/MCS-5	18	14.5	15.5	14.5	16
PDTCH/MCS-6	20	17	18	17.5	21
PDTCH/MCS-7	23.5	23.5	24	24.5	26.5[c]
PDTCH/MCS-8	28.5	29	30	30	[b]
PDTCH/MCS-9	30	32	33	35	[b]

[a] TUx = typical urban channel at x km/h; RA250x = rural area x km/h.
[b] Reference performance can't be met.
[c] Performance is specified at 30% BLER.

Acknowledged mode throughput
When error-free data delivery is necessary, the (E)GPRS network can retransmit erroneous radio blocks. There is a basic retransmission mechanism specified in Rel'97 and an advanced one (based on IR) included in Rel'99 for EPGRS. IR has been introduced in Chapter 1 and will be described in detail later on.

With the basic retransmission mechanism the achieved throughput can be calculated as a function of the BLER:

Throughput = Peak_Throughput x (1 − BLER)

where BLER is given in Figure 7.2. When IR is applied, the probability of erroneous blocks in successive retransmissions is much lower than in the initial transmission. Therefore, in order to calculate the throughput per TSL the BLER of the first and successive transmissions have to be considered. EGPRS throughput values with IR are shown in Figure 7.2. Retransmissions are done with the same MCS. No memory limitations have been considered, so IR is always possible. IR memory capable of storing 20 000 soft values (bit values before decoding) per user give performance close to the optimum. It is worthwhile mentioning the good performance of MCS-9 that in principle could always be used. In practice, the limited IR memory, the link performance under different propagation conditions and the possible delay requirements imposed by upper layers, requires the usage of the other MCS.

The effect of frequency hopping
Frequency hopping helps to combat fast fading by randomising C/I values of each burst. Randomisation of C/I has the effect of averaging the C/I values over the duration of the radio blocks so the probability of having consecutive bad C/I values is minimised. On the other hand, it increases the variance of the C/I values, therefore it is more likely that some of the bursts experience bad C/I. The effect of a burst experiencing bad C/I can be compensated by channel coding and interleaving as it was explained in Chapter 6. Nevertheless, if radio blocks are not well protected (i.e. they don't have enough coding) the randomisation of the C/I values has a negative effect, as the channel coding is not robust enough to correct the effect of some bursts having bad C/I. As a consequence, the use of frequency hopping for some of the coding schemes is not beneficial. In those cases the frequency hopping gain turns to be negative.

Figure 7.3 shows the link performance difference at 10% BLER for some selected MCSs. MCS-1 and MCS-7 (coding rate of 0.5 and 0.76, respectively) benefit from frequency hopping (FH) having a direct link level gain of 3.6 and 2.2 dB. On the other hand MCS-3 and MCS-9 (coding rate of 0.85 and 1) have a loss of 1.1 and 1.5 dB, respectively. It should be noted that MCS-9 is interleaved only over two bursts, in order to avoid bad performance with FH.

Figure 7.3 EGPRS link performance. With and without hopping. TU3 channel. 900 MHz band.

Higher frequency hopping gains can be achieved with lower BLER values. Nevertheless, when LA is applied for throughput maximisation, the normal operation point of the (M)CS never goes lower than 10% BLER. This is because LA will switch to a higher (M)CS when BLER is low.

When the users are moving faster (e.g. in a TU50 channel) fast fading process tends to be more uncorrelated having a very similar effect of FH. Therefore, the gain or loss of FH is marginal.

Ideal link adaptation
(E)GPRS network may automatically select the most appropriate (M)CS, depending on the radio link conditions, in order to maximise throughput. The link level performance will, therefore, follow the envelope of the throughput vs. CIR curves of the different (M)CSs (Figure 7.2). Figure 7.4 represents the achieved link performance with LA, in an ideal situation with optimum CIR switching points. For the same throughput there is an appreciable link level gain with EGPRS compared with GPRS that varies from a couple of dBs for 5 kbps up 15 dB for throughput values close to 20 kbps. The higher the throughput the larger the gain. Frequency hopping results tend to be some dB worse, specially in the high throughput area where high (M)CS are utilised. Figure 7.4 displays the gain of IR, which always improves link performance, in some cases over 3 dB (in no FH case). The envelope of throughput per TSL vs. C/I depends on the MCS

Figure 7.4   Throughput per TSL and IR gain vs. CIR. TU3 channel. 900 MHz band.

switching points in case that IR is not used, but when IR is used, the envelope is smoother. IR gain changes according to this envelope so that IR gain is not constant but it changes with C/I values. The IR gain is larger without frequency hopping in most of the cases. It should be noted that real RRM algorithms will have an impact on the represented link level results. In reality LA is based on the radio link condition estimation reported by the MS or measured by the BTS, which is not perfect. Therefore, a realistic LA algorithm will make non-optimum (M)CS decisions. Nevertheless the use of IR, which actually adapts the transmission to radio link conditions by increasing the coding rate on successive retransmissions, has been proved to be a quite efficient way to complement the LA algorithm. In effect the use of IR reduces LA implementation requirement and, it is better to talk about LA&IR together as an efficient mechanism to adapt the transmission to changing radio link conditions. In conclusion, the combination of LA&IR algorithm allows the network to operate very close to optimum conditions.

Noise limited performance

In this section, link level performance results for (E)GPRS with a TU3 channel in noise limited scenarios are presented. In this case there is no interference in the cell and link level performance is limited by the thermal noise, the noise bandwidth defined by the receiver filter and the noise generated by the receiver or noise factor. For a more detailed explanation see Chapter 10.

Figure 7.5 shows the throughput per TSL vs. E_s/N_0 for GPRS and EGPRS with IR and ideal LA in a TU3 channel with and without frequency hopping. Signal-to-noise ratio (E_b/N_0) can be calculated from E_s/N_0 as $E_s/N_0-4.77$ dB for 8-PSK modulation. E_b/N_0 is the same as E_s/N_0 with GMSK. A 8-PSK symbol

Figure 7.5 Throughput per TSL and IR gain vs. E_s/N_0 for GPRS LA and EGPRS LA&IR. TU3 channel. 900 MHz band.

corresponds to 3 transmitted bits, therefore $10^*\log(3)=4.77$ is subtracted in order to obtain the energy per bit (Eb). Notice how in noise limited scenarios frequency hopping has the same effects as it had in interference limited situations, i.e. frequency hopping introduces some gain for coding schemes with high protection and some losses when using coding schemes with very low protection.

From Figure 7.5 it is possible to observe how GPRS and EGPRS perform similarly for low throughputs per TSL. For Es/No larger than 7 dB EGPRS allows a much higher throughput per TSL than GPRS. Incremental redundancy combining produces similar throughput gains as in interference limited scenario. IR benefit is larger in the non-hopping case as it was explained in the previous section.

GPRS/EGPRS throughput vs. cell range

In order to calculate the coverage or throughput as a function of the distance for (E)GPRS, a noise limited scenario is considered with performance limited by thermal noise and transmitter output power. The link level results and the link budget calculation are used for this purpose.

Link budget calculation is used to map path loss to E_s/N_0 as described in Chapter 10. Using Figure 7.5, E_s/N_0 can be translated into throughput per TSL. Figure 7.6 on the right shows how EGPRS DL throughput changes with the path loss. E_s/N_0 is a function of the received signal power, symbol rate and noise figure. The received power depends for example on the transmitter output power, the transmitter and receiver antenna gains, cable losses and path loss.

Figure 7.6 Throughput per TSL for (E)GPRS with IR and GPRS in a TU3 channel. Ideal link adaptation. 900 MHz band. Ideal frequency hopping.

Path loss (L) is a function of the distance. The vehicular test model specified in [3] has been considered in this analysis. According to the reference, the path loss can be calculated according to:

$$L = 40(1 - 4 \times 10^{-3}\Delta h_b)\log_{10}(R) - 18\log_{10}(\Delta h_b) + 21\log_{10}(f) + 80 \text{ dB}$$

where R is the distance between base station and mobile station in kilometres, f is the carrier frequency in MHz and Δh_b is the base station antenna height, in metres, measured from the average rooftop level.

Once the path loss as a function of the distance is known, E_s/N_0 vs. cell range can be obtained. Finally, combining this function with the throughput per TSL obtained from link level simulations, the throughput per TSL for each coding scheme vs. cell range can be obtained.

Based on the link budget presented in Section 10.1 the path loss for AMR 5.15 voice codec (see Chapter 6) in uplink is 156.2. Assuming the vehicular path loss model the cell range, considering an antenna height of 15 m, is 8.68 km. In the same way AMR 12.2 voice codec path loss is 151.4 and the corresponding cell range is 6.47 km. The downlink (DL) throughput per TSL can be represented vs. relative ranges taking AMR 12.2 range as a reference. In the case of AMR 5.15 the coverage relative to AMR 12.2 is calculated as 8.68 × 100/6.47 = 134%.

Figure 7.6 on the left shows DL throughput per TSL vs. relative cell range for EGPRS with IR and GPRS in a TU3 channel when using ideal LA.

7.2 (E)GPRS radio resource management

The radio link control (RLC)/medium access control (MAC) layer involves a number of new procedures optimised for data transmission in packet switch mode. A description of the new RLC/MAC mechanisms that have been introduced in the standards is presented in Chapter 1. The design of the algorithms that make use of the new mechanisms is not included in the (E)GPRS specifications. Nevertheless, it is very important to understand these algorithms and their overall performance impact.

Polling and acknowledgement strategy

Polling is a mechanism in which the packet control unit (PCU) asks the mobile station to send information about the DL link conditions and about the successfully received RLC blocks. Polling is part of the (E)GPRS selective request ARQ mechanism where non correctly received RLC blocks are retransmitted. The PCU polls the mobile station by transmitting an RLC data block with a polling bit set to 1. This message is sent on the packet associated control channel (PACCH) (Section entitled *GPRS packet data logical channels*). The mobile responds to the polling by transmitting a packet DL acknowledge (ACK)/negative acknowledge (NACK) message on the PACCH sub-channel scheduled by the PCU, as displayed in Figure 7.7. A packet DL ACK/NACK message contains radio link condition information and acknowledgement bitmaps for the RLC receiving window at the MS. Note that polling can be used in both RLC acknowledged and unacknowledged RLC mode.

Polling can also be coordinated with the TBF dropping algorithm. By detecting excessive losses of DL ACK/NACK messages it is possible to detect inoperative radio link conditions.

It is really important to find an optimal frequency of the polling mechanism. The polling algorithm may be able to adapt to different radio link conditions by polling more often when high BLER is detected. Additionally it has to take into account the frequency at which RLC blocks are scheduled. If the mobile is not polled frequently the RLC window could get stalled during the transmission. On

Figure 7.7   Polling mechanism.

Figure 7.8   RLC window stalling.

the other hand if the mobile station is polled very often it will create a lot of signalling overhead in UL.

The transmitter has a window in order to address RLC blocks. In this window the RLC blocks are classified as untransmitted, pending ack, acked and nacked. An RLC block is in untransmitted status if it has not been transmitted yet for the first time. Once it is transmitted it changes to pending ack state. When a packet DL ACK/NACK message is received the PCU updates the transmitting window as follows. If the block has been positively acknowledged the block is set to acked. If the block has been negatively acknowledged the status is set to nacked. If no acknowledgement information is received for a transmitted block the status is still pending ack. Once the transmitting window is updated the window is slid up to the first unacknowledged block, see steps 1–3 in Figure 7.8.

The polling mechanism is applicable only in the DL direction, however a similar mechanism is used for UL. In the UL direction, the PCU may send packet UL ACK/NACK messages to the mobile in order to inform the MS which blocks have been successfully received. When a packet UL ACK/NACK message is received the mobile updates its transmit window. If the block is positively acknowledged the status is set to *acked*, and if the block is negatively acknowledged the status is set to nacked. A block is not negatively acknowledged until a timer expires in order to avoid retransmissions when the RLC block has already been retransmitted within the RLC acknowledgement delay. A counter is used instead of a timer in Rel'99 onward. After that, the transmit window is slid further up to the first unacknowledged RLC block. The algorithm that controls the sending of UL ACK/NACK messages is called UL acknowledgement strategy.

The effect of RLC window stalling

Stalling may occur in UL and DL directions. When a nacked block is retransmitted its status is changed from nacked to pending ack. Also when an untransmitted block is transmitted its status is changed to pending ack. If there are no blocks whose status is either nacked or untransmitted within the RLC window, meaning that all the RLC blocks have been transmitted but no acknowledgement information has been received yet, the window is said to be stalled. In this situation, new incoming RLC blocks can't be transmitted until new acknowledgements are received so that the RLC window can advance. In a stalling situation the RLC layer may retransmit pending ack blocks one after another starting from the beginning of the transmit window. Those are pre-emptive retransmissions that will help to solve the RLC window congestion.

Figure 7.8 shows graphically the stalling process. In DL the size of the RLC window is 64 RLC blocks for GPRS and 1024 (max.) for EGPRS. The risk of stalling is reduced in EGPRS by means of two improvements. First of all, the transmit window is much larger. Secondly, the retransmission mechanisms are more efficient.

Link adaptation algorithms for (E)GPRS

Introduction

In (E)GPRS the information is transmitted in packets using different channel coding schemes. The purpose of coding is to improve the transmission quality when radio link conditions are bad, since the coding introduces some redundancy in the packets in order to detect and correct bit errors that occur during the transmission.

By increasing the amount of redundancy, the error correction can be improved but at the same time the net bit rate is reduced. In order to maximise the throughput, robust coding schemes should be used in poor radio link conditions and coding schemes with less protection in good radio conditions.

Link adaptation (LA) algorithms are used to select the optimum (M)CS as a function of the radio link conditions. The selection of the coding scheme is based on channel quality estimates.

In a typical LA the receiver measures the carrier to interference ratio or the raw bit error rate and makes an estimate for the channel quality by averaging the measurements over some time interval [4]. The LA algorithm makes the decision about the coding scheme by comparing the estimated channel quality with certain threshold values in order to obtain maximum throughput. Based on the available channel quality estimates (CIR, BLER and bit error probability (BEP)) different LA algorithms are possible.

LA algorithms for DL and UL   Radio link measurements are the same in RLC acknowledged and unacknowledged modes. The same packet DL ACK/NACK and packet UL ACK/NACK messages are sent in both cases.

In DL, the PCU polls the mobile, which responds by sending the GPRS or EGPRS packet DL ACK/NACK message. These messages contain measurements with information about the link quality. The GPRS report contains the available interference levels in serving cell, the RXQUAL (received signal quality), C (received signal level) and SIGN_VAR (average variance of received signal level). Similar information is included for EGPRS, apart from mean and co-variance averaged values of BEP which are included instead of RXQUAL and SIGN_VAR values. Based on the information of radio link conditions, the network selects the optimal coding scheme to use in DL acknowledged and unacknowledged mode.

In the uplink, the network commands the mobile station with the packet UL ACK/NACK message the coding scheme to use. The selection of the coding scheme is based on the same LA algorithm used in DL using measurements performed by the base station.

Coordination with power control algorithm Both LA and power control algorithms have link quality measurements as input. The quality thresholds of both algorithms should be coordinated. Power Control should not interfere LA in order to avoid that a reduction in output power results in a change of coding scheme and a decrease in throughput. The LA algorithm usually has priority over the power control algorithm.

Quality of service requirements input to LA The cross-points or thresholds for LA can be determined based on the user quality of service (QoS) profile. The general rule in acknowledge mode is to select the (M)CS that maximises throughput per TSL. Nevertheless, for certain users with very strict delay requirements it could be better to use a lower coding scheme in order to avoid too many retransmissions that introduce more delay. If RLC unacknowledged mode is used the cross-points could be set to force the use of lower coding schemes to achieve the minimum reliability QoS requirements.

LA in GPRS

An ideal LA algorithm for GPRS (TU3 channel with ideal FH) would follow the envelope of the C/I vs. throughput curves with different Coding Schemes, see Figure 7.9. In GPRS the algorithms can be typically based on BLER and CIR estimation. Both measurements can be calculated from reported information from the MS or the BTS. If C/I is used for LA, different thresholds have to be set up for the hopping and non-hopping cases. This is because the C/I vs. through-put curve of each CS changes with the use of FH.

LA in EGPRS

As it was mentioned in previous sections the retransmission mechanism is more efficient in EGPRS due to the use of incremental redundancy (IR). LA should be coordinated with the IR mechanism for optimum performance.

Incremental redundancy IR is a physical layer performance enhancement for the acknowledged RLC mode of EGPRS. IR adjusts the code rate to actual

Figure 7.9 LA algorithm based on CIR thresholds for GPRS.

channel conditions by incrementally transmitting redundant information until decoding is successful. The automatic repeat request (ARQ) protocol takes care of requesting and retransmitting incorrectly received blocks. IR improves the reception of retransmissions by combining the information from the original failed transmission with the new retransmitted block and, in this way, increases the probability of correct reception. Both the modulation and coding schemes and the puncturing schemes may be changed in subsequent retransmissions [5]. IR is an EGPRS specific implementation of type II hybrid ARQ transmission scheme [16].

Figure 7.10 shows an example of IR operation with MCS-9. The original data are first convolutionally encoded with coding rate 1/3. Before the first transmission, the encoded data are punctured by removing 2/3 of the encoded bits. The punctured bits are transmitted over the radio interface. The effective channel coding rate is 1/1, i.e. the same number of bits as the original data is transmitted. The receiver is able to decode and recover the original data if radio link conditions are good enough. Otherwise, the network will re-send the same coded data but with a different puncturing scheme: different bits are now removed. The receiver keeps in its memory the information of the received bits before decoding (soft values). When the second transmission is received, the receiver has now two-thirds of the received encoded bits, so the effective coding rate (after combining with the first transmission) for the second transmission is 1/2. The probability of receiving correctly the original data is now higher than in the

Figure 7.10 IR transmission and combining with MCS-9.

first transmission. If the block is still incorrect the network will send again the same encoded data with a different puncturing scheme, containing the rest of the original encoded data. The effective coding rate for the third transmission is 1/3, and the probability of receiving correctly the data is much higher.

Without IR, the BLER of each transmission is potentially the same. However, with IR the BLER of retransmission after combining is much lower than BLER of the initial transmission. Therefore, the final net throughput without IR is lower than the throughput with IR. For this reason, the LA crossing points or thresholds need to take into account the use of IR.

IR is mandatory for the MS but not for the network. If the MS runs out of memory it cannot store incorrect received blocks to be combined with the new ones. This situation can be reported to the network, and the LA mechanism should act accordingly assuming that IR will not be done at the MS side.

Retransmission with a different coding scheme   EGPRS has nine different coding schemes (MCS-1–MCS-9). The coding schemes MCS-1–MCS-4 use Gaussian minimum shift keying (GMSK) modulation and MCS-5–MCS-9 use 8-PSK modulation. These coding schemes are grouped in three families as specified in Section 1.5.2. Family A consists of coding schemes MCS-1 and MCS-4, family B consists of MCS-2, MCS-5 and MCS-7 and family C consists of MCS-3, MCS-6, MCS-8 and MCS-9.

In GPRS all the retransmissions of an RLC block have to be performed with the same coding scheme as the original transmission. This fact increases the risk of stalling in the transmit window and may create additional delays. In EGPRS retransmissions of an RLC block with different coding schemes are allowed.

However, these retransmissions must be done using the MCS from the same family. MCS-6 and MCS-9 can be freely selected as they have the same RLC block size. One MCS-9 radio block contains two RLC blocks, each one of which could be sent separately in two MCS-6 radio blocks. The same principle applies to MCS-5 and MCS-7. In these cases, IR combining between two different MCSs is possible. Note that IR is always possible when retransmitting with the same MCS.

In order to allow retransmissions with other coding schemes, re-segmentation and padding can be used. However, as these procedures modify the data before the channel coding, IR combining is not possible in these cases. When using padding, dummy bits are added to an MCS-8 block to make it suitable for MCS-6, forming a MCS-6-padded block (see Figure 7.11). Padding is only possible with MCS-8. When using re-segmentation an RLC block is split in half and separately transmitted, but must be acknowledged as one. The following re-segmentations are possible (see Figure 7.11): 1 MCS-4 block into 2 MCS-1 blocks, 1 MCS-5 block or 1 MCS-7 block into 2 MCS-2 blocks, 1 MCS-6 block or 1 MCS-9 block into 2 MCS-3 blocks and 1 MCS-6-padded block into 2 MCS-3-padded blocks.

Padding and re-segmentation are useful, in principle, when there is the need to retransmit the same radio block with lower MCS due to radio link condition

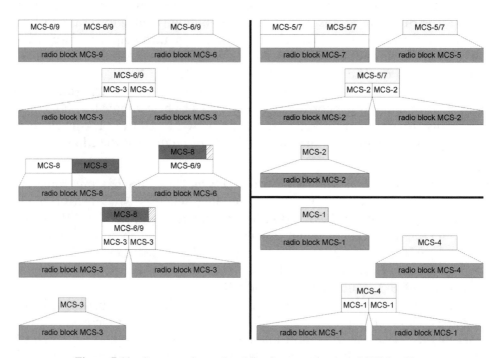

Figure 7.11   Segmentation and padding between the three MCS families.

8 PSK MEAN BEP																															
1	2	3	4	5	6	7	8	9	10	11	12	13	14	15	16	17	18	19	20	21	22	23	24	25	26	27	28	29	30	31	32
5	5	5	5	5	5	5	5	5	5	5	5	6	6	6	6	6	7	7	7	7	7	7	7	8	8	8	8	8	9	9	9
5	5	5	5	5	5	5	5	5	5	6	6	6	6	6	7	7	7	7	7	7	7	8	8	8	8	8	8	9	9	9	9
5	5	5	5	5	5	5	5	5	5	6	6	6	6	6	7	7	7	7	7	7	7	8	8	8	8	8	8	9	9	9	9
5	5	5	5	5	5	5	5	5	6	6	6	6	6	6	7	7	7	7	7	7	7	8	8	8	8	8	8	9	9	9	9
5	5	5	5	5	5	5	5	6	6	6	6	6	6	7	7	7	7	7	7	7	7	8	8	8	8	8	8	9	9	9	9
5	5	5	5	5	5	5	6	6	6	6	6	6	7	7	7	7	7	7	7	7	7	8	8	8	8	8	8	9	9	9	9
5	5	5	5	5	5	6	6	6	6	6	6	7	7	7	7	7	7	7	7	7	7	8	8	8	8	8	8	9	9	9	9
5	5	5	5	5	6	6	6	6	6	6	7	7	7	7	7	7	7	7	7	7	7	8	8	8	8	8	8	9	9	9	9

(Row labels at left: 8PSK_CV_BEP, rows 1–8)

Figure 7.12 LA look-up table for 8-PSK modulation. From EGPRS specifications [6].

changes. IR combining of some the retransmitted radio blocks is normally enough to achieve optimum performance if the packet delay is not a critical factor. Therefore, padding and re-segmentation can be avoided in most of the situations.

EGPRS link adaptation implementation example
Link adaptation can be based on BEP values estimated in the receiver. Mean (MEAN_BEP) and normalised co-variance (CV_BEP) BEP values are reported by the MS as explained before. Look-up tables (CV_BEP, MEAN_BEP), like the one in Figure 7.12 are then used to decide the coding scheme used in the transmission of the radio blocks. The LA look-up table in the figure is optimised for throughput maximisation without IR. In case IR is used a different table should be used.

(E)GPRS channel allocation

Channel allocation algorithms are responsible for assigning channels to (E)GPRS MSs. There are two issues to consider in this process: the interaction with CS and interaction with other (E)GPRS users.

Interaction with CS services
In GSM/EDGE networks, CS and (E)GPRS connections have to live together sharing the same resources (timeslots). Two different strategies are possible: sharing all the available resources between PS and CS or splitting these resources in two separate pools. In both cases, the channel allocation algorithm selects the resources to be used for each connection. A channel is considered to be a particular TSL using certain frequency parameters.

Two separate pools of resources This resource sharing method is based on splitting all the channels into two different pools, the circuit switched pool and the packet switched pool, see Figure 7.13. This means that voice and CS data

Figure 7.13   Two pool channel allocation method.

connections are allocated to channels that belongs to the CS pool, and (E)GPRS connections are allocated to channels within the PS pool, as far as there is capacity available in the pools. Upgrades and downgrades of the two pools are possible in order to adapt the size of the pools to traffic load conditions. A minimum number of channels can be reserved for each pool to ensure a certain minimum PS or CS capacity.

The main benefits of having two separate pools are:

- The PS channels remain always consecutive to each other, and thus the probability of allocating the maximum number of TSLs supported by the MS is maximised. This improves statistical multiplexing so that better utilisation of available TSLs can be achieved.
- CS services need stable TSL allocation and have continuous traffic flow. But on the other hand, PS services need frequent TSL allocations with shorter duration, and the traffic tends to be bursty. The channel allocation algorithm can be optimised for the two kinds of traffic inside each pool, and the interaction between the two pools is minimised.

The main drawback is that there is the need to do continuous re-packing of the connections towards the end of the pools, so that non-utilised channels are at the border of the two pools and can be converted to PS or CS when required.

One common pool of resources   This way of sharing the resources consists in forming a pool of channels in which CS and PS connections are allocated. Both types of connections have to compete for the use of the resources in the cell. The concept is depicted in Figure 7.14.

Figure 7.14   One pool method.

The main advantage of this method is that the pool of available channels is larger, so it gives more freedom for channel selection. There are also some drawbacks of this method. It is not possible to optimise channel allocation for PS and CS services separately. It may happen that the free resources available in the pool are fragmented into non-consecutive channels, and multislot PS and CS mobiles cannot use these TSLs efficiently.

Interaction with other EGPRS users
Once the potential (E)GPRS channels are identified there is the need to select most suitable channels for the (E)GPRS connection. The simplest way to do that is to select consecutive channels that are less utilised. Considering only best-effort users, the channel utilisation can be measured as the amount of users that are using the channel. If there are already N mobiles in one channel, a new mobile would get a portion of $1/(N + 1)$ of that TSL capacity. The channels that give the maximum capacity, i.e. the channels with fewer allocated mobiles, will be selected for the new connection. If (E)GPRS connections have different priorities the channel allocation may consider the different user priorities in order to calculate the utilised portion of channels. It is also possible to reserve absolute values of channel utilisation for guaranteed throughput services over (E)GPRS.

(E)GPRS scheduler

Once the resources, timeslots and frequencies have been assigned, the PS users can be multiplexed with other PS users sharing the same channels. Multiplexing is controlled by the scheduling mechanism in the UL and DL direction.

The scheduler assigns turns to each user depending on certain criteria. It is possible to see the scheduler as a queuing system for each TSL as shown in

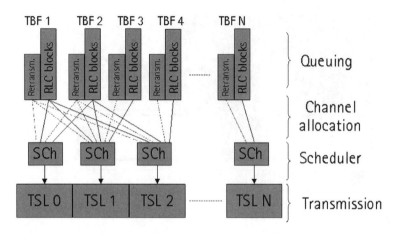

Figure 7.15 The scheduler assigns turns to the different users for each TSL.

Figure 7.15. Scheduling algorithms for best-effort and interactive traffic classes, and for guaranteeing throughput for streaming services will be presented next.

Scheduling algorithms

Round robin The simplest scheduling algorithm is the round robin, in which each user is assigned a transmission turn with the same priority. When all the users have transmitted once, the first user starts to transmit again. Round robin can be applied to each TSL. More elaborate scheduling mechanisms with similar effect to the previously described are also possible.

Weighted round robin Weighted round robin does not share equally the scheduling turns between the users because the users have different priorities. The number of assigned turns is proportional to a weighting factor or priority. The weights assigned to each user are fixed. Weighted round robin is useful when (E)GPRS subscribers are divided in different classes (e.g. gold, silver and bronze), with different subscription characteristics and different service levels.

Scheduling algorithms to guarantee throughput Streaming services need guaranteed throughput through the air interface. There are two possibilities in order to guarantee throughput: dedicate timeslots (similar to CS) or share the timeslots with other services. When the timeslots are shared, the scheduler assigns transmission turns to guarantee a minimum throughput. The rest of the transmission turns can be used for other users or extra traffic over the guaranteed throughput. See Figure 7.16.

Figure 7.16   Use of shared channels for streaming services.

Power control

Power control can also be used with packet switched services. Power control for (E)GPRS is more challenging than for speech due to the bursty nature of Packet Switched services. (E)GPRS power control can be supplied for uplink and downlink.

Uplink power control
The purpose of UL power control is mainly to save battery from the mobile station. However, it helps also to reduce interference. When the radio conditions are good, the MS transmits with lower output power. The algorithm is described in Chapter 1. It is performed autonomously by the mobile with the help of some parameters provided by the network. No interference is taken into account in the algorithm, only path loss is considered.

The operator can modify α and Γ_{CH} parameters described in Section entitled *Power control procedure* in 1.4. Γ_{CH} determines the minimum mobile output power, and α determines how much a change in the received level (C) is translated into a change of MS output power (it changes the slope). Figure 7.17 shows mobile station output power vs. DL received signal level for different values of α and Γ_{CH}. Note that by setting different parameter values it is possible to control the minimum level when UL power control starts to operate, as well as how aggressively the power control is applied when this minimum level is surpassed.

The mobile station shall use the same output power on all four bursts within one radio block. When accessing a cell on the PRACH or RACH (random access channels described in Section entitled *GPRS packet data logical channels* in 1.4) and before receiving the first power control parameters during packet transfer, the mobile shall use the output power defined by PMAX that is the maximum MS transmitted power.

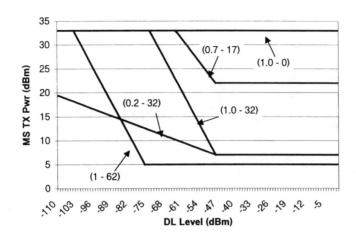

Figure 7.17 Mobile station output power (MS TX Pwr) vs. DL received signal level (DL level) for different (alpha–gamma) values.

Downlink power control

The objective of downlink power control (DL PC) is to optimise the base station output power maintaining the same quality in the ongoing connections. The DL PC algorithm decreases the interference in the system. In a mixed scenario with packet and circuit switched services DL PC results in a reduction of the interference generated from PS to the speech services.

Figure 7.18 shows how the DL PC algorithm is composed of two steps mainly. Firstly, the mobile performs quality measurements in downlink and secondly, depending on these measurements, the base station output power is updated. Depending on the kind of measurements and the used criteria to change the output power, several DL PC algorithms can be defined.

BTS transmitter power can be written as

BTS PWR = BCCH PWR − P0 − Delta

where BCCH PWR is the BCCH channel transmission power, P0 is a constant value that introduces negative power offset and Delta is a component that

Figure 7.18 Downlink power operation.

depends on radio link conditions. Different P0 can be assigned to individual users. When dynamic UL RLC block allocation is in use (see Section entitled *Medium access control and radio link control layer* in 1.4) the maximum value of Delta is limited to 10 dB. This is to prevent uplink state flag (USF) being lost by another MS sharing the same TSL with the user receiving the radio block. This is critical as the USF has one-to-many transmission nature and any user sharing the same TSL need to decode the USF in order to transmit in the UL direction. In case of static UL RLC block allocation, USFs are not needed, and therefore Delta values are not limited.

DL PC should act outside the LA work window in order to avoid that a decrease in output power results in a change of coding scheme and a decrease in throughput. If only a subset of (M)CS is utilised, e.g. CS-1–2, the DL PC work window is larger and more gain can be achieved. The implementation of all modulation and coding schemes in EGPRS reduces significantly the DL PC work window and the interference reduction is very limited unless a certain data capacity is sacrificed. In any case, DL PC is always convenient when mixing voice and data services on the same hopping layer, in order to reduce interference levels generated to speech services. P0 can be used for this purpose. A reduction of 3 dB in (E)GPRS transmitted power doesn't have a big impact on average data capacity, but it definitely has a major impact on speech quality, see Section 7.5.

7.3 GPRS system capacity

Introduction

In this chapter, performance of GPRS systems is evaluated by network level simulations. The focus is on downlink performance since in most applications and scenarios the data flow is clearly asymmetrical, thus the data capacity bottleneck will be in DL. Typically, uplink traffic may consist mostly of TCP acknowledgement messages, although occasionally certain UL capacity is required in order to send user data.

Performance indicators (throughput, delay) are measured between BSC (PCU) and MS, i.e. they are not end-to-end values. The analysis is focused on the characterisation of data transport capabilities of the GPRS base station subsystem (BSS) according to release Rel'97 specifications. End-to-end values depend heavily on issues like higher protocol layers (including the application level), GPRS core network elements, and the quality of the fixed line connection in between the two entities. These issues are out of the scope of this analysis and will be handled separately in Section 7.7.

GPRS/EGPRS simulation model

PCU-to-MS throughput and delay

Figure 7.19 GPRS/EGPRS functionality in the simulator.

Modelling issues and performance measures

A basic description of the simulator is described in Appendix E. Below, the GPRS-specific features and modelling are described. Figure 7.19 describes schematically the GPRS/EGPRS simulation model. All the necessary details affecting the analysed performance values are taken into account in the simulations.

Radio link modelling

Only downlink user traffic is simulated. Downlink packet ACK/NACK messages are modelled on uplink. These blocks experience interference that correlates with the downlink CIR, according to

CIR_uplink = CIR_downlink + $N(2, 4)$

where all the units are expressed in dB, and $N(2, 4)$ is a random normally distributed value with mean of 2 and variance of 4. This relationship is based on separate simulations with both up- and downlink traffic assuming 90/10 asymmetry.

Link level results use two-phase mapping tables according to principles in [7]. First, burst-level CIR values are mapped on to raw BER values (two tables are needed, one for GMSK and one for 8-PSK). Next, the arithmetic mean and standard deviation of raw BER samples over the interleaving period are calculated. These values are then mapped on to block error probability (BLEP) according to used coding scheme. Once the BLEP is calculated, a random experiment is done to decide whether the block had an error.

Traffic model
All GPRS sessions arrive at the network according to Poisson process and end after every data block has been positively acknowledged. The traffic pattern generated for all GPRS users is FTP-like traffic (file of 120 kbytes download) and follows the guidelines described [3]. Note that layers above LLC are not implemented in the simulator.

Radio resource management model
The channel allocation algorithm selects the least occupied consecutive slots from the same transceiver (TRX). If all the slots are full (maximum number of users in the same TSL is limited to 6), the TBF is blocked and a new TBF establishment procedure is started after 5 s. Round robin type of scheduling is applied, so each active TBF transmits a fixed amount of time after which the transmission turn is given to the next TBF in the queue. Scheduling is done on slot basis, i.e. every slot in each TRX has its own queue. Network polling is modelled in such a way that every MS has a polling counter, which is initially set to zero. Every time a RLC block is retransmitted as a result of negative acknowledgement the polling counter is increased by 3 and in all other cases (normal transmission or retransmission due to stalling) the polling counter is increased by 1. When the value of the counter reaches 18, polling bit is set true in the next RLC block and the counter is reset to zero.

Cell re-selection is triggered by power budget algorithm with penalty timer and hysteresis (see values in Table 7.E.3 in Appendix E). If the MS is in packet idle mode, it just attaches itself to the new cell. In packet transfer mode, current TBF is released and TBF establishment attempt is made to the new cell. If at least one channel could be allocated, MS can continue data transmission after 1 s (estimated processing and signalling delay). The unacknowledged LLC PDUs transmitted in the old cell are copied and retransmitted in the new cell. If channel allocation fails, the MS will experience a delay of 5 s after which it returns to the packet idle mode.

Link adaptation is based on C/I and uses all four GPRS coding schemes. The algorithm exploits the minimum C/I values optimised for coding schemes 2–4. Frequency hopping is random radio frequency (synthesised) hopping. Mobile allocation index offset (MAIO) management (see Chapter 5) is used with 1/1 reuse scheme. Power control is not used.

MAC layer modelling
TBF establishment procedure (downlink) is triggered whenever new data arrives at the PCU addressing a certain MS. A downlink TBF establishment delay (defined in Table 7.E.3, Appendix E) is taken into account before the actual data transmission can take place.

TBF blocking occurs when PCU is unable to allocate radio resources for the use of MS. A timer *T3190* is started and re-establishment of the TBF cannot be started until the timer expires (expiry limit of the timer *T3190* is presented in Table 7.E.3).

TBF release procedure is entered whenever there are no LLC data in the PCU's buffer waiting for transmission. Since delayed TBF release procedure is used in DL, data transmission can proceed without any additional delays whenever new data are available. Before the actual release, all non-acknowledged blocks inside the transmitting window must be correctly received and the expiry limit of the delayed TBF release timer must be reached. This timer is started whenever there is no packet Ack/Nack message with final block indicator (FBI) bit set to true is received by the PCU and it is reset if new data arrive for the specific MS. The expiry limit of the release timer is presented in Table 7.E.3, Appendix E.

TBF dropping is based on lost pollings in the uplink direction. Whenever the network receives a valid RLC/MAC control message from the MS, it shall reset counter *N3105*. The network shall increment the counter for each radio block (allocated to that MS with the relative reserved block period (RRBP) field) for which no RLC/MAC control message is received. If *N3105* reaches *N3105max*, the network releases the downlink TBF. *N3105max* is a parameter defined in Table E.3.

RLC/LLC layer modelling
All the relevant features of the RLC protocol are taken into account in the simulations. These include active transmitting and receiving windows, selective ARQ with block sequence numbers, final block indicators, polling bits and downlink packet acknowledgement message with measurement reports. The RLC layer operates in acknowledged mode, LLC in unacknowledged mode.

Upper layer modelling
Layers above LLC (TCP/IP, RTP/IP, ...) are not explicitly modelled in the analysis of this section. The effect of upper layers will be considered in Section 7.7.

Simulation parameters
Some of the most important GPRS-specific simulation parameters and their default values are presented in Appendix E. Note that the general simulation parameters and network set-up presented in Appendix E are also utilised in these simulations.

GPRS performance in a separate non-hopping band

In this section, performance analysis of the GPRS system in a separate non-hopping band is presented. The analysis is done with two different LA algorithms. The first algorithm uses only the two most robust coding schemes (CS-1 and CS-2) and the second algorithm all four GPRS coding schemes. Only CS-1 and CS-2 are used in most of the initial GPRS networks but it doesn't achieve the full benefits of GPRS.

Performance of the GPRS system is analysed with different reuse patterns using one TRX with eight TSLs available (except in the trunking gain analysis). All simulated users are best-effort GPRS users, i.e. only GPRS traffic is simulated and there is no prioritisation between different users. The case with mixed CS and PS traffic in the same TRX traffic is analysed separately in Section 7.5.

First, basic GPRS capacity and performance figures are presented starting with the behaviour of user-specific mean TBF throughput and mean gross LLC frame delay curves. These are both good indicators of the quality experienced by an average GPRS user. Next, 10th percentile throughput and 90th percentile delay are shown, which are good network level QoS indicators. This basic performance analysis concludes with the spectrum efficiency values achieved with a GPRS system. This analysis includes both the effect of different reuse patterns and different user-specific QoS criteria.

After evaluating the general GPRS performance, the effects of two special cases related to GPRS deployment are discussed. First, GPRS performance on BCCH layer is studied. Next TSL capacity analysis is presented. In the last part of this section the trunking gain analysis shows how the traffic handling capacity of each individual TRX increases when more TRXs are added to the cell.

Mean throughput and mean delay
Figure 7.20 shows the evolution of mean throughput and delay of different reuse patterns and LA. Throughput is measured as the average throughput delivered to the LLC layer during the duration of the TBF. Delay refers to LLC frame delay: the time from when the LLC frame starts to be transmitted by the PCU until it is entirely received by the MS. As it was stated in the previous section, RLC operates in acknowledge mode (retransmissions are possible). Throughput values range from 90% to about 20% of the maximum theoretical values of the highest coding schemes used in the LA algorithm.

Best throughput values are achieved with loose reuses, as expected. The performance levels experienced with reuse patterns 2/6, 3/9 and 4/12 are quite close to each other while 1/3 has clearly lower but still satisfactory performance. However, it should be noted that this comparison does not take into account the amount of spectrum needed for each reuse pattern.

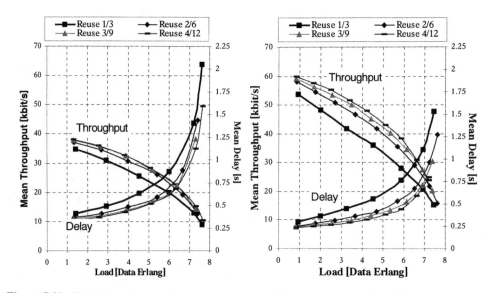

Figure 7.20 Evolution of mean throughput and mean delay in a non-hopping GPRS network with different reuse patterns. Left: GPRS with CS-1–2. Right: GPRS with CS-1–4. Three TSL capable MS. One TRX. Macro-cell environment.

As the load increases there are two main reasons for throughput reduction: increasing interference levels and TSL sharing between users. Both effects will be analysed later on by means of TSL capacity and throughput reduction concepts, respectively. For 1/3 reuse and 6 Data Erlangs, 50% of the throughput reduction is due to interference and another 50% is due to TSL multiplexing. The same 6 Data Erlangs with 3/9 reuse has 15 and 85% throughput reduction due to interference and TSL sharing, respectively.

As the load (Data Erlangs) gets close to the absolute maximum (8 Erlangs for 1 TRX), the dominant performance degradation effect is the TBF multiplexing over the same timeslots and performance is similar regardless of the frequency reuse.

When comparing the performance of the used LA schemes, it is seen that in terms of mean throughput values the additional coding schemes CS-3 and CS-4 bring around 50% higher throughputs. The gain is smaller when the reuse gets tighter. This is due to higher interference levels, which decrease the usefulness of less robust coding schemes. Still, even with small reuse distance and high load the users close to the base station can utilise the higher coding schemes effectively.

10th percentile throughput and 90th percentile delay
In the previous section, only the mean values were compared. Mean values are proportional to the overall spectrum efficiency of the whole network, but they do not tell about the distribution of quality among users. This section focuses on user quality at the lower bound of the distributions. Figure 7.21 shows the minimum throughput values achieved by 90% of the users (10th percentile

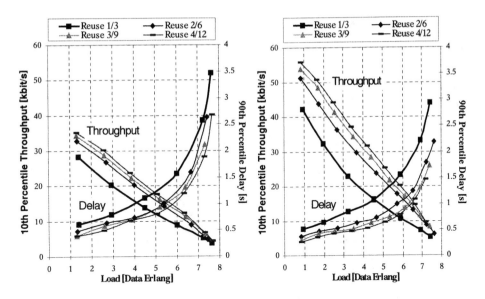

Figure 7.21 Evolution of 10th percentile throughput and 90th percentile LLC frame delay in a non-hopping GPRS network with different reuse patterns. Left: GPRS with CS-1–2. Right: GPRS with CS-1–4. Three TSL capable MS. One TRX. Macro-cell environment.

throughput), and maximum delays for 90% of the transmitted LLC frames (90th percentile delay).

Note also that the decrease in throughput is now more noticeable than in the mean value curves. This is because interference levels are affecting directly low throughput users. The average quality is not influenced that much due to the high quality experienced by the majority of the users, but these percentile figures are immediately affected.

Minimum throughput values are up to around 50% lower than the average values. Note that a simple round robin scheduler is used, therefore scheduler-based throughput equalisation techniques are not applied to users with bad throughputs. With 50% load (4 data Erlangs) 90% of the users achieve 23 and 33 kbit/s net throughput in reuse 4/12 with CS-1–CS-2 and CS-1–CS-4, respectively

Spectrum efficiency and TBF blocking
Previous sections considered the user perceived throughput and delay. In this one, the spectrum efficiency of the network is shown. The GPRS system traffic is increased until it is fully loaded. At this saturation point, TBF blocking starts to grow exponentially according to queuing theory and additional increase in offered load does not any more increase the spectrum efficiency. Figure 7.22 shows the evolution of spectrum efficiency and TBF blocking as a function of network load.

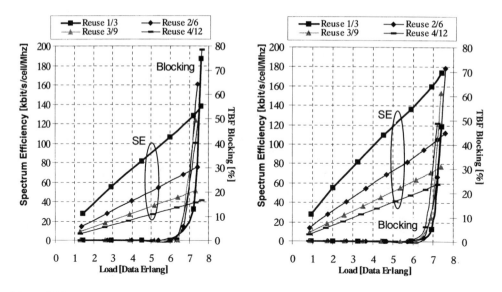

Figure 7.22 Evolution of spectrum efficiency and TBF blocking in a non-hopping GPRS network with different reuse patterns. Left: GPRS with CS-1–2. Right: GPRS with CS-1–4. Three TSL capable MS. One TRX. Macro-cell environment.

The tighter the reuse is, the better spectrum efficiency can be achieved. But pure load based comparison is not always sensible, since it doesn't take any quality requirement into account. This spectrum efficiency and quality trade-off is tackled in a later section.

From Figure 7.22 it can also be seen that TBF blocking starts to take place with the load close to 8 Data Erlangs (absolute maximum limit). The increase of blocking is very rapid and the network will reach the state of congestion. Reaching this congestion point should be avoided when dimensioning of the GPRS network.

Performance in the BCCH layer

Earlier sections only considered non-BCCH TRXs. In this section, the performance of GPRS deployment in the BCCH layer is analysed. There are seven TSLs available for GPRS as TSL 0 is reserved for BCCH. The analysis is performed with two different reuse patterns (3/9 and 4/12) and the achieved results are compared against the results with the same reuse patterns in the non-BCCH layer (Figure 7.23). The non-BCCH layer has only seven TSLs available to allow a fair comparison with BCCH. The analysis is performed with the LA algorithm using all four available GPRS coding schemes. Note that reuse 3/9 is very aggressive for standard BCCH deployments. From the radio bearer quality point of view, the BCCH layer differs from the non-BCCH layer considerably. The most important characteristic of the BCCH layer is that all the downlink bursts in the BCCH frequency must be transmitted constantly and with full power.

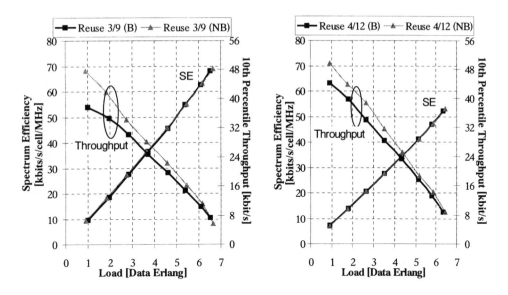

Figure 7.23 Evolution of spectrum efficiency and 10th percentile throughput in a non-hopping GPRS network with different reuse patterns in both BCCH ('B') and non-BCCH ('NB') layers using CS-1–4. Left: reuse 3/9. Right: reuse 4/12. Three TSL capable MS. Both graphs are with one TRX and seven TSLs available.

In the BCCH layer the TSL capacity, see Section 5.2, is constant and independent of GPRS load. Therefore, the throughput variation is only due to user multiplexing over the same TSL. The difference in performance with the non-BCCH layer is due to the lower interference levels at low network loads in the non-BCCH layer. As the load increases and the network becomes fully loaded, the interference levels are similar and throughputs are equivalent.

TSL capacity
TSL capacity analysis is useful for GPRS performance characterisation and network dimensioning with different number of allocated TSLs and different number of available TSLs for GPRS (see Section entitled *TSL capacity* in 5.2). TSL capacity tells how much data the network can transmit with a fully utilised TSL. TSL capacity depends on interference levels and RLC protocol performance. It doesn't depend on HW dimensioning. Figure 7.24 shows TSL capacity for non-BCCH layer with different reuses. The figure on the left-hand side shows TSL capacity not taking into account pre-emptive retransmissions. Pre-emptive retransmissions are pending ACK RLC blocks that may be transmitted during TBF release phase and RLC protocol stalling situation. If those retransmissions are not accounted for, the TSL capacity depends only on interference levels. As the network load increases, the TSL capacity decreases due to higher interference levels. When pre-emptive retransmissions are considered the TSL capacity

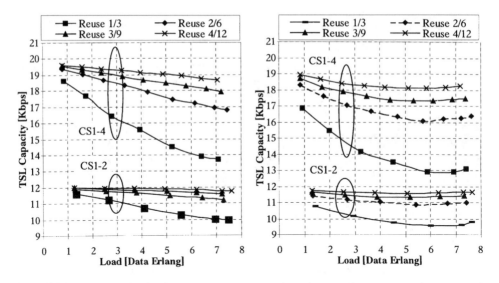

Figure 7.24 TSL capacity for frequency reuses. Non-BCCH layer. Left: no pre-emptive retransmissions. Right: with pre-emptive retransmissions. Three TSL capable MS.

curve changes, see left figure. The effect is more evident for higher interference levels and low Data Erlang values. For the traffic model considered (FTP), the dominant effect causing the TSL capacity reduction is the RLC window stalling. When the network is heavy loaded (Data Erlangs approaching 8), the network multiplexes users on the same TSL, thus reducing the number of pre-emptive retransmissions as they have lower priority than any user data. Simulation assumes three TSL MS, however if lower MS TSL capability was used the probability of RLC protocol stalling would be lower (as throughput is lower). Figure 7.25 shows TSL capacity in the BCCH layer. If pre-emptive retransmissions are not considered, the TSL capacity is constant due to the continuous transmission in the DL direction.

Trunking efficiency analysis
There is a trunking gain when more than one TRX is dedicated for GPRS. Figure 7.26 shows simulation results in a non-hopping GPRS network with 1/3 reuse pattern dedicating one, two or three TRXs in each sector exclusively for GPRS. LA with CS-1–4 is utilised in the analysis. The figure represents blocking probability and throughput reduction factors as a function of TSL utilisation (see Sections entitled *Throughput reduction factor* and *TSL utilisation* in 5.2). It is a measure of how user throughput is affected by TSL sharing among several users. TSL utilisation is a measure of how many of the available TSLs for (E)GPRS are being utilised, on average, to carry data traffic. It varies between 0% (no data traffic) up to 100% (full utilisation of the available resources).

Figure 7.26 shows TBF blocking probability and SE for the same network

Figure 7.25 TSL capacity for different frequency reuses. BCCH layer. Left: no pre-emptive retransmissions. Right: with pre-emptive retransmissions. Three TSL capable MS.

with different number of TRXs. By increasing the number of TRXs it is possible to reach higher TSL utilisation for the same TBF blocking probability. This effect is equivalent to the one experienced by speech traffic. Basically, the existing hardware can be used more effectively when the number of available channels increase and the results show that GPRS service is no exception.

Throughput reduction factors are also affected by increasing the number of TRXs. By doing so the throughput reduction factors increase and, therefore, it is possible to achieve the same average throughput with higher TSL utilisation. This means that SE can be increased while maintaining the same average throughput. Table 7.3 summarises SE values achieved with different reduction

Figure 7.26 Evolution of spectrum efficiency and throughput reduction factors in a non-hopping macro cell GPRS network with reuse pattern of 1/3 and with different number of TRXs and CS-1–4. Three TSL MS. Three TSL capable MS. One TRX.

Table 7.3   EGPRS maximum SE [kbps/Cell/MHz] with different
number of TRXs. Three TSL mobile station

Reduction fact.	0.9	0.8	0.7	0.5
1 TRX	52.5	68.4	75.6	96.6
2 TRX	68.4	80.4	87.5	101.5
3 TRX	80.5	89.6	95.2	103.6

factors. Table 7.3 shows, for example, that for a reduction factor of 10% (this defines network quality) the SE is 50% higher with three TRXs compared with one TRX. This is purely a trunking efficiency gain.

Simulation results have shown that GPRS throughput reduction factors are the same for any frequency reuse including fractional reuses (e.g. 1/1 hopping). Only in the case where interference levels are very high the reduction factor curve changes.

Reduction factors and TBF blocking from Figure 7.26 will change with different number of allocated TSLs. Trunking efficiency depends on the MS TSL capabilities (Nu) and total number of TSLs available for GPRS (Ns). Doubling Ns has the same effect as dividing Nu by two, from the trunking efficiency point of view. In effect trunking efficiency depends exclusively on the rate (N) of the two factors N = Ns/Nu. For example, one TRX with one TSL allocation has the same trunking efficiency as three TRXs with three TSL allocations. The later assumes an ideal channel allocation algorithm that is able to share evenly the load over the available TSL. A properly designed channel allocation algorithm should get close to this situation.

In order to improve trunking efficiency one TSL allocation should be preferred, but the price paid is a lower user throughput. TBF blocking curves and reduction factor can be computed theoretically under a number of assumptions. This analysis is presented in Appendix B. Theoretical results are close to the simulated ones although they tend to be more optimistic. There are two main effects not accounted for in the theoretical analysis when comparing with simulation results:

- In simulations, load is not always equally shared among all the available TSLs. This is due to channel allocation limitations (allocated TSLs have to be consecutive); and, specially, because periodical channel re-allocation of the existing TBFs is not performed. This has the effect of worsening the throughput reduction factors in simulations up to 80% TSL utilisation.
- When interference levels are so high that the TSL capacity can't be considered constant and independent of TSL assignment dynamics, the theoretical analysis is less exact. In this case, the shape of the trunking efficiency curves changes. This is the case of frequency hopping at high EFL levels as mentioned before.

GPRS performance in a separate band with RF hopping

This section presents spectrum efficiency values for GPRS traffic allocated in the hopping layer. The evolution of the most important network quality indicators is also presented. The LA CIR thresholds have been adjusted for hopping channels. A comparison of the non-hopping and hopping networks will be presented first. Then, a soft-capacity study of GPRS in the hopping layer will be performed.

Comparison between FH and non-FH deployments

In this section comparison between frequency hopping and non-hopping networks is presented. In order to be able to produce a fair comparison, the analysis is performed with one TRX configuration and with the same band-width. The selected non-hopping configuration is reuse 3/9 and the corresponding FH configuration is 1/1 with nine frequencies. Note also that the comparison is performed with both link adaptation algorithms.

From Figure 7.27 it can be seen that FH does not bring any quality or capacity gain compared with the non-hopping case. It actually degrades the performance, especially in the CS-1–4 case. This behaviour is different to that with speech traffic, where significant capacity gains are achieved with FH. The main reasons for this difference are:

- Very limited link level gains achieved with frequency hopping when retransmissions and LA are applied. See Section 7.1. In practice, frequency hopping gains are more noticeable with very low BLER, but very low BLER doesn't make a difference in terms of acknowledged mode throughput.
- Highest CS (CS-3 and CS-4) has lower channel coding, so the gain achieved with speech channels becomes a FH loss with CS-3 and CS-4. The lowest CSs (CS-1 and CS-2) have a certain frequency diversity gain (although limited), but interference diversity gain turns negative. Co-channel interference levels in a fractionally reused network can be very high (e.g. with 1/1 reuse the same frequencies are reused on every site). The limited interleaving depth (four bursts) and limited coding are not enough to cope with the strong co-channel burst hits. For the case of GPRS with CS-1–4, FH results are specially degraded. CS-3 and CS-4 suffer the most with FH, thus limiting significantly the SE.

With 1/1 hopping reuse it is possible to introduce up to three TRXs and still avoid co-channel interference from the adjacent cells. By doing this it is possible to eliminate HW limitations and make the network purely interference limited. In this case, it is possible to achieve better spectrum efficiency than the reference 3/9 reuse.

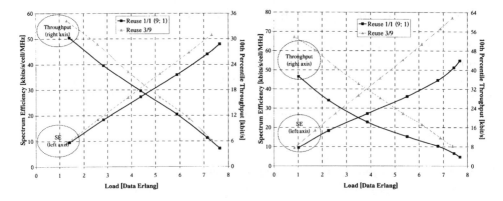

Figure 7.27 Evolution of spectrum efficiency and 10th percentile throughput in both non-hopping and RF hopping macro cell GPRS networks. Left: GPRS with CS-1–2. Right: GPRS with CS-1–4. Three TSL capable terminals. One TRX.

In any case it can be stated that GPRS can be introduced into the hopping network without major problems. Although it is recommended to introduce it first in the BCCH layer and utilise the hopping layer whenever it is required.

Reuse 1/1 hopping analysis
In this chapter a QoS related soft-capacity analysis is performed. The used FH scheme is reuse 1/1 and the number of hopping frequencies is fixed to 18.

Figure 7.32 shows TSL capacity vs. effective frequency load (EFL). The analysis is performed with one, two and three TRXs. The same TSL capacity is achieved for the same EFL values, independently of the number of TRXs as soon as there is low TSL sharing among different users. The later is due to the high probability of RLC protocol stalling that generates a lot of pre-emptive retransmissions. This effect, of increased TSL capacity when TSL sharing increases is due to pre-emptive retransmissions and has been also found in non-hopping cases (see Figures 7.24 and 7.25), although with 1/1 hopping reuse the effect is much more evident. Taking an EFL level of 6%, it corresponds to an average user throughput of $3 \times 12 = 36$ kbps with a three TSL MS. If there is any HW limitation, throughput values should be corrected by the reduction factor. Figure 7.28 represents reduction factor values for one, two and three TRXs. Note also that this analysis is performed with LA utilising all GPRS coding schemes. Figure 7.29 represents the 10th percentile throughput values and 90th percentile delay in pure soft-capacity situation.

GPRS spectrum efficiency with QoS criterion

Quality of service criterion plays an important role in capacity evaluations. Typically, spectrum efficiency increases as a function of input load until the

Figure 7.28 TSL capacity and reduction factors for GPRS in the frequency hopping layer. All GPRS coding schemes.

network is fully loaded. At the same time, the quality of calls naturally decreases. It is sensible to assume that even for best-effort services, there is certain minimum QoS that must be fulfilled (this is required already by some higher level protocols like TCP). When simulating spectrum efficiency for a packet data network, offered load is increased until the point where this minimum QoS level is maintained. Since user perceived quality, not average network level quality, is the indicator to be measured, the QoS criteria must be user-specific. QoS criteria are specified in terms of minimum throughput and maximum delay achieved by 90% of the throughput and delay samples. Figure 7.30 represents GPRS SE figures for different reuses and minimum QoS criteria for three TSL GPRS mobile stations. Four different reuse patterns are considered without hopping, in addi-

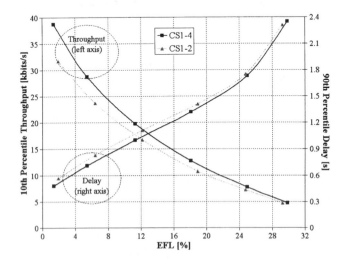

Figure 7.29 Tenth percentile throughput and 90th percentile delay vs. EFL for GPRS in a RF-hopping macro cell network with reuse 1/1. Three TSL capable MS.

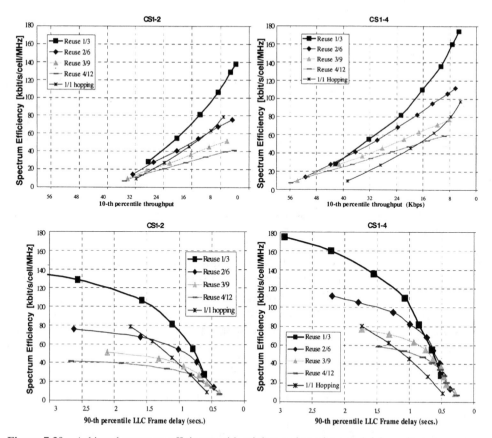

Figure 7.30   Achieved spectrum efficiency with minimum throughput and delay criteria in a macro cell GPRS network with different reuse patterns. Left figures: GPRS with CS-1–2. Right figures: GPRS with CS-1–4. One TRX.

tion to 1/1 reuse with frequency hopping. The SE values are limited by a combination of HW and interference blocking, except in the 1/1 hopping case where there are no HW limitations so that performance is purely interference limited. Hopping performance is degraded also due to RLC protocol stalling, especially in the GPRS CS-1–4 case. Stalling increases artificially the load (up to 30%) and the interference levels due to pre-emptive retransmissions. With lower user throughputs stalling is less frequent so that SE improves appreciably.

As it can be seen in Figure 7.30, the best spectrum efficiency can be achieved with low user data rates and very tight frequency reuse (reuse 1/3). These kinds of assumptions are very suitable for best-effort data. However, reuse pattern will not be able to fulfil high quality criteria (high throughput and low delay) in some locations at the cell border. With tighter QoS criteria, i.e. assuming a higher percentage of users above the minimum quality, the spectrum efficiency will be reduced and looser reuse (like 2/6 or 3/9) may give better spectrum efficiency.

Reuse partitioning principle to increase spectral efficiency and QoS provisioning

From the previous section it has been seen that tight reuses provide the highest spectral efficiency, but they suffer from high interference levels at the edge of the cell. Those calls with bad radio link conditions can be moved to a different layer (e.g. BCCH layer) where the reuse is not so aggressive and throughput at the cell border is much better. In this way two contradictory targets can be met for data services: high spectral efficiency together with high QoS. Very aggressive deployments are also thinkable, like utilising just one frequency for the high capacity layer. By making sure that users with good radio link conditions use this layer, spectral efficiency can be boosted. In order to implement these schemes, the reuse partitioning principles presented in Section 6.2 can be used.

7.4 EGPRS system capacity

Introduction

In this section, performance of the EGPRS (Rel'99 specifications) system is evaluated by network level simulations. As in the GPRS related analysis presented earlier, focus is once again on downlink performance. The performance evaluations are done both from the capacity (spectrum efficiency) and quality (throughput, delay) points of view. Analysis concentrating on the evolution of timeslot capacity as a function of EFL is also presented.

Modelling issues and performance measures

The modelling of EGPRS service is naturally based on the modelling of GPRS service. In addition to the features already supported in GPRS, EGPRS introduces many new enhancements. The EGPRS specific modelling issues that are taken into account in the simulations are described below.

RRM algorithms
Link adaptation is similar to that described in Section entitled *LA in EGPRS* in 7.2.

RLC layer
Incremental redundancy with receiver memory limitations is implemented. During IR operation, if the incorrectly received RLC block does not fit into MS memory it will be discarded and it can't be soft-combined with the old data. MS IR memory considered is such that it is possible to store 61 440 soft values (received raw bits before de-coding) per TBF. This is more than enough in order to avoid IR memory problems. Block splitting and block re-segmentation inside the same MCS family are modelled as well as header errors in EGPRS

RLC blocks. RLC window size is 384 blocks, which is the maximum value for three-slot EGPRS MS.

EGPRS performance with link adaptation in a separate non-hopping band

In this section performance analysis of the EGPRS system in a separate non-hopping band is presented. Performance of the EGPRS system is analysed with different reuse patterns and with one TRX configuration. All users are best-effort EGPRS users, i.e. only EGPRS traffic is simulated and there is no prioritisation between different users.

This section is organised in the same way as the corresponding GPRS section. The evolutions of mean net throughput and mean gross LLC frame delay are presented first. Then the evolutions of 10th percentile throughput and 90th percentile delay are studied. After these quality related sections the spectrum efficiency values achieved with the EGPRS system will be presented. The analysis includes also the comparison of different reuse patterns with different QoS criteria.

Mean throughput and mean delay
The achieved mean throughput values and delays are depicted in Figure 7.31. The achieved values are significantly better than for GPRS. Once again, 1/3 reuse has the worst performance in terms of throughput and delay due to the higher interference levels, while the difference between other reuses is not so large. With low load it is possible to get quite close to the theoretical maximum bit rates with all reuses. Even with high loads very good performance values can be achieved compared with GPRS. With the same load levels, average throughput values are 2.5–3 times better than with GPRS CS-1–4 and 3–4 times better than with GPRS CS-1–2.

Figure 7.31 Throughput and delay performance of EGPRS in a non-hopping macro cell network with different reuse patterns. Three TSL capable MS. One TRX.

10th percentile throughput and 90th percentile delay
Figure 7.31 shows minimum throughput and maximum delay achieved in 90% of the samples (10th percentile throughput and 90th percentile delay, respectively). Note also the relatively larger degradation with 1/3 with respect to other reuses, compared with GPRS results. The absolute throughput and delay levels achieved with certain loads are significantly better than with GPRS. Compared with GPRS with CS-1–2, the percentile throughput values are 2–3.5 times higher. When comparing with GPRS with CS-1–4, gains are of the order of 1.5–2.5 times higher. EGPRS gains are lower with tighter reuse schemes since the higher coding schemes provided by the EGPRS bearer are less used.

Spectrum efficiency and TBF blocking
Figure 7.32 shows the spectrum efficiency growth of an EGPRS network as a function of network load. TBF blocking is shown in the same figure in order to see the level of saturation in the network. With EGPRS, the achieved levels of spectrum efficiency are two- to threefold compared with the same GPRS cases. And as with GPRS, the TBF blocking starts to occur as the load approaches the absolute maximum (8 Erlangs). Although reuse 1/3 has the worst throughput performance, it achieves the highest SE due to the lower number of utilised frequencies.

The simulated SE figures in a fully loaded situation are very much in line with the ones estimated in field trials and presented in Section 7.6.

Figure 7.32 Spectrum efficiency and TBF blocking of EGPRS in a non-hopping macro cell network with different reuse patterns.

Performance in the BCCH layer

EGPRS has a similar behaviour in the BCCH layer compared with GPRS. Due to the continuous transmission, lower throughput values are achieved when the network load is low compared with the same reuse but utilising a non-BCCH TRX. Figure 7.33 shows 10th percentile throughput values and SE with different network loads for different reuses. Seven TSLs are available for both cases. SE is practically the same for both BCCH and non-BCCH cases, the difference is only found in percentile values, that defines the minimum quality.

TSL capacity

TSL capacity achieved with EGPRS is presented in Figure 7.34. Due to the longer RLC window and better retransmission techniques the probability of stalling is much lower than with GPRS. Therefore, pre-emptive transmissions do not influence TSL capacity values very much. TSL capacity values have a larger dynamic range compared with GPRS. For example, reuse 1/3 TSL capacity is almost halved at very high loads. Nevertheless, TSL capacity remains almost constant in the BCCH layer and therefore BCCH is very suitable for achieving high throughput values. By allocating absolute scheduling priority to a user the throughput reduction becomes 1 and throughput depends exclusively on number of allocated TSLs and the TSL capacity. With a four TSL capable terminal allocated in DL and BCCH reuse of 5/15, high priority users may experience throughput close to $4 \times 48 = 192$ kbps on average, independently of BCCH traffic load.

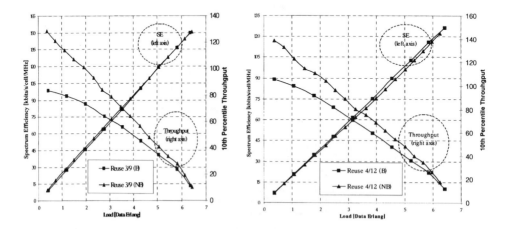

Figure 7.33 Evolution of spectrum efficiency and 10th percentile throughput in a non-hopping GPRS network with different reuse patterns in both BCCH ('B') and non-BCCH ('NB') layers using CS-1–4. Left: reuse 3/9. Right: reuse 4/12. Three TSL capable MS. Both graphs are with one TRX and seven TSLs available.

Figure 7.34 TSL capacity for BCCH and non-BCCH layer.

Trunking efficiency analysis

The same trunking efficiency analysis presented in Section entitled *GPRS performance in a separate non-hopping band* in 7.3 is applicable to EGPRS. Figure 7.35 on the right shows EGPRS reduction factors for one TRX with different reuses. Reuses 2/6, 3/9 and 4/12 have very similar reduction factors. Comparing with GPRS, the reduction factors are also similar to the GPRS reduction factors (found in Figure 7.26).

Reuse 1/3 has lower reduction factors as it has a much larger soft-blocking component. Reduction factors can be seen in Figure 7.35. Reduction factors with two and three TRXs for reuse 1/3 are included in Figure 7.35.

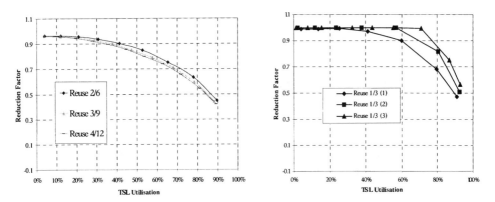

Figure 7.35 EGPRS reduction factor for different reuses. Left: one TRX. Right: one, two and three TRXs with 1/3 reuse. Three TSL capable MS.

EGPRS performance in a separate band with RF hopping

This section presents simulated spectrum efficiency values for the EGPRS network with different reuse schemes in the RF hopping situation. The evolution of the most important network quality indicators is also presented. As in the corresponding GPRS section, Section entitled *GPRS performance in a separate band with RF-hopping* in 7.3, the simulations were done using only non-BCCH hopping TRXs with pure EGPRS traffic.

This section is organised so that a comparison of non-hopping and hopping EGPRS networks is presented first. Finally, a soft-capacity analysis of the EGPRS system is presented.

Comparison between FH and non-FH deployments

In this section a comparison between frequency hopping and non-hopping networks is presented. In order to be able to produce a fair comparison, the analysis is performed with one TRX configuration and with the same bandwidth. The selected non-hopping configuration is reuse 3/9 and the corresponding FH configuration is 1/1 with nine frequencies. Figure 7.36 shows that FH does not bring any quality or capacity gain with one TRX.

With nine frequencies and 1/1 hopping reuse it is possible to increase the number of TRXs up to three (controlling the intra-site co-channel interference, without such control it can go up to nine). In this way HW limitation effects can be minimised, thus networks become pure soft-capacity limited, and better spectrum efficiency than the reference 3/9 case can be achieved.

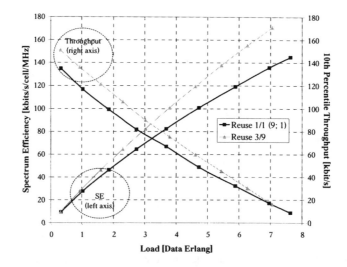

Figure 7.36 Evolution of spectrum efficiency and 10th percentile throughput in both non-hopping and RF hopping macro cell EGPRS networks. Three TSL capable MS. One TRX.

Figure 7.37 Timeslot capacity as a function of EFL for EGPRS in a RF hopping macro cell network with different reuse patterns. Three TSL capable MS. (X, Y) in legend means X hopping frequencies and Y TRXs.

Reuse 1/1 hopping analysis

TSL capacity is represented in Figure 7.37 as a function of EFL. Results are shown for 9 and 18 frequencies. Note that TSL capacity is the same as soon as the network is loaded with the same EFL. In EGPRS TSL capacity figures are not affected by pre-emptive re-transmissions in the same way as in GPRS case, as probability of RLC protocol stalling is much lower. Reduction factors for one and three TRXs are also shown on the right-hand figure. Reuse 1/1 with one TRX has a similar reduction factor to that in the non-hopping cases. In case of three TRXs and nine frequencies, the higher soft-blocking component changes the shape of the reduction factor.

Spectrum efficiency with QoS criterion

Figures 7.38 and 7.39 show spectrum efficiency with minimum QoS criteria for EGPRS traffic with three TSL mobile stations. Different frequency reuses without frequency hopping are considered. Reuse 1/1 results with frequency hopping are also included in the figure. In these simulations, reuse 1/1 is not HW limited (TSLs are not shared by more than one user). Compared with GPRS, the achieved spectrum efficiency for the same quality is twofold with low quality criteria. EGPRS spectrum efficiency increases with tighter QoS criteria. For example, EGPRS SE is 10 times higher compared with GPRS CS-1–4 with a 10th percentile throughput of 48 kbps. The SE gain of EGPRS is due to two main reasons. First, the much better link performance of EGPRS in acknowledge mode (see Section entitled *Interference limited performance* in 7.1) makes it possible to achieve same user throughputs with less C/I values. Secondly, the EGPRS has threefold larger peak throughput than GPRS.

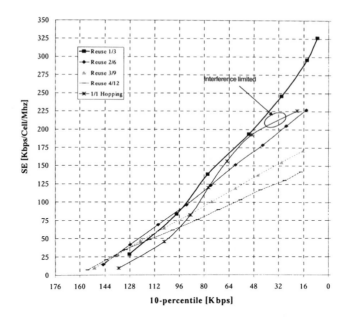

Figure 7.38 Achieved spectrum efficiency values with minimum throughput criteria in a non-hopping macro cell EGPRS network with different reuse patterns. One TRX. Three TSL capable MS.

Figure 7.39 Achieved spectrum efficiency values with maximum delay criteria in a non-hopping macro cell EGPRS network with different reuse patterns. One TRX. Three TSL capable MS.

Reuse 1/1 with frequency hopping is purely interference limited. In this case, network load is increased until the interference level is such that minimum QoS criteria can't be met. This represents pure soft-capacity performance of EGPRS. Reuse 1/3 without hopping suffers from a combination of HW and interference limitations. But still the achieved spectral efficiency for 1/3 non-hopping is better than the pure soft-capacity limited 1/1 hopping case, although the difference is not so dramatic as in the GPRS case. The reason for this is the better performance of EGPRS in FH networks, as described in Sections entitled *GPRS performance in a separate band with RF-hopping* in 7.3 and *EGPRS performance with link adaptation in a separate non-hopping band* in 7.4.

Reuse 1/3 is very suitable for low throughput and delay requirements but for more restrictive throughput or delay requirements looser reuses are more suitable. Properly designed RRM should follow the envelope of Figure 7.38, maximising the SE for different QoS requirements. This can be accomplished, for example, with reuse partitioning techniques described in Section entitled *Reuse partitioning principle to increase spectral efficiency and QoS provisioning* in 7.3.

Spectral efficiency for different TSL capabilities
All the previous simulation results correspond to three TSL capable terminals. Nevertheless, by increasing terminal TSL capability and therefore the number of allocated TSLs per connection there is more potential data bandwidth allocated to each connection. This is translated into higher throughputs and better spectral efficiency under interference limited scenarios for a minimum quality criterion. On the other hand, if the number of allocated TSLs is reduced there will be a contrary effect.

SE analysis for different allocated TSLs can be performed with existing results. Figure 7.34 can be used to compute TSL capacity values for different network loads (Data Erlangs) and frequency reuses. The number of TRXs available for (E)GPRS can be taken into account considering reduction factors from Figure 7.35. In the case of FH, Figure 7.37 can be used for evaluation of both TSL capacity and reduction factors.

The average throughput values can be estimated as

User throughput = TSL capacity × Number of TSLs × reduction factors.

Changing the number of allocated TSLs affects trucking efficiency, and thus it affects reduction factors. In any case the average user throughputs will be better with higher number of allocated TSLs. Figure 7.40 shows simulated and ideally estimated average throughput values for one TRX and reuse 3/9. Estimated values (on the right figure) use reduction factors from Appendix B, which assumes that data load is evenly distributed over the TRX. TSL capacity is calculated from Figure 7.37. This figure also shows estimated average throughput values for different number of allocated TSLs.

Figure 7.40 Left: simulated average user throughputs for three and four allocated TSLs. Right: estimated average user throughput for different number of allocated TSLs. One TRX. Reuse 3/9.

A higher number of allocated TSLs is directly translated into a higher spectrum efficiency with a minimum quality criterion. Figure 7.40 (left) shows the SE values achieved with minimum throughput criteria (10th percentile throughput) for one, three and four allocated TSLs. The SE gain is more noticeable when the minimum quality is high, in the interference limited area. With lower minimum throughput the gain of allocating more TSLs disappears as the network becomes HW limited. The SE is always better with more allocated TSLs therefore, form the performance point of view, it is always better to allocate the maximum TSL capability supported by the MS.

Figure 7.41 (right) shows SE vs average and 10th percentile user throughput with 4 TSL capable mobile stations. 4 TSL is the highest DL configuration

Figure 7.41 Left: EGPRS SE vs. 10th percentile user throughput for 1, 3 and 4 TSL capable MS. 1 TRX. Right: EGPRS SE vs. average and 10th percentile user throughput with 4 TSL capable MS. 1 TRX. Reuse 1/3.

possible in Rel'99 simple mobile station without the need of duplicating the radio parts (see chapter 1). Rel'5 introduces enhancements that allows the support of 6 TSL in DL with simple terminals (see chapter 2). The most spectral efficient reuse is considered (reuse 1/3), therefore the figure represents the maximum EGPRS downlink capacity with simple MS design. The highest SE is achieved with an average datarate of 64 kbps. In order to double the average throughput (up to 128 Kbps) the SE is reduced by 30%. In this case the minimum throughput achieved by 90% of the user is 56 kbps. On the other hand at the highest spectral efficiency the minimum throughput is much lower (less then 16 kbps).

Throughput distribution analysis

This section will analyse how the throughput experienced by EGPRS MSs varies across the network. The analysis is made on the TBF level, which means that the information generated for each individual TBF is collected and analysed. Figure 7.42, on the left, shows a typical throughput snapshot (corresponding to a certain instant) which describes the throughput variation as a function of MS location. Figure 7.42, on the right, shows the corresponding average throughput distributions in the form of probability density function. The simulation case has been selected in such a way that its load corresponds to the level of 3 Data Erlangs. The achieved mean throughput is in this case about 100 kbit/s and the 10th percentile throughput about 50 kbit/s. The quality achieved in certain locations does not only depend on the distance to the serving BTS. The load of the serving BTS might cause queuing and therefore despite excellent radio link quality the throughput experienced by a certain MS can occasionally be much less than the maximum, even when located very close to the BTS. However, the throughput close to the cell centre is naturally still on average

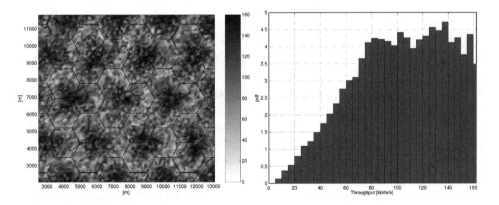

Figure 7.42 EGPRS throughput map and distribution in a macro cell network. Reuse 1/3. One TRX. Three TSL capable MS.

better than in the cell border area. No throughput compensation techniques have been considered for those results. Clever scheduling and channel allocation design may distribute more equally the throughput by allocating more scheduling turn to users at the border of the cells. Note that even with tight reuses like 1/3 a high percentage of users will have high throughputs. At the cell edge, some users may experience low throughput. Those users can be handed over to a different layer with not so tight reuse (e.g. BCCH layer or hopping layer) to improve the throughput if necessary.

Effect of traffic burstiness

Earlier results assumed the FTP traffic model where the users access the system to download a file with a fixed length. With some other traffic patterns EGPRS traffic can be very bursty. For busty traffic there are two main KPIs to consider: LLC frame delay and TBF throughput. LLC frame delay is applicable to short packets. TBF throughput is still a good quality measure for long files download corresponding to long TBF duration. System level-wise, bursty traffic has a tendency to increase protocol and signalling overhead. Some of the effects are:

- The number of established TBFs is higher thus generating more signalling in (P)CCCH channels. Signalling load in common control channel is studied in Chapter 11.
- A noticeable delay due to TBF establishment procedure is experienced for short and infrequent packets. A typical TBF establishment delay is on the order of 250 ms. Delayed TBF release functionality (see Section entitled *GPRS and EGPRS enhancements in Rel'4* in 1.5) helps to reduce this effect.
- During the TBF release procedures, the network may generate too many pending ACK block transmissions (pre-emptive retransmissions) that increase network load, but most of them are unnecessary (they do not contribute that much to SE).

Less protocol efficiency with bursty traffic is traduced into an increase of Data Erlangs for the same cell throughput. This, in practice, will generate higher interference levels and will reduce the spectral efficiency. Figure 7.43 on the left shows SE vs. 10th percentile throughput with reuse 3/9 and two traffic models: FTP and combined. Only TBF longer than 5 s have been considered for throughput statistics calculation. The combined traffic model is a combination of WWW (30%), FTP (60%) and email download model (10%). SE is slightly lower with combined traffic as it was foreseen. Figure 7.43 shows also how the amount of unnecessary retransmissions generated mostly during the TBF release phase increases in the case of combined traffic. The figure on the right shows average LLC frame delay. It can be seen that the average delay is worse for combined traffic. As the network starts to be fully loaded the main reason for delay is network congestion and there is no difference between the two models.

Figure 7.43 FTP and combined traffic performance. Reuse 3/9. Three TSL capable MS.

The performance with bursty traffic can be improved by reducing the radio round trip time to the minimum. By locating RLC/MAC protocol stack as close as possible to the BTS, radio round trip time can be minimised. For short and infrequent packets the minimum delay depends basically on the overhead delay introduced by the TBF establishment and release procedures, which may be of the order of 500 ms (although this value depends very much on network implementation).

7.5 Mixed voice and data traffic capacity

In previous sections data capacity figures assume (E)GPRS traffic only. Therefore, they represent absolute performance of (E)GPRS systems. In practical deployments different kind of users (voice and data) may be using the same frequencies and the same HW resources. Different services will interact with each other affecting the performance of each other. The interaction may be controlled to a certain degree by radio resource management (RRM). Different priorities can be applied to different services that have a direct impact on capacity figures. The most important case, since it will be the most widely deployed in initial deployments, is the one in which the (E)GPRS data users use spare resources (timeslots) not utilised by GSM speech users. The average number of free TSLs that can be potentially used by (E)GPRS can be calculated from the average voice Erlangs per cell and the total number of TSLs available on each cell. Table 7.4 presents the number of free TSLs for different cell configurations. It is assumed that one TSL is reserved for signalling with one and two TRX configurations, while two TSLs are required for the other configurations (signalling capacity dimensioning in mixed cases is covered in Chapter 11). By adding an extra TRX, best-effort cell data capacity can be improved. It is also possible

Table 7.4 Available TSLs for best-effort (E)GPRS traffic

Number of TRX	Voice Erlangs at 2% blocking	Free TSLs	Share of TSL available for (E)GPRS
1	2.9	4.1	58%
2	9.0	6.0	40%
3	14.9	7.1	32%
4	21.9	8.1	27%
5	29.2	8.8	23%
6	36.5	9.5	21%

to dedicate some TSLs for (E)GPRS use exclusively. In this way a minimum cell data capacity can be guaranteed.

In order to calculate data capacity, the key figure is the amount of (E)GPRS traffic that can be transmitted in a fully utilised TSL. This concept is called TSL capacity and has been described in previous sections. Maximum data capacity per cell can be expressed as

Cell data capacity (kbps/cell) = Free TSL × TSL capacity.

TSL capacity depends mainly on the total interference level generated by voice and data services. In practice, it is impossible to fully utilise the free TSL value with (E)GPRS. That would mean 100% TBF blocking and very poor quality. CS and PS dynamics are interacting between each other so that TBF blocking probability and reduction factors depend on the share of Data Erlangs and Voice Erlangs.

Mixed traffic scenarios

There are two main scenarios of mixed traffic: BCCH and hopping layer with very different behaviour.

BCCH layer
In the BCCH layer, the TSL capacity is quite independent of network load and traffic mix (percentage of data and speech users) in the DL direction. There is continuous transmission with constant power (DL PC is not allowed), over BCCH frequencies in order to enable the MS to measure neighbour cells. As a result DL interference levels are constant and independent of traffic load. Typical performance figures for TSL capacity that can be obtained from (E)GPRS traffic only simulations, are 12, 19 and 43 kbps for BCCH reuse 4/12 with GPRS CS-1–2, GPRS CS-1–4 and EGPRS, respectively. An additional characteristic of the BCCH layer is that speech performance in BCCH will not be degraded by (E)GPRS service interference.

Hopping layer

In the hopping layer, the interference level depends on how much the network is loaded (EFL operation point of the network). It can also be very dependent on the traffic mix as data services tend to generate more interference than voice. This is because DL PC is not as efficient for (E)GPRS as it is for voice services, and DTX is not applicable for data services. There are two key questions that that will be answered next:

- How many Data Erlangs can be allocated without degrading speech capacity?
- What is the TSL capacity in mixed traffic cases?

The simplest case to consider is when downlink power control (DL PC) and DTX are not used in voice services, and also DL PC is not used by (E)GPRS. In this case, generated interference levels are equivalent. Therefore, it is possible to translate Data Erlangs into Voice Erlangs. Soft-capacity analysis for data presented in this chapter and for voice, presented in Chapter 6, are fully applicable by considering total EFL as the addition of voice EFL and data EFL. In other words, a Total Erlang is considered as the addition of Voice Erlangs and Data Erlangs. Cell data capacity can be calculated considering Data Erlangs and TSL capacity. TSL capacity vs. EFL results from previous sections can be used to calculate TSL capacity. For example, suppose that hopping layer is loaded with 8 Voice Erlangs and 4 Data Erlangs. If the number of hopping frequencies is 24, the network is loaded with 6.25% EFL. Based on Figure 7.37 the TSL capacity is 38 kbps giving a total cell data capacity of $4 \times 38 = 152$ kbps.

If DL PC and DTX are used by voice users, a direct conversion between Data Erlangs and Voice Erlangs is not possible. Figure 7.44 shows voice quality vs. EFL curves for AMR and EFR codecs with a quality limit of 2% outage of samples with FER > 4.2%. The EFL curve without power control and DTX would correspond to the speech performance with almost 100% (E)GPRS traffic. The maximum data EFL that can be reached is 4.8%. On the other hand, EFL without power control would correspond to the performance close to 100% voice traffic. The maximum voice EFL that can be reached with DL PC is around 7%, which correspond with a capacity gain of 45% that is in line with the field trial results presented in Section entitled *Power control* in 6.1. An extra 30% gain can be achieved with DL DTX activated (see Section entitled *Discontinuous transmission* in 6.1), therefore the maximum voice EFL is 9.1%. This is roughly 2 times more Erlangs than in almost 100% (E)GPRS case. From this analysis it can be extracted the following simple rule of thumb: *1 Data Erlang is equivalent to 2 Voice Erlangs in terms of interference generation*. The same exercise can be done with AMR voice users, in this case the rule of thumb from simulation results is the same. Note that for EFL calculations, the activation of DTX doesn't change the Voice Erlang values. Depending on the minimum quality

Figure 7.44 Percentage of samples with bad quality (FER > 4.2% and 2% outage) vs. EFL for AMR and EFR.

target and different network topology, different translation factors can be found, so only field trial data will help to fine-tune the translation factor values. In order to calculate TSL capacity, Voice Erlangs need to be translated into the equivalent Data Erlangs. Consider, for example, 8 Voice Erlangs of EFR traffic, 4 Data Erlangs and 24 hopping frequencies. Eight Voice Erlangs can be converted into 4 equivalent Data Erlangs. Four Data Erlangs plus 4 equivalent Data Erlangs corresponds to 4.2% EFL for data capacity calculations. TSL capacity calculated from Figure 7.37, gives around 41 kbps corresponding to total cell data capacity of 4 × 41 = 164 kbps.

If DL PC is used by (E)GPRS and DL PC and DTX are activated for voice users, the analysis requires (E)GPRS DL PC performance analysis. Figure 7.45 represents EFL vs. speech quality (EFR codec) for 45 and 55% of Voice Erlangs and Data Erlangs, respectively. Three different simulation results are presented: No GPRS DL PC, GPRS DL PC with P0 = 0 and GPRS DL PC with P0 = 3 dB. P0 represents a constant BTS transmitter power reduction applied to all GPRS connections. The use of P0 is useful to control interference generated by (E)GPRS traffic. Considering a quality limit of 2% samples with 4.2% FER, just reducing the GPRS transmitted power 3 dB (P0 = 3 dB), there is a capacity gain of 18%. This means that it is possible to introduce 18% more speech and Data Erlangs (where 45% of the load is speech). P0 gains are directly applicable to EGPRS as well. Three decibels of transmitted power reduction doesn't have a major impact on average throughput (see Figure 7.45). Apart from a fixed offset, the DL PC algorithm can be applied to control transmitted power depending on

Figure 7.45 Voice quality (left) and average throughput (right) per connection with 55% GPRS (CS-1–2) and 45% EFR traffic. DTX not activated. Three TSL capable MS.

radio link conditions. This brings additional gains that depend on the modulation and coding schemes implemented. GPRS CS1-2 will have the best gains while the gain for EGPRS with MCS-1–9 is marginal.

7.6 (E)GPRS performance estimation based on real network measurements

The performance of (E)GPRS may differ in different networks due the different topology and propagation environment. Simulation results presented in previous sections show the performance of a regular network with macro-cell propagation characteristics. This section presents a methodology to estimate (E)GPRS performance in a real network based on measurement reports sent back to the network by the MS. This methodology has been used in a field trial carried out in a major European capital (Network 9, in Appendix F). Results provide accurate (E)GPRS performance figures expected in that city.

Any MS with an established voice connection reports periodically (every 480 ms), the received signal level C from the serving cell and the six strongest neighbouring cells (N_1, N_2, ..., N_6). The MS may be detecting and measuring more than six cells but only the six strongest are reported. This will change with the introduction of enhanced measurement report in Rel'99, which largely expands the terminals' reporting capabilities. The reported cells are ordered from the strongest to the weakest. The relative level of signal between serving and reported cell makes it possible to estimate the C/I values, in case the reported cell is a co-channel interferer. Therefore, this information can be utilised to envisage C/I distributions with a given frequency reuse. In order to do this, the cells are re-planned with the desired frequency reuse and co-channel interfering cells are defined. Measurement reports are collected from the trial area and stored in a database. The trial area should cover a sufficient number of cells. Each measurement report is processed as follows:

- If one (or more) of the six reported cells is a co-channel interferer, the user will experience a throughput that corresponds to $C/I = C/N_i$. Where N_i is sum of co-channel interferer levels.

- If there is no co-channel interferer in the six strongest reported cells, it can be said that C/I will be, in the worse case the one associated with the weakest neighbour (N_6). Noting that throughput vs. C/I curves of (E)GPRS tend to saturate for a certain C/I value, $[C/I]_{sat}$, there are two cases to consider:

- If $C/N_6 > [C/I]_{sat}$, the MS connection would experience the maximum throughput corresponding to $C/I = [C/I]_{sat}$.

- If $C/N_6 < [C/I]_{sat}$, the MS connection would experience, at least, the throughput associated with $C/I = C/N_6$. In this case it is impossible to know the exact throughput and only worse case calculations can be done. In order to produce more realistic results, C/I values are compensated with a factor calculated from a static simulator for each reuse. This factor is deduced by calculating the C/I with the information provided by the six strongest cells, and then comparing with the calculated C/I values considering all cells.

- If the signal level from the serving cell is too low, the radio conditions can be noise limited. For each measurement report, the C/N (signal-to-noise ratio) is also calculated as $C/N = C - \text{noise level}$. The noise level is assumed to be -111 dBm. If C/N is larger than C/I, the link is noise limited and throughput value is calculated with C/N value using throughput vs. C/N link level curves.

Figure 7.46 shows TSL capacity distributions. More than 13 million measurement reports have been collected and processed from around 145 cells from one BSC. Micro-cells have been excluded from the analysis. Results correspond to a fully loaded network as continuous co-channel interference is assumed in the above described procedure. The procedure also assumes full power transmission, therefore, DL PC can't be taken into account. Frequency hopping can't be included in the methodology, so non-hopping C/I vs. throughput mapping tables (that can be found in Section 7.1) have been utilised. This provides an accurate analysis in the case of BCCH and worst case scenario for the non-BCCH TRXs.

TSL capacity can be considered as average user throughput if there is only one user per TSL. This can be achieved with proper HW dimensioning, configuring the available parameters (if they are available to control the maximum number of users per TSL) or by allocating the highest relative priority in the scheduler to some users. TSL capacity together with SE figures have been collected in Table 7.5.

Figure 7.46 EGPRS and GPRS cumulative throughput distributions.

Timing advance (TA) information is also included in measurement reports. This information can be considered in the analysis to calculate user throughput figures as a function of the distance from the BTS. Figure 7.47 shows the results. User distance distribution can be found in Figure 7.48.

Figure 7.47 Variation of throughput vs. distance with EGPRS and GPRS.

Table 7.5 (E)GPRS performance estimated figures from field trial data

Reuse	EGPRS			GPRS (CS-1–4)		
	Minimum throughput/TSL with 90% location probability	TSL capacity	kbps/MHz/cell	Minimum throughput/TSL with 90% location probability	TSL capacity	kps/MHz/cell
3	6	35.4	472	4	14.9	199
6	16	43.1	287	10	17.3	115
9	26	46	204	14	18.2	81
12	32	47.6	159	16	18.8	63

Figure 7.48 User distance distribution.

7.7 Application performance over (E)GPRS

The basic (E)GPRS service provided by the BSS is the delivery of LLC data packets to the MS that are coming from the PS core network. Therefore, all performance figures presented in previous sections are measured at LLC layer level. Nevertheless, QoS experienced by the end users is measured at application level. BSS plays only a part of end-to-end QoS provisioning, according to the QoS architecture described in Chapter 3. The BSS role in the end-to-end QoS picture is called radio access bearer service (following UMTS terminology). The performance analysis presented in previous sections describes the performance of the basic service provided by (E)GPRS bearers. It describes the performance up to LLC layer. The impact of the (E)GPRS bearer over the quality experienced in the application layers is studied in this section.

The most used Internet application protocols are depicted in Figure 7.49, together with their associated protocol stack at the MS. (E)GPRS BSS provides data link (Layer 2) relay service from the serving GPRS support node (SGSN) to the MS. Internet applications use one of two possible transport protocols: TCP or user datagram protocol (UDP). TCP is a connection oriented protocol with reliable data transmission, while UDP is not connection oriented and does not provide a reliable data transmission service. WWW, FTP and email applications use TCP protocol, while wireless application protocol (WAP), real-time transport protocol (RTP), Domain Names Server (DNS) and simple network management protocol (SNMP) services use UDP.

Figure 7.49   (E)GPRS user plane protocol stack.

TCP performance

TCP is a protocol originally designed and optimised for fixed PS networks. It may not be optimal for mobile networks without a certain degree of optimisation. The main characteristics of TCP to be considered in (E)GPRS networks are:

- The application data are split into segments of appropriate size. Typical maximum size is nearly 1500 octets (Ethernet maximum packet size with IPv4), although some applications may limit the packet size to only 576 octets.
- It ensures reliable data transfer. It uses an acknowledgement scheme where the sender receives ACKs to previously sent packets. The sender will retransmit the packets that are considered to be not acknowledged.
- It tries to adapt to different speeds by using all the available bandwidth. Flow control mechanism is used so the sender and received bit rate can be adjusted due to network conditions.

A well-known problem of TCP protocol is the amount of overhead to be transmitted over the radio interface. In (E)GPRS networks, the bandwidth is much more scarce and expensive than in fixed networks. Therefore, the protocol overhead should be minimised. The TCP protocol headers, together with IP headers produce an overhead of 40 octets on each TCP/IP packet (IPv4). In order to reduce this overhead, it is better to increase the length of the IP segments to the maximum possible (1500 octets). It is also possible to apply data compression techniques on the subnetwork dependent convergence protocol (SNDCP) layer [8] already in Rel'99. Compression techniques can be applied to the TCP/IP and RTP/IP headers as well as the data part of the IP packets. Van Jacobson's header compression [9] algorithm can reduce the TCP/IP header length to 4 octets (for most of the packets). Degermark *et al.*'s header compressing [10] can be applied to RTP/IP headers, improving the efficiency for real-time data applications. The data compression algorithm is V.42 bis [11].

Most of the problems with TCP over (E)GPRS and, in general, over wireless networks, come from the existing algorithms to adapt data rates to different speeds. Those protocols have been optimised for fixed networks and don't consider the special characteristics of wireless networks. The most important effect of (E)GPRS is the high impact on the overall round trip time (RTT). RTT can be considered as the total time a small packet takes to travel from the service content server through the mobile network (core network and BSS) to the MS and back, see Figure 7.50. The RTT of a (E)GPRS network can change between 0.5 and more than 1 s (depending on the user throughput) while typical RTT of fixed networks can be 1 ms for local area network (LAN), 20 ms for integrated services digital network (ISDN), 20–200 ms for analogue modem and 125 ms for average Internet.

Long round trip delay makes initial data rates very slow due to TCP long connection establishment. TCP uses three-way handshake for connection establishment, see Figure 7.51. Therefore, the time required for establishment procedure is 1.5 × RTT which gives 1.5 s, assuming 1 s (E)GPRS RTT. In TCP, the data rate is given by the length of a congestion window (CWND), which indicates the amount of packets that can be in transfer without causing congestion. The CWND is adjusted following the 'slow start' mechanism depicted in Figure 7.52. The CWND length is exponentially increased, starting from 1, as acknowledgements arrive at the sender. This procedure is repeated until the saturation point is reached. Saturation is reached when packet losses are detected by the TCP layer or a maximum window length (advertised window), indicated by the receiver, is reached. The TCP layer detects packet losses when multiple ACK (typically three) arrive for the same segment (meaning that the subsequent segments are lost) or when retransmission timer expires. In Figure 7.52, saturation has been detected with three duplicate ACK after the congestion window has increased it size up to 16 segments. After saturation is reached CWND stays constant until a segment loss is detected. Then, CWND is reduced to half of its previous value, so routers are able to drain their queues, and TCP enters the *congestion avoidance* phase. In this phase, CWND increases linearly until saturation is reached again.

Figure 7.50 Elements affecting the round trip time.

Figure 7.51   TCP signalling chart.

The impact of 'slow start' is very dependent on RTT. With long RTT, like in (E)GPRS, it takes longer to reach the nominal achievable throughput. Therefore, short TCP connections are not transmitted with the maximum achievable throughput. Optimisation of TCP parameters and optional features can be used to optimise TCP performance for wireless networks [12,13]. Some of the recommendations proposed are: use a maximum segment size of 1460 bytes, use local MTU path discovery option, use an initial CNWD of two segments and do not use delayed acknowledged mode during the 'slow start' phase. The TCP data rates can be increased in the order of 30% with simple TCP optimisation.

Figure 7.52   TCP slow start.

(E)GPRS cell re-selection delays have also an impact on TCP performance. The typical outage time is a random value that ranges from 0.5 to 2 s. During this time, TCP timers can expire and cause the CNWD window to be reduced, generating unnecessary retransmissions of TCP segments. In general, fast moving (E)GPRS terminals can generate problems for the TCP layer due to fast changing radio link conditions and cell re-selection outages. GERAN Rel'4 introduces the network assisted cell change (NACC), which reduces the cell reselection outage down to 500 ms (max.). Additionally, with the use of the network controlled cell re-selection (NCCR) feature, it is also possible to optimise the cell re-selection performance, so the impact to short outage in upper layers can be reduced.

It is worthwhile to comment that WAP, an optimised protocol for wireless network, uses UDP as a transport protocol when Internet protocol are available, thus avoiding most of the problems associated with TCP. But unfortunately, the majority of the Internet services do not use WAP, but they use instead hypertext transfer protocol (HTTP), which utilises TCP connections. HTTP 1.1 has some advantages over HTTP 1.0 as it reduces the number of TCP connections (openings and closings) and reuses the same TCP connections for multiple objects.

UDP performance

UDP transport protocol is less problematic than TCP, as it doesn't require retransmissions and protocol overhead is much lower. The application layer handles things like possible required retransmissions and re-ordering of the datagrams. In most data applications, (E)GPRS works in RLC ACK mode, thus guaranteeing reliable transmission over the air interface.

Some streaming and conversational services, like voice over IP (VoIP), uses RTP over UDP. In this case the (E)GPRS network operates in unacknowledged mode. RRM management is then responsible to maintain the negotiated reliability QoS parameter. The most straightforward mechanism to control reliability is the link adaptation (LA) algorithm, that would select the highest (M)CS possible while being under a maximum BLER. It is worthwhile to note that certain streaming services, that don't have so tight delay requirements, allow RLC to operate in ACK mode.

Performance enhancement proxy

A performance enhancement proxy (PEP) [14] is an entity in the network acting on behalf of an end system or a user. See Figure 7.53. The objective of a PEP is to improve the end-user application performance by adapting protocols/information to a bearer. There are two main types of PEP: transport layer and application layer PEP, although a PEP may operate in more than one layer at the same time.

Transport layer PEP introduces enhancements to overcome the inherit problem of TCP, especially 'slow start'. There are different mechanisms at

Figure 7.53 Performance enhancement proxy (PEP).

this level. Some of them are: TCP ACK handling (TCP ACK manipulation), modified TCP stacks and fine-tuning of TCP parameters.

Application layer PEP solutions include solutions like: cacheing, information compression and optimisation of TCP by the application (fewer TCP connections).

7.8 Some GPRS performance measurements in field trials

TSL capacity figures have been measured in a live network (Network 10, in Appendix E) with data collected from the network management system during the busy hour. Results are presented in Figure 7.54. GRPS traffic is allocated in BCCH layer. CS1-2 are only implemented in the network and Automatic Link Adaptation is in use for all GPRS MSs. TSL capacity figures are close to the maximum (12 Kbps) in all cells. This is in line with network level simulations results from figure 7.24, where the TSL capacity was already very close to 12 Kbps with reuse 12. In Network 10 radio link conditions are good, so almost all

Figure 7.54 TSL capacity measured in a GPRS network.

the time the CS-2 is in use. Control RLC blocks have been excluded from TSL capacity calculations, as they tend to disturb the measurements due to the low GPRS traffic in the network.

Application-level round trip time

Application level round trip time can be measured using PING application. Ping measures the delay from when the MS sends a short packet, it is received by the application server and finally the MS receives the server response back. Round trip times change depending on GPRS network implementation and optimisation, typically varying between 1 and 2 s. If the application server is another MS, the round trip time is longer, of the order of 2.5–5 s.

FTP download performance

FTP is an application protocol that runs on top of TCP, therefore it is affected by TCP procedures. Figure 7.55 shows application level throughput as a function of C/I measured in a field trial. At good radio link conditions the throughput per TSL is around 10 kbps, slightly worse than CS-2 peak throughput (12 kbps). This represents a 16% overhead introduced by upper layers and GPRS procedures. Note that this figure may depend on how well optimised the TCP/IP protocol is. At C/I of 10 dB, the application throughput per TSL is around 6.7 kbps, while having a look to link performance, the LLC level throughput per TSL is around 10 kbps. Now the overhead is around 33%. No data compression was used at SNDCP later. Similar throughput can be achieved with HTTP for WWW browsing.

Figure 7.55   Measured application level average throughput as a function of C/I. No FH.

WAP browsing

WAP browsing worked very well in GPRS trials. User satisfaction was very good even for very bad radio link conditions (C/I = 3 dB).

Real-time services

Different services like low-resolution movie, slide show and audio have been tested under different C/I conditions. The services have been tested over different radio link condition in RLC acknowledge mode. Low-resolution movie (moving images plus voice optimised for 30 kbps) has good quality with C/I greater than 20 dB. For worse radio link condition quality degrades very much and becomes unusable with C/I equal to 6. If the movie is optimised for 18 kbps better quality can be achieved down to 10 dB. Real-time slide show (static images plus voice) at 30 kbps works well for C/I higher than 10 dB, although at 20 kbps the quality is better and it is maintained down to 8 dB. Audio MP3 monaural at 16 kbps doesn't show any degradation down to 10 dB. Although it can be used with lower quality with lower C/I.

RLC acknowledge mode proved to be very efficient for real-time services over UDP. There is roughly a 4 dB link level gain for the same application quality with the introduction of LA. Most of the applications have buffers that can cope with the delay introduced by retransmissions.

Table 7.6 shows achieved quality for several real-time services from the quality test performed during a field trial. Mean opinion score (MOS) values are averaged from four different persons.

Table 7.6   Real-time services MOS as a function of link quality measured in laboratory trial. GPRS CS-1–2. Three TSL MS capability

		C/I [dB]					
Service	Speed	20	16	12	10	8	6
Low resolution movie	30 kbps	4	1.75	2	2	1.75	1
Low resolution movie	18 kbps	5	4.06	4	3	2.37	1
Slide show	30 kbps	4.75	4.56	3.98	3.56	3.43	2.06
Slide show	20 kbps	4.75	4.56	3.98	3.56	3.43	2.06
Audio MP3/Monoural	16 kbps	5	5	5	5	3	2

MOS class	Remark	Classification
5	Excellent	Complete relaxation possible, no effort required, no changes from the original
4	Good	Attention necessary, no appreciable effort needed
3	Fair	Moderate effort needed
2	Poor	Considerable effort needed
1	Bad	No meaning understood with any feasible effort

References

[1] 3GPP 45.005, Release 1999: 'Digital Cellular Telecommunications System (Phase 2 +)'; Radio Transmission and Reception.

[2] 3GPP TS GSM 05.05 version 8.8.0 Release 1999: 'Digital Cellular Telecommunications System (Phase 2 +)'; Radio Transmission and Reception.

[3] Universal Mobile Telecommunications System (UMTS): Selection Procedures for the Choice of Radio Transmission Technologies of the UMTS (UMTS 30.03 version 3.2.0). ETSI Technical Report 101 112 (1998-04).

[4] Pons J., Dunlop J., 'Bit Error Rate Based Link Adaptation for GSM', Proceedings of the 1998 9th IEEE International Symposium on Personal, Indoor and Mobile Radio Communications, PIMRC, Part 3, pp. 1530–1534.

[5] 3GPP 45.03, Release 1999: 'Digital Cellular Telecommunications System (Phase 2 +)'; Channel Coding.

[6] 3GPP 45.008 Release 1999: 'Digital Cellular Telecommunications System (Phase 2 +)'; Radio Subsystem Link Control.

[7] Hämäläinen S., Slanina P., Hartman M., Lappeteläinen A., Holma H., Salonaho O., 'A Novel Interface between Link and System Level Simulations', Proc. ACTS Mobile Telecommunications Summit, Aalborg, Denmark, October 1997, pp. 599–604.

[8] 'Subnetwork Dependent Convergence Protocol (SNDCP)', 3GPP 3G TS 24.065 version 3.1.0.

[9] RFC-1144, Jacobson V.: 'Compressing TCP/IP Headers for Low-speed Serial Links'.

[10] RFC-2507, Degermark M., Nordgren B., Pink S.: 'IP Header Compression'.

[11] ITU-T, Recommendation V.42 bis: 'Data Compression Procedures for Data Circuit-terminating Equipment (DCE) using Error Correcting Procedures'.

[12] Xylomenos G., Polyzos G., Mähönen P., Saaranen M., 'TCP Performance Issues over Wireless Links', IEEE Communication Magazine, Vol. 39, No. 4, April 2001.

[13] 'TCP over 2.5G and 3G Wireless Networks', IETF Internet-Draft, October 2001.

[14] Border J., Kojo M., Griner J., Montenegro G., Shelby Z., 'Performance Enhancements Proxies Intended to Mitigate Link-related Degradations', IETF Informational Document – RFC 3135, June 2001.

8

Dynamic Frequency and Channel Allocation

Matti Salmenkaita

Dynamic channel allocation is a wide definition for functionality that aims to optimise the usage of radio resources by applying intelligence in the channel selection process. Dynamic channel allocation (DCA) schemes can be based on many different variables and different applications provide varying degrees of freedom in the channel selection. The DCA schemes are commonly categorised as traffic adaptive (TA-DCA) or interference adaptive (IA-DCA) [1].

Traffic adaptive DCA algorithms allow dynamic usage of a common channel pool thus avoiding the lack of flexibility that is characteristic for a fixed channel allocation. The primary aim of the TA-DCA schemes is to adapt to the fluctuations in the traffic distribution thus minimising the call blocking probability due to the lack of available channels. The TA-DCA algorithms avoid interference between cells by considering the channels used in the nearby cells and by utilising some form of pre-planned compatibility matrix or exclusion zone to govern the process. Examples of TA-DCA schemes can be found in [2,3,4].

Interference adaptive DCA algorithms are similarly able to minimise the blocking probability by allowing dynamic usage of a common channel pool. However, instead of a fixed compatibility matrix the IA-DCA schemes utilise real-time interference measurements to avoid interference. The usage of real-time measurements allows higher flexibility and adaptability in the channel selection process and therefore generally leads to better performance with the expense of increased complexity [5,6,7].

This chapter presents a dynamic channel allocation scheme called dynamic frequency and channel assignment (DFCA[1]). DFCA is a centralised IA-DCA scheme based on the estimation of carrier to interference ratio (CIR) on possible channels. The aim is to optimise the distribution of interference taking into account the unique channel quality requirements of different users and services. DFCA utilises cyclic frequency hopping for maximum frequency diversity gain. With adequate base transceiver station (BTS) support DFCA can freely choose the mobile allocation (MA) list, mobile allocation index offset (MAIO), timeslot and the training sequence code that will be used for each connection regardless of the physical transceiver. The basic requirement for DFCA is that the air interface is synchronised, as synchronisation provides timeslot alignment and time division multiple access (TDMA) frame number control that are needed for timeslot level CIR estimation with cyclic frequency hopping.

8.1 Air interface synchronisation

GSM synchronisation basics

The basic time unit in the global system for mobile communications (GSM) air interface is a TDMA frame consisting of eight timeslots. The placement of logical channels in timeslots as well as the current frequency selection in the frequency hopping sequence are determined by the TDMA frame number. In an unsynchronised network the frame clocks in base stations are initialised independently of each other leading to random frame alignment and numbering. This is illustrated in Figure 8.1. The air interface can be considered synchronised when the TDMA frame alignment and frame numbers of all BTSs are known and controlled. An example of this is presented in Figure 8.2.

Figure 8.1   Random TDMA frame alignment.

[1] Nokia proprietary functionality

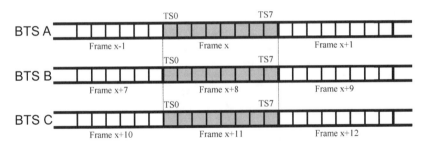

Figure 8.2 Synchronised TDMA frame alignment.

Implementation of synchronisation

Typically the base stations that are physically located in the same BTS cabinet are using the same frame clock source and therefore they are always synchronised relative to each other. It is also quite straightforward to chain a common clock signal to all the BTS cabinets in the same site, and this way achieve local synchronisation even if the BTSs are located in different cabinets.

Area level synchronisation is more difficult to achieve, as the common master clock signal must be provided in all BTS sites which may be located several kilometres apart. A common solution to this problem is to use the stable and globally available clock signal from the global positioning system (GPS) satellite system. A GPS receiver in every BTS site provides a frame clock signal that is then distributed to all BTS cabinets in the site. This frame clock is calculated from the GPS clock signal according to the same rules in all the sites leading to common TDMA frame alignment and numbering in all synchronised base stations.

TDMA frame number considerations

A common TDMA frame number is not desirable at area level as it leads to co-incident broadcast control channel (BCCH) and traffic channel (TCH) multi-frames in all the base stations. For BCCH this results in co-incident transmission of synchronisation bursts in every 10 TDMA frames. The synchronisation burst carries the base station identification code (BSIC) of a cell and it is needed together with the frequency correction burst for a mobile station to identify the neighbouring cell and to synchronise with it. Synchronisation with a common TDMA frame number leads to a situation where in 9 out of 10 TDMA frames there is no synchronisation burst transmission from any base station and in 1 out of 10 TDMA frames all the base stations transmit the synchronisation burst simultaneously due to co-incident BCCH multiframes.

The mobile stations in active mode attempt to search for the neighbours only during idle frames that occur once within each TCH multiframe of 26 TDMA

Figure 8.3 Relation of the BCCH and TCH multiframes.

frames. On the other hand, frequency correction burst or synchronisation burst transmission occurs 10 times within one BCCH multiframe of 51 TDMA frames as illustrated in Figure 8.3. Thus, in a synchronised network with a common TDMA frame number only 10/51 = 19.2% of the TCH idle frames can be used to search for the neighbouring cells. Because of this, the mobile stations would need significantly longer time to identify the neighbouring base stations leading to decreased BSIC decoding performance.

The second problem with the common frame number is that the slow associated control channel (SACCH) transmissions that always occur in the same instant within the TCH multiframe would occur simultaneously all over the synchronised area. Since SACCH is always transmitted regardless of the voice activity, it would not benefit from discontinuous transmission (DTX) at all.

These problems can be avoided by applying different frame number offsets throughout the synchronised area. Each base station can have a fixed frame number offset that is added to the global frame number. This is illustrated in Figure 8.2 where the BTSs B and C have relative frame number offsets of 8 and 11, respectively, towards BTS A. It is also possible to apply the frame number offsets on the site level so that all base stations in one site use the same frame number.

Synchronisation accuracy

Synchronisation has to be accurate enough to ensure that proper timeslot alignment is maintained as network features relying on synchronisation such as DFCA assume nearly perfect timeslot alignment so that there is no interference between different timeslots. However, the guard period that is located at the end of the bursts provides some room for overlapping without detrimental effects (Figure 8.4).

All bursts, except the access burst, have a guard period lasting 8.25 bit periods that corresponds to 30.46 μs and about 9.1 km in propagation distance. Thus,

overlapping within the guard period   **partial burst overlapping**

Figure 8.4   Burst overlapping.

assuming an ideal situation with the mobile station located right next to the serving base station, the interfering transmission can originate up to 9.1 km away without burst overlapping between the adjacent timeslots. This distance is referred to here as the synchronisation radius (D_0). Generally, the synchronisation radius is the maximum propagation distance difference between the serving and the interfering signal. If the propagation distance difference exceeds D_0, then partial burst overlapping starts to occur. The synchronisation radius is illustrated in Figure 8.5.

Typically the high serving cell signal level close to the base station ensures that the CIR is sufficiently high to avoid quality degradation. The most sensitive location is the cell border area where the serving cell signal level is the lowest. Two different cases can be identified depending on the relative location of the mobile station. These are illustrated in Figure 8.6.

In both cases the propagation distance for the serving cell signal C is denoted as d_c and for the interfering signal I as d_i. Additionally in the uplink cases the propagation distance of the uplink signal that is being interfered is denoted as d_{c1}. In the uplink the timing advance is used to compensate the signal propagation delay, so ideally from a timing point of view d_c and d_{c1} appear to be zero. Consequently, propagation delay difference as observed in the base station corresponds to $d_i - d_c$ which is also the propagation distance difference in the downlink direction. So, ideally the uplink and the downlink situations are similar. However, the granularity of the uplink timing advance introduces an additional error source that doesn't exist in the downlink direction.

Figure 8.5   Propagation distances.

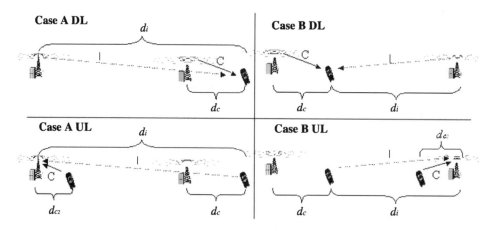

Figure 8.6 The two studied cases considering both downlink and uplink.

In order to avoid burst overlapping between successive timeslots, the propagation distance difference $(d_i - d_c)$ must be below the synchronisation radius D_0. In this analysis the impact of the interfering base stations outside the synchronisation radius is considered, so in the worst case the d_i can be expressed as

$$D_0 = d_i - d_c \Rightarrow d_i = D_0 + d_c$$

In the ideal case D_0 is about 9.1 km as presented before. However, multipath propagation introduces delayed signal components that effectively reduce D_0. The delay and magnitude of these signal components can be estimated by studying the commonly used multipath channel models presented in [8]. It is assumed that the components with more than 10 dB of attenuation can be excluded. Typical delays after which the multipath components have been attenuated by more than 10 dB are presented in Table 8.1 for typical urban (TU), rural area (RA) and hilly terrain (HT) models. For each model also the corresponding D_0 is presented.

At this stage the impact of the multipath propagation is estimated, but the accuracy of the synchronisation is not yet accounted for. The synchronisation accuracy depends on the stability of the timing source, the way the synchronisation signal is distributed and the calibration of the used equipment. In practice,

Table 8.1 Impact of multipath channels on D_0

Channel model	No multipath	Typical urban (TU)	Rural area (RA)	Hilly terrain (HT)
Approx. delay for > 10 dB attenuation	0 μs	5 μs	0.4 μs	16 μs
D_0	9.1 km	7.6 km	9.0 km	4.3 km

Figure 8.7 D_0 as a function of the synchronisation error.

with GPS based synchronisation accuracy of the order of 2 μs is achievable. The granularity of the uplink timing advance setting is 1 bit period corresponding to approximately 3.69 μs. This granularity introduces an additional error source that is only applicable in the uplink direction. Therefore, for the uplink an additional error of 3.69 μs must be added. Figure 8.7 presents D_0 as a function of the synchronisation error for different multipath channels.

Burst overlapping may occur when an interfering transmission originates from a base station that is located more than $D_0 + d_c$ away from the mobile station. However, if the D_0 is high enough the propagation loss has already attenuated the interfering signal sufficiently so that the interference impact is negligible. The worst case situation is when the mobile station is located in the cell border where the serving cell signal strength is the lowest. Figure 8.8 presents the cell border CIR as a function of the cell radius d_c assuming 2 and 6 μs synchronisation errors and the worst case situation where $d_i = D_0 + d_c$ and burst overlapping starts to occur.

Figure 8.8 shows that in typical urban and rural environments the cell radius can be up to several kilometres without significant interference from overlapping bursts. Typically urban area cells are so small that the interference from the cells outside the synchronisation radius is negligible. Since the rural cells are usually much larger, the overlapping bursts may cause some degradation of the part of the burst that is overlapped. However, small overlapping is not likely to cause significant quality degradation as the training sequence is located in the middle of the burst and the channel coding is in most cases able to correct the errors. Hilly terrain is the most restrictive as the strong delayed signal components shrink the synchronisation radius.

Figure 8.8   Cell border CIR as a function of cell radius of d_c.

8.2 DFCA concept

DFCA is based on the timeslot alignment and TDMA frame number control provided by network level synchronisation. The timeslot alignment ensures that the GSM air interface timeslots are co-incident throughout the network. This makes it possible to take all the interference considerations to the timeslot level.

As GSM uses the combination of frequency division multiple access (FDMA) and time division multiple access (TDMA), the radio channel is determined by the frequency and the timeslot. When a channel assignment needs to be performed as a result of a newly initiated connection or handover, DFCA will evaluate all the possible channels and then choose the most suitable one in terms of CIR for the assignment. For this reason, an estimate of the CIR is determined for each available radio channel.

CIR estimation

The estimation of CIR is based on the downlink signal level measurements of the serving cell and the neighbouring cells for which the BCCH frequencies are measured. These measurements are continuously performed and reported to the network by the mobile stations in active mode. This is a standard GSM functionality mainly used to provide path loss information for handover decisions. If one of the reported neighbouring cells would use the same frequency as the serving cell, the potential CIR can then be calculated as follows:

$$CIR_n = RXLEV_{servcell} + Pr_{servcell} - RXLEV_{neigbcelln}$$

where RXLEV is the GSM parameter corresponding to the measured signal level expressed in dBm and Pr is the current power control power reduction in the serving cell.

With legacy handsets the maximum number of reported neighbouring cell RXLEV values is limited to six. However, the GSM standard release 99 compatible handsets are able to report more by using enhanced measurement reporting (EMR). In most cases reporting only six neighbours is sufficient for handover evaluation purposes but for interference estimation the potential CIR has to be determined for all the surrounding cells that may be significant interference sources. Typically, in an urban environment, there may be more than six such cells. The CIR estimate towards such unreported cells can be obtained by collecting long-term neighbouring cell RXLEV measurement statistics from all the mobile stations in a cell. From these statistics a CIR with a given outage probability can be determined for all the surrounding cells. This way the network is able to have a CIR estimate for all the potentially interfering cells even if some of them were not included in the latest measurement report. The directly measured CIR is available for the neighbouring cells that are included in the latest measurement report and for the rest of the surrounding cells the statistical CIR estimation can be used. Since the downlink path loss is generally equal to the uplink path loss, the CIR statistics can also be used to determine an estimate of the uplink CIR.

The method described above provides means to determine a potential co-channel CIR towards each of the reported neighbouring cells, but the CIR is only realised when the same or adjacent frequency is being used in a neighbouring cell in the same timeslot. In case of adjacent channel interference the resulting CIR is scaled up by a fixed 18 dB offset as this is typical adjacent channel attenuation in GSM. In order to determine the real CIR the network must examine the timeslots and frequencies used by the active connections in the potentially interfering cells.

Power control typically leads to reduced transmission powers. The current transmission power reduction of the interfering connection is taken into account by scaling up the estimated CIR accordingly. By processing this information a matrix showing the estimated CIR for all the possible frequency and timeslot combinations can be generated.

Combination with frequency hopping

The air interface synchronisation makes it possible to control the frame numbers in each of the cells. This allows utilisation of cyclic frequency hopping while still keeping the interference relations fixed in the network. The frequency band dedicated to DFCA operation is divided into one or more MA lists. A frequency in the non-hopping case corresponds to a combination of a MA list and a MAIO in the cyclic frequency hopping case. Two connections using the same MA list and MAIO are co-channel interferers to each other if the cells have the same frame number. This property is illustrated in Figure 8.9. Offset in the frame number means similar offset (modulo MA length) in the MAIO of the interfering connection.

Figure 8.9 Fixed interference relations with cyclic hopping.

In case of cyclic hopping, a radio channel can be uniquely defined by the used MA list, MAIO and the timeslot. Thus, the radio channels are identified by these variables in the CIR matrix as presented in Figure 8.10. The CIR matrix contains the estimated CIR for each available radio channel. Some timeslots may not always be available as the availability of a timeslot depends on base station hardware resources that might be already in use by other connections.

MA, MAIO	0	1	2	3	4	5	6	7
1, 0	5	24	Not available	13	23	Not available	45	Not available
1, 1	32	8		26	14		9	
1, 2	11	6		18	34		33	
1, 3	27	13		22	6		14	
2, 0	36	21		16	13		37	
2, 1	25	10		14	24		17	
2, 2	9	7		20	31		22	
2, 3	21	9		11	15		24	

Time Slot

Figure 8.10 Example CIR matrix.

Radio channel selection

In the channel assignment the CIR matrix is searched for a radio channel that provides a CIR that fulfils the CIR requirement of the user. However, the new assignment may cause excessive interference to some ongoing connections in the neighbouring cells. To avoid this, a similar CIR evaluation is performed for all the ongoing connections sharing the same or adjacent radio channel. For each possible radio channel and each affected user the margin between the predicted CIR and the user CIR target is calculated. The channel allocation algorithm chooses the channel taking into account the estimated CIR for the new connection as well as for all the affected existing connections. The effect of a new allocation on other users is thus taken into account. The expressions presented below show how the minimum CIR margin is determined for each candidate radio channel. Finally the radio channel with the highest positive CIR margin is selected for channel assignment. The evaluation of the radio channels based on the CIR margin is illustrated in Figure 8.11.

$$CIR_{margin_{C_{own}}} = min(CIR_{est_{C_{own-DL}}} - CIR_{target_{C_{own-DL}}}, CIR_{est_{C_{own-UL}}} - CIR_{target_{C_{own-UL}}})$$

$$CIR_{margin_{C_{other_n}}} = min(CIR_{est_{C_{other-DL_n}}} - CIR_{target_{C_{other-DL_n}}}, CIR_{est_{C_{other-UL_n}}} - CIR_{target_{C_{other-UL_n}}})$$

$$CIR_{margin_{min_C}} = min(CIR_{margin_{C_{own}}}, CIR_{margin_{C_{other_n}}})$$

Figure 8.11 Example of DFCA radio channel selection.

As a result of the above procedure the channel with the most CIR margin above the CIR targets will be chosen. Thus, the DFCA algorithm aims to maximise the CIR while maintaining the relative CIR margins between users with different CIR requirements. As the network load increases, the provided CIRs will gradually decrease. The CIR differentiation between user classes is achieved because the CIR margin is calculated based on user class specific CIR targets. Optionally, a soft blocking CIR limit can be defined for each user class. Whenever the channel assignment would lead to an estimated CIR to fall below the soft blocking CIR limit, the assignment is rejected. This procedure ensures that the new or any of the existing connections will not be exposed to intolerable interference.

Information exchange

As described before, the DFCA algorithm needs to have access to mobile station (MS) measurement reports and real-time channel occupancy and transmission power information in the neighbouring cells. Due to these requirements it is natural to locate the DFCA algorithm in the network element that handles the radio resource management. In the standard base station subsystem (BSS) architecture this network element is the base station controller (BSC). One BSC can only handle a limited number of base stations that typically cover one geographical area. Normally each of the radio resource management (RRM) functions is only aware of the channel occupancy situation within the area it controls. The same applies also for the availability of the MS measurement reports. To allow transparent DFCA operation across the border areas the RRM functions have to exchange channel occupancy, MS measurement report and current transmission power information.

DFCA frequency hopping modes

The optimum DFCA operation requires a DFCA frequency hopping mode where a base station has to be able to use any DFCA MA list and MAIO in any of the timeslots. This means that the base station must support timeslot level synthesised frequency hopping and wideband combining. Wideband combining that is required for synthesised frequency hopping may not be feasible in some base stations due to increased RF losses and older base station hardware may not support DFCA hopping mode. However, DFCA can be used in the RRM even if the DFCA hopping mode is not available. The following operating modes are possible:

- DFCA frequency hopping
- synthesised frequency hopping (RF hopping) with DFCA RRM
- baseband frequency hopping (BB hopping) with DFCA RRM

DFCA hopping mode provides full radio channel selection freedom for the DFCA algorithm and therefore the optimum performance. The synthesised frequency hopping with DFCA RRM consists of a BTS that is configured for synthesised hopping with a fixed DFCA MA list and fixed pre-planned MAIOs for each transceiver (TRX). Because of the fixed MA list and the fixed MAIOs the DFCA algorithm has limited freedom to choose the most suitable channel for each connection. Each timeslot in every TRX has already a fixed MA list and MAIO that cannot be dynamically changed. Thus, the DFCA algorithm can only select the most suitable timeslot from all the free timeslots. In low load situation the number of free timeslots is high so DFCA is able to optimise the performance quite effectively. However, in high load situation only a few time-slots may be available severely restricting the possibilities to find a suitable radio channel and therefore degrading the performance. The baseband frequency hopping with DFCA is very similar to the synthesised hopping case. Also with baseband hopping the MA list and the TRX MAIOs are fixed. The only addi-tional restriction with baseband hopping is that the MA list must contain as many frequencies as there are DFCA TRXs in the base station. If the number of the DFCA TRXs is low, this will limit the frequency diversity gain and therefore cause further performance degradation.

With both synthesised frequency hopping and baseband frequency hopping a basic frequency planning must be performed. In synthesised hopping the MAIO reuse must be planned and in the baseband hopping the allocation of DFCA MA lists to different BB hopping BTSs must be decided. The quality of this under-lying frequency plan determines maximum achievable quality during the busy hours.

If the share of DFCA hopping capable BTSs is sufficiently high and they are equally distributed among the RF and/or BB hopping BTSs, then the full channel selection freedom of the DFCA hopping capable BTSs provides very effective adaptation to the surrounding RF/BB hopping BTSs. This decreases the impor-tance of the frequency planning of the RF/BB hopping BTSs and at the same time enables increased robustness also during the busy hours maintaining the DFCA gains.

8.3 Application of DFCA for circuit switched (CS) services

Multitude of CS services

Optimising the network performance has become an increasingly difficult task as the variety of circuit switched services has increased. Nowadays there are basic full-rate and half-rate services, advanced multi-rate (AMR) codecs for both full-rate and half-rate channels as well as high speed circuit switched data (HSCSD) services with the option of utilising multiple timeslots for one connection. There are some differences in channel coding between all these services and as a result the robustness against interference varies significantly. For example, an AMR connection can tolerate almost 6 dB higher interference level than a connection using enhance full-rate (EFR) codec while still maintaining comparable performance in terms of FER or Mean Opinion Score (MOS). Half-rate (HR) channels have generally less room for strong channel coding and the interleaving depth of HR channels is four instead of eight that applies for full-rate (FR) channels. Thus, FR channels in general tolerate higher interference level than HR channels. These link performance issues have been covered in more detail in Chapter 6.

The minimum CIR a connection can tolerate sets the lower limit for the frequency reuse distance. However, with a mixture of services in the network the least robust of the used channel types determines the minimum tolerable reuse. Adaptive power control helps to project the gains of the robust connections to the more vulnerable ones by aggressively reducing the transmission powers of the robust connections. However, simulations have shown that power control alone is insufficient for balancing the performances of different services.

For better balancing of performances the different CIR requirements have to be taken into account already in the channel assignment. One possible solution is the usage of the reuse partitioning concept introduced in Chapter 6, where the frequency band has been divided into two separate pools that are then allocated to TRXs providing two different layers with different frequency reuse factors. The robust AMR connections can be primarily assigned to the underlay with low reuse factor and high spectral efficiency while the less robust connections are assigned to the overlay with conventional reuse factor where sufficiently high CIR can be guaranteed. This approach is reasonably powerful but it suffers from lack of flexibility as increase in the AMR capable handset penetration requires constant review of the frequency band split between underlay and overlay bands and a change in frequency bands generally requires adjustments in the frequency plan. Since legacy mobiles cannot be guaranteed sufficient quality on the underlay, the overlay must be conservatively dimensioned so that instances where a legacy handset needs to be assigned to underlay due to overlay blocking are minimised.

The DFCA way

DFCA is ideal for managing users with different CIR requirements. With DFCA different CIR targets can be assigned according to the robustness of the connection. Examples are presented in Table 8.2.

With DFCA, each connection is automatically assigned a channel that aims to satisfy the interference requirements associated with the connection type.

Table 8.2 Example CIR targets

Connection type	Target CIR	Soft blocking threshold
FR/EFR	15 dB	10 dB
HR	18 dB	13 dB
AMR FR	9 dB	4 dB
AMR HR	15 dB	10 dB
HSCSD	18 dB	Not used

Furthermore, the same requirements are evaluated every time a new potentially interfering channel assignment is being done thus aiming to keep the CIR above the target during the whole connection. An ongoing connection is never allowed to fall below the soft blocking limit as a result of the establishment of an interfering connection. Thus, DFCA automatically optimises the usage of the frequency resource without need for network planning changes as a consequence of growing AMR penetration and changes in the geographical traffic distribution. Growing AMR usage can simply be seen as increasing network capacity as higher effective frequency loads (EFLs) can be achieved while still maintaining sufficient quality for both EFR and AMR connections.

8.4 Simulations

The performance of DFCA has been evaluated in different scenarios by network level simulations. The used simulation tool and the simulation parameters are described in Appendix D. A complete DFCA RRM algorithm was implemented into the simulator. In all scenarios the DFCA performance is compared with a reference case that utilises random frequency hopping with 1/1 reuse strategy and fractional loading taking advantage of both power control and discontinuous transmission. Only the traffic on the non-BCCH TRXs is considered. The connection quality is studied by observing the frame erasure rate (FER) in 2-s intervals. FER over 4.2% is considered an indication of unsatisfactory quality during the past 2 s. The network performance is benchmarked by calculating the percentage of these bad quality instances. In the graphs this percentage is called an 'outage'.

Most of the simulations have been performed in an ideal regular three-sectorised network layout. In order to verify the DFCA performance in a more typical scenario, a real city centre network layout was implemented into the simulator with varying antenna orientations, cell sizes and cell traffics.

Performance in ideal network layout

The ideal network layout consists of 75 cells in regular three-sectorised configuration. The traffic density is even and the cells have equal number of transceivers.

EFR

In the first simulation case the performance of DFCA is studied with all EFR
handsets. The limited robustness of EFR channel coding requires low effective
frequency loads. In standard interference limited scenarios downlink is the
limiting link mainly due to application of uplink diversity reception and the
regularity of the network layout. In this scenario DFCA is able to give a significant boost to downlink performance and with 2% outage level an overall
network capacity gain of 66% can be observed (Figure 8.12).

AMR

In the second case the DFCA performance is studied assuming that all the
handsets in the network are AMR capable (Figure 8.13). Both AMR FR and
AMR HR channels are considered. In high load situation usage of AMR HR
is preferred due to the efficient usage of hardware resources. In the reference
case some connections in marginal conditions have to be kept in the FR
channel due to higher robustness of the FR channel whereas in DFCA the
best performance in high load situation is achieved with full AMR HR usage.
Fixed codecs with 5.9 kbit/s bit rate are used for both AMR FR and AMR
HR. At 2% outage level the DFCA gain relative to the reference case performance is 35%.

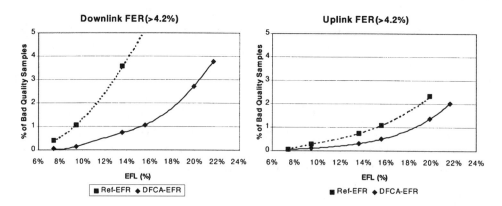

Figure 8.12   DFCA performance with EFR users.

Figure 8.13   DFCA performance with AMR users.

Partial AMR penetration

Since the AMR support requirement for handsets has only been introduced in a relatively recent release of the GSM specifications (Rel'98), it is common that the share of AMR capable end users is growing gradually as old handsets are replaced with newer ones. In this case full AMR gains are not available until 100% AMR capable handset penetration is achieved. The challenge for an operator is how to extract the full benefit of the AMR capable handsets while still providing good quality of service for the legacy handsets.

The following simulations have been performed to examine this problem and to evaluate the impact of the CIR provisioning capability of DFCA. The DFCA algorithm can be defined to allow the assignment of radio channels with lower CIR for the AMR capable handsets while providing sufficiently high CIR for the legacy handsets. Naturally, in both the reference case and the DFCA case more aggressive power control settings are applied for the AMR connections. The quality experienced by the EFR users is examined separately as this is typically limiting the maximum tolerable network load. Total FER performance comprising of combined EFR and AMR users' FER statistics is also presented. All AMR connections are AMR FR.

In the first examined case the AMR capable handset penetration has reached 25% (Figure 8.14). In the reference case the quality experienced by the EFR users has improved from the pure EFR case presented in Figure 8.12 due to the lower power levels used by the interfering AMR users, but the quality gap between the EFR users and the AMR users is significant. It is therefore clear that the performance experienced by the EFR users is the limiting factor that defines the maximum achievable capacity.

DFCA CIR provisioning proves to be very effective in equalising the qualities experienced by these two user groups. With DFCA controlled channel assign-

Figure 8.14 DFCA performance with 25% AMR penetration.

ment the quality gap disappears as the EFR users experience virtually identical quality with the AMR users. At 2% outage level a DFCA gain of 56% is achieved when the quality experienced by the EFR users is considered.

Figure 8.15 presents the situation with 75% AMR penetration. The larger share of AMR capable handsets is reflected in better overall performance as higher loads can now be achieved. However, the significant quality gap between the EFR and AMR users remains in the reference case. The results indicate a 43% DFCA gain when the quality of the EFR users is considered.

Narrowband performance
Narrowband deployment is presented in Chapter 9, where general narrowband specific issues are presented in detail. In this section the impact of narrowband deployment on DFCA performance is presented.

Figure 8.15 DFCA performance with 75% AMR penetration.

In order to evaluate the behaviour of DFCA in narrowband cases, two narrowband scenarios are studied. The BCCH TRXs are modelled in the simulations, but only the quality of the traffic on the hopping TRXs is considered. In these simulations the frequency band available for DFCA is very limited. The small number of frequencies causes three detrimental effects on the DFCA performance.

Firstly, the small number of frequencies available for DFCA limits the number of available channels. This degrades the DFCA performance, as the probability of finding a suitable channel from a small pool of channels is smaller. This can be viewed as a consequence of degraded trunking efficiency in the channel selection.

Adjacent channel interference coming from the BCCH frequency band can also have a significant negative impact of DFCA performance if the number of frequencies in the DFCA MA list is small. With DFCA that is using cyclic frequency hopping the adjacent channel interference coming from the BCCH band impacts regularly every nth burst where n is the number of frequencies in the DFCA MA list. Since BCCH is always transmitted with full power in the downlink direction, the impact of this interference can be significant for some users especially if the DFCA MA list is short. Furthermore, DFCA cannot take this interference contribution into account in the channel selection so in some cases this additional uncounted interference may be critical for the connection quality.

The small number of frequencies can also limit the possibilities of dividing the DFCA frequency band into multiple MA lists. With only one DFCA MA list the frequencies in the list are consecutive. Adjacent frequencies have typically significant fast fading correlation that degrades the frequency diversity gain especially when cyclic frequency hopping is used. The fast fading correlation between the frequencies is modelled in the simulation tool as presented in Appendix E.

3.6 MHz band   The first evaluated narrowband case is the extreme case of 3.6 MHz total frequency band. Out of this band 12 carriers have been allocated for BCCH and the remaining 6 carrier frequencies have been allocated to DFCA. The DFCA band is directly adjacent to the BCCH band.

The results in Figure 8.16 show about 26% capacity gain, but the gain reduction is obvious when compared to the wideband DFCA case where 66% gain is achieved.

4.8 MHz band   In this case the total available frequency band is 4.8 MHz corresponding to 24 carrier frequencies. Again 12 of these are allocated to the BCCH. The remaining 12 form the DFCA frequency band that is again adjacent to the BCCH band.

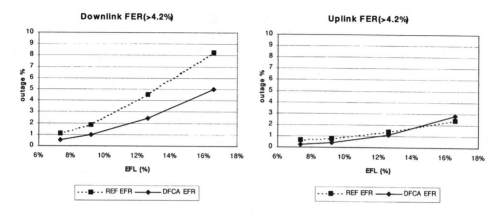

Figure 8.16 DFCA performance in 3.6 MHz narrowband case.

The doubling of the DFCA frequency band provides a significant improvement in DFCA performance. Although the overall performance is still slightly below the performance of the ideal wideband case, the DFCA gain is already 55% at the 2% outage level (Figure 8.17).

It can be concluded that although the DFCA performance degrades in the extremely narrowband case, the gains have returned to normal level in the 4.8 MHz case.

Impact of MS measurement errors
The CIR estimation that is done within the DFCA algorithm is based on the serving cell and neighbouring cell signal level measurements that are performed and reported by the mobiles. The inaccuracies in these measurements can cause errors in the DFCA CIR estimations and therefore degrade the performance. Each GSM handset must satisfy the minimum signal level measurement accuracy requirements that are stated in the GSM specifications [9].

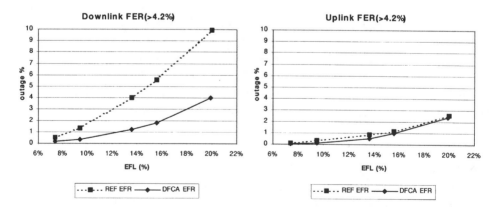

Figure 8.17 DFCA performance in 5 MHz narrowband case.

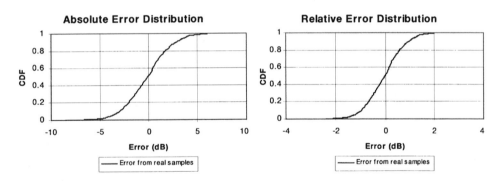

Figure 8.18 Modelling of MS measurement errors.

The absolute signal level accuracy requirement is ± 6 dB in both normal and extreme temperature conditions. If the signals are in the same frequency band and at least 14 dB over the MS reference sensitivity level then the relative signal level accuracy requirement is ± 2 dB. In these simulations the occurrence of these errors is assumed to be normally distributed with 99% probability of the error being within the above mentioned limits. The cumulative density functions of the errors are presented in Figure 8.18.

The simulations were run assuming EFR users in a regular network layout. Figure 8.19 shows that DFCA is very robust against the impact of these measurement errors. In downlink performance no difference can be noticed and in uplink the impact is hardly noticeable.

DFCA never relies on the absolute reported signal level values. Instead, it always calculates the potential CIRs from each measurement report and uses them for the background CIR statistics and also as a direct input in the CIR estimation. This direct CIR calculation makes DFCA immune to the absolute MS measurement errors. Only relative error has effect. But this effect is limited as it is averaged out in the CIR statistics that are used to determine the statistical

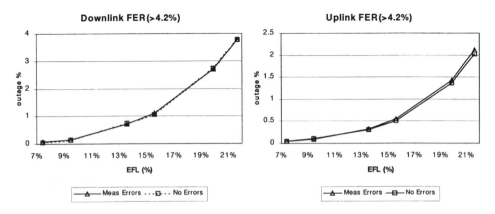

Figure 8.19 DFCA EFR performance with measurement errors.

CIR values. The relative error impacts only the part of the CIR estimation that uses the potential CIR values that have been determined directly from the latest measurement reports.

Impact of the number of reported neighbours
Since the DFCA CIR estimation is based on the reported serving cell and neighbouring cell signal levels, the number of the neighbouring cells that are included in the measurement reports has some impact on the DFCA performance (Figure 8.20). In all the previous simulations it was assumed that each measurement report contains the signal levels of six strongest neighbours that corresponds to the capabilities of legacy handsets. However, this may not always be the case. Sometimes high BCCH interference level may not allow successful BSIC decoding in which case the neighbour cannot be reported. Also, the newer handsets are able to utilise enhanced measurement reporting that allows reporting of many more neighbours in one report and by using interleaving of the reported cells even more can be reported over successive reports.

The results show that the number of reported neighbouring cells has virtually no impact in lower network loads. In higher loads the impact is clearly distinguishable, but not dramatic. The results show that about 5% extra gain can be achieved if the measurement reports contain 10 neighbours instead of 6.

In low load situations the DFCA algorithm keeps the interfering connections in far away cells as this maximises the CIR. With higher loads the interfering connections must be allowed in closer neighbouring cells. For DFCA the importance of the latest downlink measurement reports is significant only when there is a possibility that an interfering connection may exist in the reported neighbours. In low load situation this is not the case and therefore the impact of the number of reported neighbours is very small. In low load situation the inter-

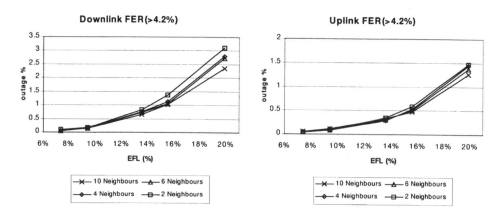

Figure 8.20 Impact of the number of reported neighbours.

ference estimation relies heavily on the background interference matrix (BIM) that is relatively insensitive to the number of neighbours included in each downlink measurement report.

Performance in typical network layout

In the previously presented simulation results DFCA performance has been evaluated in ideal network layout with even traffic distribution, equal cell sizes and standard antenna orientations. This kind of approach is justifiable in order to be able to get easily comparable and reproducible results in a highly controlled standard environment. However, real networks have always a varying degree of irregularities. The performance of DFCA has also been benchmarked in a typical scenario since it was expected that the highly dynamic nature of DFCA should enable good adaptation to the irregularities.

The typical scenario is created by importing the macro layer network layout of a real network in a city centre and its surroundings. The modelled network is the same as Network 1 in Appendix F. An uneven traffic distribution is created in order to achieve the cell loads that were observed in the real network. As a result the cell traffic is highly variable. The number of TRXs per cell ranges from two to six according to the real network configuration. A map of the modelled city centre and the cell dominance areas of the whole area are presented in Figure 8.21.

EFR performance
At first the performance with EFR users is examined. A comparison with the same case in the ideal network layout scenario in Figure 8.22 reveals that the irregularities cause a clear relative degradation in uplink performance.

Figure 8.21   Irregular network layout with cell dominance areas.

Figure 8.22   DFCA performance with EFR users.

However, with a small margin downlink is still the limiting link. The DFCA gain has been clearly improved. At 2% outage level the DFCA capacity gain is about 72%.

AMR
The AMR simulations in the irregular network scenario have been performed with the same assumptions used in the corresponding ideal network scenario presented is Section entitled *AMR* in 8.4. In the case of AMR users the irregular network is clearly uplink limited. When the uplink performance is studied at 2% outage level a capacity gain of 62% can be observed (Figure 8.23).

AMR narrowband performance (3.6 MHz)
The impact of narrowband deployment in the typical network scenario has been studied with the same assumptions that have been used in all AMR

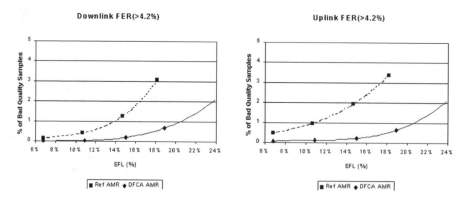

Figure 8.23   DFCA performance with AMR users.

Figure 8.24   DFCA performance with 3.6 MHz band and AMR users.

simulations in this chapter. The used frequency band is 3.6 MHz out of which 2.4 MHz is used by the BCCH layer. This leaves only 1.2 MHz available for DFCA operation.

As with the previous wideband results, also now the uplink is the limiting link in the reference case. However, at 2% outage level DFCA case is downlink limited. DFCA gains of 20 and 45% can be observed for downlink and uplink, respectively (Figure 8.24).

Impact of frequency hopping modes
As explained in Section entitled *DFCA frequency hopping modes* in 8.2, a complete DFCA implementation requires that the base station is capable of selecting the radio transmission and reception frequency independently for every timeslot in each DFCA TRX. This allows the DFCA algorithm to freely choose the radio channel for each connection. In DFCA the radio channel is defined by the selection of the timeslot, MA list and MAIO.

However, DFCA can be used together with conventional synthesised and baseband frequency hopping modes. With these hopping modes each TRX is hopping over a fixed MA list using a fixed MAIO in all the timeslots. Therefore, the radio channel selection is limited to the fixed radio channels that have been pre-assigned to the available timeslots. As a consequence, the radio channel selection freedom available for the DFCA algorithm is directly dependent on the number of free timeslots in the cell. In low load situation the freedom is the greatest but as the load and timeslot usage increases, the radio channel selection freedom decreases.

In this simulation the performance of the synthesised frequency hopping mode with DFCA RRM is compared with the performance of the DFCA frequency hopping mode that provides full radio channel selection freedom. Also some intermediate cases are studied where 25, 50 and 75% of the base

Figure 8.25   Performance with partial DFCA hopping capability.

stations are DFCA frequency hopping capable while the rest use synthesised hopping with DFCA RRM. All these cases are compared with the reference case that uses fractionally loaded 1/1 reuse strategy with random synthesised frequency hopping with conventional RRM (Figure 8.25).

As expected, a reduction in the share of the DFCA frequency hopping capable base stations leads to a reduction in the DFCA performance. However, if at least 50% of the base stations are capable of DFCA frequency hopping then only a small performance degradation can be observed indicating that the surrounding DFCA frequency hopping capable base stations are able to adapt efficiently to the limitations of the synthesised hopping base stations. Therefore, it is evident that almost optimum performance can be achieved if at least 50% of the base stations have DFCA hopping capability and they are equally distributed among the synthesised hopping base stations.

In all but the peak load situations also the more heavily restricted DFCA cases provide significant quality gains. This is clearly visible especially in uplink, where the 100% synthesised hopping with DFCA RRM as well as the case with 25% of DFCA hopping capable base stations provide over 100% reduction in the FER outage. This is almost equal to the quality gain provided by the cases with a higher share of DFCA hopping capable base stations and proves that the channel selection freedom in low loads is sufficient for good DFCA performance.

Summary of the simulation results

A summary of the simulation results is presented in Table 8.3.

Table 8.3 Summary of the simulation results

Simulated scenario	Gain in ideal network (limiting link in parentheses)	Gain in typical network (limiting link in parentheses)
EFR	66% (DL)	72% (DL)
AMR	35% (DL)	62% (UL)
75% EFR/25% AMR	56% (DL)	N/A
25% EFR/75% AMR	43% (DL)	N/A
3.6 MHz EFR	26% (DL)	N/A
4.8 MHz EFR	55% (DL)	N/A
3.6 MHz AMR	N/A	20% (DL)

8.5 DFCA for packet switched (PS) services

Packet switched data services can be used by a multitude of different applications that can have very different quality of service (QoS) requirements. The most important of these is the throughput. Some of the services such as applications using streaming data can be very demanding requiring a high and guaranteed throughput whereas some other services such as the short message service or background email download may not have specific throughput requirements. The complete QoS framework has been presented in Chapter 3.

The user throughput depends on the interference level on the used radio channels as well as on the amount of radio resources assigned to the user. The radio resources are defined by the number of timeslots and the amount of scheduling turns within these timeslots. The radio resources can be efficiently managed by conventional means presented in Chapter 6, but without DFCA the interference level on the radio channels cannot be accurately controlled. For example, in a random frequency hopping network the frequency load defines an average level of interference, but large random deviations remain depending on the timeslot and the location of the user. This makes the estimation of the interference levels prior to the channel assignment inaccurate. DFCA provides means to estimate and control the level of interference in timeslot level. Thus, with DFCA all the variables that define the achievable throughput can be kept in close control.

The optimum network configuration is very much dependent on the QoS requirements of the carried traffic. For example, a background type of traffic doesn't require guaranteed throughput so the most spectral efficient network configuration is one with very tight frequency reuse. Smaller reuse distance implies that with the same frequency band more TRXs per cell are installed. The 1/3 reuse strategy allows four times TRXs per cell compared with the 4/12 reuse. On the other hand 1/3 reuse also implies that the interference level in the network is much higher than in the 4/12 reuse. The high interference leads to reduced throughput per timeslot. However, this is compensated for by the fact that in the 1/3 reuse case there are four times more timeslots available in each cell. If the mobiles support several timeslots in the downlink direction, reasonable throughputs can be achieved even in the highly interfered tight frequency reuse case. The high number of TRXs per cell in this case allows a high number of users per cell thus leading to high spectral efficiency.

On the other hand, if the service requires high throughput the limited number of timeslots that can be used for one connection start to become the throughput-limiting factor. In such a case the high throughput per timeslot becomes essential so that the required throughput can be achieved. Figure 8.26 presents the spectral efficiency in terms of kbit/s/cell/MHz as a function of the frequency reuse scheme for a number of QoS criteria and three timeslots (TSLs) supported by the MS. The assumptions behind these results and more detailed results have been presented in Section entitled *EGPRS performance with link adaptation in a separate non-hopping band* in 7.4.

The results clearly indicate that if low throughputs are sufficient then extremely tight frequency reuse factors provide the best spectral efficiency. On the other hand, such tight reuses are completely unsuitable for services that require very high throughputs. The challenge for the operator is how to combine these two conflicting requirements.

Typically, the options are limited as conventional frequency planning doesn't allow reuse differentiation and normally the speech traffic uses the same TRXs

Figure 8.26 EDGE spectral efficiency with different QoS criteria (10th percentile throughput) in different frequency reuse schemes. Number of TSLs supported by the MS is three.

as the PS traffic. Theoretically it would be possible to split the frequency band into two or more parts and apply a reuse-partitioning scheme where different TRXs would have different frequency reuse factors. This way some TRXs could be deployed with just a few frequencies in order to form a layer with very tight frequency reuse. These TRXs would then be optimised to handle low throughput background traffic with high spectral efficiency. However, this kind of approach is not very flexible as fixed separate frequency bands are needed.

DFCA allows a flexible and efficient management of radio resources also for the PS services. The full radio channel selection freedom and the ability to estimate the CIR based on the user's downlink neighbouring cell measurements provide means to optimise the radio channel selection so that the user throughputs and the spectral efficiency can be maximised. Furthermore, DFCA can also guarantee that the desired interference level is maintained during the connection regardless of the user's location and movement as long as the network is not coverage limited. This makes it possible to guarantee higher throughputs without out need for excessive safety margins that degrade the spectral efficiency.

References

[1] Beck R., Panzer H., 'Strategies for Handover and Dynamic Channel Allocation in Microcellular Mobile Radio Systems', IEEE Veh. Technol. Conf., 1989, pp. 178–185.
[2] Cimini L. J., Foschini G. J., Chih-Lin I, Miljanic Z., 'Call Blocking Performance of Distributed Algorithms for Dynamic Channel Allocation in Microcells', IEEE Trans. Commun., Vol. 42, No. 8, August 1994, pp. 2600–2607.
[3] Jordan S., Khan A., 'A Performance Bound on Dynamic Channel Allocation in Cellular Systems: Equal Load', IEEE Trans. Veh. Technol., Vol. 43, No. 2, May 1994, pp. 333–344.
[4] Baiocchi A., Priscoli F. D., Grilli F., Sestini F., 'The Geometric Dynamic Channel Allocation as a Practical Strategy in Mobile Networks with Bursty User Mobility', IEEE Trans. Veh. Technol., Vol. 44, No. 1, February 1995, pp. 14–23.
[5] Cheng M., Chuang J., 'Performance Evolution of Distributed Measurements-based Dynamic Channel Assignment in Local Wireless Communications', IEEE J. Selected Areas Commun., Vol. 14, No. 4, May 1996, pp. 698–710.
[6] Borgonovo F., Capone A., Molinaro A., 'The Impact of Signal Strength Measures on the Efficiency of Dynamic Channel Allocation Techniques', IEEE Int. Conf. Commun., Vol. 3, 1998, pp. 1400–1404.
[7] Verdone R., Zanella A., Zuliani L., 'Dynamic Channel Allocation Schemes in Mobile Radio Systems with Frequency Hopping', 12th IEEE International Symposium on Personal, Indoor and Mobile Radio Communications, Vol. 2, September/October 2001, pp. 157–162.
[8] GSM 05.05, 'Radio Transmission and Reception', 3GPP, release 1999.
[9] GSM 05.08, 'Radio Subsystem Link Control', 3GPP, release 1999.

9

Narrowband Deployment

Angel Rivada, Timo Halonen and Juan Melero

Bandwidth is a scarce resource. Some networks have to be initially deployed with a very limited amount of frequency spectrum. A potential limitation in these narrowband deployment scenarios is the number of frequencies required for the broadcast control channel (BCCH) layer. Traditionally, global system for mobile communications (GSM) networks have been deployed with loose BCCH reuses due to the permanent full-power transmission of BCCH transceivers (TRXs) in downlink (DL). In narrowband scenarios, this could lead to limit the number of available frequencies for the hopping layer deployment, effectively limiting its performance and the number of additional deployed TRXs per cell. Therefore, in narrowband cases, the tightening of the BCCH reuses is critical to maximise the overall system performance. This chapter assumes Radio Frequency (RF) hopping design resulting in two effective layers with separated frequency band, the BCCH and the hopping layer.

In this chapter the factors that differentiate narrowband networks from wideband scenarios are identified. Additionally, deployment guidelines and solutions are presented in order to achieve the maximum spectral efficiency with the scarce spectrum resources available. BCCH deployment schemes will be analysed and functionality to tighten the BCCH reuse will be described. Finally, the common BCCH functionality and a specific adaptive multi-rate (AMR) analysis are presented as tools to maximise the performance of narrowband deployment scenarios.

9.1 What is a narrowband network?

Networks having an allocated spectrum smaller than 5 MHz per link (10 MHz full-duplex transmission with 5 MHz per link) are considered to be narrowband networks in the analysis performed in this chapter. Extreme examples of these scenarios can be found across the US, where in some cases GSM has been initially deployed with 3.6 MHz or even less bandwidth. In other countries in Europe and Asia, narrowband scenarios are less usual due to the different spectrum licensing policies of the local regulatory bodies. However, in some specific situations other issues limit the number of available frequencies. There are some typical scenarios where the deployment bandwidth is limited:

- frequency spectrum re-farming, e.g. technology migration from existing time division multiple access (TDMA) IS-136 or code division multiple access (CDMA) IS-95 to GSM/enhanced data rates for global evolution (EDGE);
- narrow licensed frequency spectrum due to specific regulations of the country, i.e. 5 MHz allocations in US Personal Communication System (PCS) 1900 band;
- micro cell deployment (band splitting).

An efficient BCCH deployment is critical in the narrowband scenarios. Due to the nature of the GSM BCCH channel, GSM operators have traditionally reserved 15 frequencies (3 MHz) or more for the BCCH TRX deployment. This means that in the case of 3.6 MHz deployment (18 frequencies), only three frequencies would remain for additional transceivers. Not only does this limit the number of additional TRXs deployed but as well seriously impacts the performance of the 'hopping layer' since the frequency diversity gains would be substantially limited.

Frequency spectrum re-farming. Technology migration

The term spectrum re-farming is used to describe reorganisation of certain frequency spectrum. Reasons behind the re-farming are often related to a radio technology migration (e.g. from TDMA IS-136 or IS-95 to GSM/ EDGE). However, re-farming is a time-consuming process and it does not happen over night, mostly due to the fact that the terminal migration is a lengthy process. In most of the cases, wireless operators have to maintain both the old and new radio technology systems using a common frequency spectrum. Therefore, operators have to utilise minimum amount of spectrum for the initial deployment of the new radio technology in order to cope with existing customers in the old system during the migration period. In these circumstances, the deployment of GSM/EDGE requires specific attention.

In the case a cellular operator decides to perform a technology migration from IS-95 (cdmaOne) or IS-136 (US-TDMA), with a limited available bandwidth, to GSM/EDGE, this will take place using a phased approach. In the first phase at least a minimum amount of frequencies have to be re-farmed in order to deploy the GSM BCCH TRX layer. The second phase provides already enough capacity for the majority of the traffic to be migrated to GSM/EDGE and in the last phase the whole frequency spectrum is re-farmed. Phases one and two are usually going to be GSM narrowband scenarios. Figure 9.1 illustrates an extreme example of IS-95 migration to GSM/EDGE. In this case IS-95 is deployed in 5 MHz per link with one or two carriers being deployed per sector. In the first phase only the BCCH layer is deployed providing sufficient capacity for initial introduction of GSM/EDGE. During this first phase the existing traffic is shared between the BCCH TRX and the IS-95 carriers. When the traffic carried by the GSM/EDGE system is such that the second IS-95 carrier becomes redundant, the second phase starts with the migration of the spectrum initially assigned to the second IS-95 carrier. Finally, when the terminal migration is completed, the remaining spectrum is migrated to GSM/EDGE.

In these technology migration scenarios, the potential cross system interference has to be taken into account. Appendix D includes an analysis of the cross system interference issues to consider when different radio technologies are deployed within the same band.

Narrow licensed frequency spectrum

Most European GSM operators have large spectrum allocations with dual band availability (900 and 1800 MHz). Also most of the Asian operators have deployed the 1800 band to get more frequency spectrum. However, there are still operators facing GSM spectrum limitations. In America, especially in the US, nation-wide operators may only be licensed 5 MHz in some of their markets (in the US the spectrum is not licensed nation-wide).

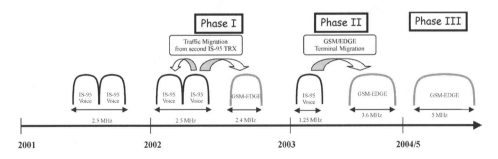

Figure 9.1 Frequency band re-farming example from IS-95 to GSM/EDGE.

Figure 9.2   US PCS 1900 licensed bandwidth allocations.

Figure 9.2 shows the fragmented US PCS 1900 licensed bandwidth allocation.

Figure 9.3 illustrates the distribution of population according to the bandwidth availability of each market covered by the top five US operators. It shows that almost 20% of markets have to deploy cellular networks under narrowband conditions (10 MHz = 5 MHz in uplink and 5 MHz in downlink).

Micro cell deployment

Another situation where many operators have to cope with narrow spectrum availability is in traffic dense urban areas, where it is necessary to deploy dense microcellular networks. In this case, frequency spectrum is often split into two separate bands, one for macro cells and another one for micro cells, both requiring separate BCCH allocations. A similar situation occurs in case of indoor

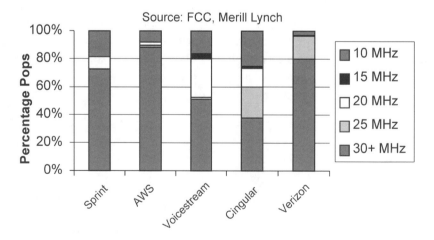

Figure 9.3   Spectrum distribution of top five US wireless operators (this allocation assumes an agreement is reached regarding Nextwave 1900 band spectrum, with such spectrum re-assigned according to 2001 spectrum re-auction results).

systems deployment. Therefore, even with relatively large amount of spectrum allocated, the operator has to deal sometimes with the defined narrowband scenario.

In order to overcome the potential performance problems associated with narrowband deployment, several solutions are described in the following sections.

9.2 BCCH band allocation

In narrowband scenarios, the BCCH deployment will determine the frequency hopping layer efficiency and capacity. This efficiency depends on how much uncorrelated the fading profiles of the bursts that make up a speech frame are. By using frequency hopping, those bursts use different frequencies and are affected by different fading conditions (frequency diversity). The higher the number of frequencies available to hop over, and the higher the frequency separation among them (to reduce the partial correlation of the fading profile), the higher the gain of frequency diversity [1]. The use of hopping also randomises the interference signals received, having an averaging effect (interference diversity). Again, this effect is more efficient when there are more frequencies to hop over. These concepts are described in detail in Chapter 5.

In narrowband cases, the number of frequencies in the hopping layer limits the overall hopping layer performance. One approach to make them more efficient from the frequency diversity point of view would be to increase the frequency spacing as much as possible in order to increase the effective hopping bandwidth. This could be done by interleaving the BCCH and hopping frequencies. However, the BCCH TRXs transmit at full power in the DL and this would create adjacent interference in the hopping layer that might degrade the performance of the hopping layer even more. Another option would be to use separate bands to isolate the effect of adjacent channel interference from BCCH, but this would suffer limited frequency diversity gains. These configurations and some others (with and without guardbands) are analysed in the next sections for different bandwidths.

BCCH reuse for narrowband scenarios

As previously mentioned, traditionally, practical GSM BCCH deployments have used a reuse of 15 or higher since the available spectrum usually did not impose limitations in the BCCH reuse to be deployed. However, in narrowband scenarios the need to tighten the BCCH reuse in order to release more frequencies for the hopping layer forces the careful analysis of the most efficient BCCH deployment strategy. With efficient network planning (site locations, antenna locations and height, downtilts, beamwidth, etc.), a reuse of 12 frequencies for BCCH is achievable. In this chapter, narrowband networks with BCCH reuse of 12 have been considered. Some methods to even lower the BCCH reuse without compromising the performance will be covered along the chapter.

Using 12 frequencies for BCCH leaves only 6 frequencies for hopping transceivers in the case of 3.6 MHz bandwidth per link and 13 hopping frequencies in the 5 MHz case (one less in the case 100 kHz guardband is required at both sides of the 5 MHz licensed band).

Narrowband BCCH deployment strategies

Initially, two basic strategies can be used to allocate the available band between BCCH and hopping layers: either select a group of contiguous frequencies for BCCH and another one for hopping frequencies (herewith called 'non-staggered'); or interleave hopping frequencies with BCCH ones ('staggered' strategy). Both criteria are shown in Figure 9.4.

Impact on frequency diversity gain
Figure 9.5 displays the link performance impact of the different allocation strategies. Curves for non-hopping and uncorrelated hopping frequency configurations are also plotted. The 5 MHz case has a frequency diversity gain over 1 dB higher than that of the 3.6 MHz case.

Additionally, the effect of the correlation of the hopping frequency fading profiles is also displayed in Figure 9.5. When using the non-staggered strategy the hopping frequencies are closer to each other and their fading profiles are more correlated than in the staggered case, where there exist more separation

Figure 9.4 Basic BCCH band allocation strategies.

Figure 9.5   Impact on frequency diversity gain based on simulations.

among the hopping frequencies. The effect is that both strategies experience different diversity gains. For 3.6 MHz, the frequency diversity gain at 1% frame error rate (FER) is around 0.6 dB higher when the staggered configuration is used instead of the non-staggered one. For 5 MHz, the difference rises to 0.8 dB at 1% FER. In this last case the staggered configuration reaches the maximum frequency diversity associated with 13 frequencies, i.e. the uncorrelated frequency diversity case.

System level performance: influence of adjacent channel interference
Figure 9.6 presents the overall system performance associated with the different configurations presented. In the *x*-axis effective frequency load (EFL) is used as a measure of traffic load (as presented in Chapter 5). It can be seen that the non-staggered strategy outperforms the staggered one, despite the higher diversity gain of the latter one. The staggered case suffers from the permanent adjacent channel interference coming from BCCH frequencies in DL (always transmitting at full power). Simulations show that the power control gain is limited by continuous adjacent channel interference. This figure also shows that the performance with 13 hopping frequencies is better than the one with 6 hopping frequencies due to higher diversity gain.

Need of guardband

According to the analysis from the previous section the best performance would be achieved with an allocation strategy that provides high diversity gain without suffering degradation due to adjacent channel interference from BCCH. One way to reduce this interference is the use of guardbands between BCCH and hopping frequencies. In the non-staggered case, using only a guard channel between the BCCH band and hopping band might reduce the adjacent

Figure 9.6 Network performance for different allocation strategies and bandwidths. Typical urban 3 km/h (TU3) and 1500 m inter-site distance, DL performance, hopping layer simulations.

interference coming from the BCCH frequency adjacent to the hopping pool of frequencies and therefore this could potentially improve the performance. However, this implies the loss of one hopping frequency with the consequent degradation in diversity gain and capacity. This and other strategies, which are displayed in Figure 9.7, are studied in this section.

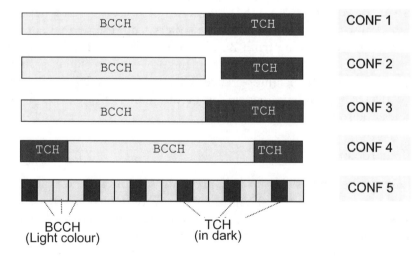

Figure 9.7 BCCH allocation strategies.

Figure 9.8 Network performance for different BCCH allocation strategies. TU3 and 3000 m inter-site distance, DL performance simulations.

Figure 9.8 displays the results for the studied configurations. In this case the staggered case (CONF 5) has an equivalent performance to that with one list with contiguous frequencies (CONF 1, non-staggered). This is due to the larger inter-site distances used, which makes the network to be more coverage limited and the effectiveness of power control is reduced as higher power has to be transmitted to cover the cell area. From the displayed results, it is clear that guardbands between BCCH and hopping layer degrade the overall performance. Finally, the best performing case is CONF 4, even though its gain is only significant in the 3.6 MHz case. This configuration uses one Mobile Allocation (MA) list with frequencies taken from two sub-bands: one at the beginning of the total available band, and another one from the end of such band, which leads to an increase in the effective hopping band (high frequency diversity) with a low impact of the BCCH adjacent interference in the hopping layer.

The main conclusion from this study is that the non-staggered configuration outperforms the staggered one, not requiring a protection guardband between the BCCH and hopping bands. CONF 4 yields the best performance, especially in very narrow deployment scenarios, when frequency diversity degradation is more significant.

9.3 BCCH underlay

As mentioned before, in narrowband scenarios the reduced number of frequencies to be deployed in the hopping layer limits the system performance and capacity. Therefore, the tightening of the BCCH reuse is critical. This section presents a functionality capable of tightening BCCH reuses without degrading the overall performance, hence releasing frequencies to be deployed in the hopping layer and increasing the system performance and capacity. This solution is called *BCCH underlay*, and belongs to the reuse partitioning category presented in Chapter 6.

Traditionally, the limiting factor for the minimum BCCH reuse has been the performance of the traffic channels carried in the BCCH carrier (usually from timeslot (TSL) 1 to 7). As BCCH reuses tighten, the interference distribution will worsen and the performance of these traffic channels will degrade below the acceptable level. On the other hand, common signalling channels carried by TSL 0 of the BCCH carrier show a more robust link performance due to the coding and retransmission schemes they use (see Chapter 11). *BCCH underlay* functionality will deploy a channel allocation strategy that ensures the BCCH traffic channels' performance is adequate. Under these circumstances the new limiting factor for the minimum BCCH reuse will be the common control channels' performance and this will allow the deployment of tighter BCCH reuses.

Description of BCCH underlay concept

BCCH underlay is a non-standardised functionality based on the reuse partitioning concept. Its basic target is to improve both coverage and capacity of radio networks. It ensures that the BCCH carrier is used only by connections which performance can be guaranteed, while the other hopping TRXs not only cover the whole cell area, but increase the effective coverage due to their frequency hopping diversity gain.

Figure 9.9 illustrates the concept. The effective BCCH coverage area can be modified in order to ensure adequate performance of the calls allocated in the

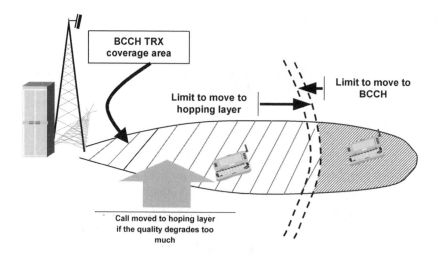

Figure 9.9   BCCH underlay principle.

BCCH carrier while maintaining a high absorption. Therefore, the cell consists of two different logical layers: the BCCH TRX layer with a limited effective coverage area, and a hopping layer that serves the whole cell, providing extended coverage at the edge of the cells.

Finally, the signalling needed to set up and maintain the calls is properly established at the border of the cells thanks to the robustness of signalling channels due to the coding and retransmission schemes used by these channels.

BCCH underlay simulation and trial results

In order to understand the benefits of *BCCH underlay*, an analysis is performed for the scenario of 3.6 MHz, when only 18 frequencies are available. Two cases have been covered: a standard deployment with 12 BCCH and 6 hopping frequencies, and a *BCCH underlay* deployment with 9 BCCH and 9 hopping frequencies. Figure 9.10 presents the results. Note that the EFL used in the *x*-axis corresponds to the total EFL, taking into account both the traffic served by BCCH and hopping TRXs. Only DL performance is displayed.

On its left-hand side, Figure 9.10 displays the performance of the BCCH layer when *BCCH underlay* is used. The BCCH band allocation strategy used is the non-staggered one. Since in this strategy BCCH frequencies are separated from hopping frequencies, the performance of BCCH in DL is almost traffic load independent, especially for the standard case. The *BCCH underlay* case shows an improved BCCH performance even if the number of BCCH frequencies deployed is lower. The right-hand side of Figure 9.10 shows the overall performance benefits of *BCCH underlay*, including both BCCH and hopping layers. The use of *BCCH underlay* provides a capacity gain of 20%.

Figure 9.10   Performance of *BCCH underlay* for 3.6 MHz. TU3 and 1500 m inter-site distance, DL performance simulations.

The gains of *BCCH underlay* are due to the higher number of frequencies available to hop over. The gain of *BCCH underlay* is therefore more important for very narrowband deployment cases, such as the ones used in migration scenarios.

This feature has been tested in trials in Network 6 (see Appendix F). The improvement in performance in terms of drop call rate (DCR) is displayed in Figure 9.11. The network had 22 frequencies available with around two TRXs per cell. The benchmark case did not have hopping in use. By using *BCCH underlay*, BCCH and hopping layers were defined, using both frequencies. The gains over the benchmark case are partly due to the use of hopping. However, it can be seen that the overall performance improves with the use of tighter BCCH reuses, due to the use of the *BCCH underlay* functionality.

According to the performance results presented in this section, *BCCH underlay* is an adequate solution to improve the performance of narrowband deployment scenarios, especially for extreme cases with very few frequencies available to hop over (3.6 MHz), which are likely in migration scenarios. For bandwidths higher than 5 MHz, other solutions, such as the reuse partitioning techniques presented in Chapter 6 provide better performance.

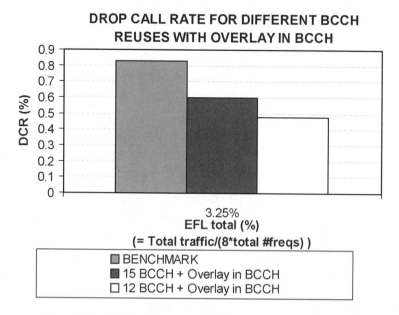

Figure 9.11 DCR performance of BCCH underlay in Network 6.

9.4 Common BCCH

Many operators have access to two or more frequency bands (900/1800 in Europe and Asia Pacific (APAC), and 800/1900 in America). For those operators, and especially if one of the licensed frequency bands is narrow, common BCCH is a competitive solution.

In standard deployment of GSM/EDGE in multiple frequency bands, each band would require a separate BCCH layer thus limiting the overall performance of the system since the BCCH layer is less spectral efficient than the hopping layer. Common BCCH functionality is introduced in the European Telecommunications Standards Institute (ETSI) GSM specifications [2]. The solution consists of the deployment of a single BCCH layer for both frequency bands. The main advantages of the common BCCH functionality are improved trunking gain, higher spectral efficiency, better quality due to decreased handover rate and better neighbour reporting performance. Common BCCH can be configured in only one of the bands of operation when resources across all bands are co-located and synchronised.

9.5 AMR and transmit diversity techniques to tighten BCCH reuse

The robustness of the codecs used with AMR allows GSM networks to work under more interfered radio conditions. Therefore, when the AMR terminal penetration is 100%, tighter BCCH reuses can be deployed, as long as the performance of common control channels does not degrade the overall performance below the required minimum quality. However, most realistic deployments have to count either on high percentages of legacy GSM mobiles (non-AMR capable), or roaming legacy GSM terminals. Chapter 6 provides simulation results analysing the AMR performance of the BCCH layer using different reuses.

Additionally, as explained in detail in Chapter 10, there are some transmit diversity techniques that can improve the DL performance without requiring terminal modifications. These techniques are delay diversity, phase hopping and antenna hopping, and provide similar diversity gains to the ones achieved with frequency hopping. Since diversity gains from different functionality are not cumulative, the gains of these transmit diversity techniques are greater in non-hopping cases, such as the BCCH, reaching 4 dB link level gain in some cases. Both traffic and signalling channel performance can be enhanced and tighter BCCH reuses can be deployed. Finally, some of these techniques may provide some improvement when frequency hopping is used, which may have some relevance when very few hopping frequencies are available, such as in narrowband scenarios, where the frequency diversity gain is limited by the short number of hopping frequencies and the small effective hopping band.

References

[1] Klingenbrunn Thomas, Mogensen P., 'Modelling Frequency Correlation of Fast Fading in Frequency Hopping GSM Link Simulations', IEEE Veh. Technol. Conf., 1999, pp. 2398–2402.
[2] GSM Recommendations 03.26, 'Multiband Operation of GSM 900/1800 by a Single Operator', ETSI, 1998.

10

Link Performance Enhancements

Riku Pirhonen and Matti Salmenkaita

Link performance sets the baseline for network level performance. The basic performance is determined by the GERAN[1] specifications that define the physical channel structure and transmission techniques like channel coding and interleaving format. Apart from the standardised solutions, it is still possible to introduce additional link performance enhancements that are fully transparent from the specification point of view. This chapter contains a review of those techniques together with their associated performance gains.

Section 10.1 presents an example of the basic GERAN link budget calculation without the introduction of any link performance improvement technique. Section 10.2 is an overview of the available techniques, focusing mainly in the uplink (UL) direction. Section 10.3 focuses more in the downlink (DL) direction which is usually the more demanding one, in terms of data throughput and receiver capabilities. Finally in Section 10.5 macrodiversity techniques applied to GERAN are described, system level gain figures are provided and comparison with alternative system level techniques like dynamic frequency and channel allocation (DFCA) is performed.

10.1 GERAN *link budgets*

The exact range of a cellular base station is set by the performance of the

---

[1] GERAN Rel'5 notation is adopted in this section, although results are fully applicable to previous GSM releases.

particular base station and mobile equipment transmitters and receivers, the service in question and the propagation and interference environment. This chapter will discuss the link budget in terms of maximum allowed path loss, which in turn can be used to evaluate practical ranges for a given environment.

Minimum performance requirements

Minimum performance requirements are defined for the mobile and the base station as part of the standardisation process. For GERAN these are described in the 3rd Generation Partnership Project (3GPP) GERAN specification series [1]. The minimum performance is specified for different services, radio channel

Table 10.1   Minimum sensitivity requirements for TU50 (Ideal Frequency Hopping) channel on the GSM 850 and 900 band[a]

Service	Error rate			Receiver sensitivity level − 102 dBm	
	FER	Class 1b RBER[b]	Class II	Base station	Mobile
TCH/AFS12.2	2.4%	1.5%	–	− 104 dBm	− 102 dBm
TCH/AFS10.2	0.85%	0.15%	–	− 104 dBm	− 102 dBm
TCH/AFS7.95	0.045%	0.032%	–	− 104 dBm	− 102 dBm
TCH/AFS7.40	0.069%	0.016%	–	− 104 dBm	− 102 dBm
TCH/AFS6.7	0.017%	0.022%	–	− 104 dBm	− 102 dBm
TCH/AFS5.9	< 0.01%	0.001%	–	− 104 dBm	− 102 dBm
TCH/AFS5.15	< 0.01%	0.001%	–	− 104 dBm	− 102 dBm
TCH/AFS4.75	< 0.01%	0.001%	–	− 104 dBm	− 102 dBm
TCH/AHS7.95	20%	2.3%	5%	− 104 dBm	− 102 dBm
TCH/AHS7.40	16%	1.4%	5.3%	− 104 dBm	− 102 dBm
TCH/AHS6.7	9.2%	1.1%	5.8%	− 104 dBm	− 102 dBm
TCH/AHS5.9	5.7%	0.51%	6%	− 104 dBm	− 102 dBm
TCH/AHS5.15	2.5%	0.51%	6.3%	− 104 dBm	− 102 dBm
TCH/AHS4.75	1.2%	0.17%	6.4%	− 104 dBm	− 102 dBm
PDTCH/MCS-1	BLER ≤ 10%			− 103 dBm	
PDTCH/MCS-2	BLER ≤ 10%			− 101 dBm	
PDTCH/MCS-3	BLER ≤ 10%			− 96.5 dBm	
PDTCH/MCS-4	BLER ≤ 10%			− 91 dBm	
PDTCH/MCS-5	BLER ≤ 10%			− 97 dBm	− 94 dBm
PDTCH/MCS-6	BLER ≤ 10%			− 94.5 dBm	− 91.5 dBm
PDTCH/MCS-7	BLER ≤ 10%			− 88.5 dBm	− 84 dBm
PDTCH/MCS-8	BTS BLER ≤ 10%, MS ≤ 30%			− 84 dBm	− 83 dBm
PDTCH/MCS-9	BTS BLER ≤ 10%, MS ≤ 30%			− 80 dBm	− 78.5 dBm

[a] Handset value for 'small handset' category.
[b] RBER = residual bit error rate.

conditions, frequency bands, mobile speeds, frequency hopping operation mode and various equipment classes. Furthermore, in the type approval test process there are individual tests defined for channel noise, co-channel interference and adjacent channel interference cases for mobiles and base stations. The performance requirements are jointly approved in the standardisation process by the operators, regulators and equipment manufacturers. This minimum required performance as such is not necessarily the point where a network is operated. The test cases are defined so that the piece of equipment that passes the tests operates according to the baseline on which operators base their network deployment and end users can trust that this equipment purchased for use in the cellular system will perform as expected.

The various test parameters stated above make a large set of test case combinations. As an example, Table 10.1 shows sensitivity requirements for the environment described with the typical urban (TU) channel model. These specific test cases for adaptive multi-rate codec (AMR) and packet data assumes use of frequency hopping, mobile speed of 50 km/h and global system for mobile communications (GSM) 850 or 900 MHz band.

All enhanced general packet radio system (EGPRS) mobiles have to support incremental redundancy (IR), which is a type II hybrid automatic repeat request (H-ARQ) retransmission scheme designed to improve data throughput for non-delay critical data. The performance of this feature is tested in MCS-9 mode in static channel at -97 dBm noise level. In these conditions the long-term throughput per timeslot, when measured between the logical link control (LLC) and radio link control (RLC)/medium access control (MAC) layers, has to exceed 20 kbit/s.

The practical base stations typically exceed the requirements by a pretty large margin while mobile handset performance tends to be closer to the specified requirements. The performance numbers stated in various sources for any product are usually given as 'typical' or 'guaranteed', the latter taking into account possible variations in performance introduced in the manufacturing and installation processes.

Link budget

In order to define the cell range a link budget for the connection between the base station and the mobile needs to be calculated for uplink and downlink directions. Link budgets take into account the transmission power, losses at the transmitter and receiver end, antenna gains and propagation path loss. In order to calculate a link budget for a fading radio connection and define the slow fading margin, one needs to define the desired percentage of the area where service is to be available at given probability as well as some basic environment related factors such as path loss exponent and area correction factors specific for the propagation model.

Table 10.2 Receiver sensitivity

A	Background noise	kT	− 174.0 dBmW/Hz	
B	Noise bandwidth	B	52.6 dB Hz	180 kHz
C	Noise factor	NF	5 dB	Estimate
D	Noise[a]	N_0	− 116.4 dBm	D = A + B + C
E	Signal to noise ratio	E_b/N_0	6 dB	Specific for the service
F	Receiver sensitivity		− 110.4 dBm	F = D + E

[a] At the receiver filter output.

The maximum allowed path loss is defined by the receiver sensitivity, transmitter power and additional gains and losses that the transmission path has. In interference-free conditions the receiver sensitivity is limited by the natural thermal noise, which can be calculated from the receiver temperature T and Boltzmann's constant k. The receiver filter defines the receiver bandwidth and thus the noise bandwidth B. The noise generated by the receiver itself can be factored in by the noise factor, NF. These terms define the noise level in the receiver. By calculations, measurements or simulations one can define what signal-to-noise ratio is required for a particular service to provide desired bit error probability. Table 10.2 summarises these parameters and shows an example calculation for receiver sensitivity.

A more detailed link budget calculation will take into account the specific base station configuration and actual environment. The detailed base station configuration factors in losses due to connectors and cables, impacts of duplexer and combiner units depending on the number of transceiver units per sector, gains from directive antennas and so on. Table 10.3 shows a more detailed link budget calculation for two AMR codec modes as well as three packet data rates. Data rates are given as per time slot, so e.g. a multislot mobile capable of supporting 3 time slots would be able to provide three times to data rates in this table.

Table 10.3 gives the maximum allowed path loss in dB. For any particular environment the actual range can then be calculated with a propagation model tuned for that environment. Differences in path loss and area correction factors can introduce significant differences in the cell ranges, so in this table it is not feasible to compare any ranges in terms of kilometres.

Traditional voice service has required a balanced link budget that provides the same range for up- and downlink directions. In a balanced link budget mobile and base station can keep up the connection to the same point and once one direction starts to loose the connection the other direction would not need to keep it up either. From Table 10.3 it can be seen that the basic link budget has better performance in the downlink direction. In modern deployments uplink performance is further improved by 4–5 dB with help of diversity reception that will be discussed later.

Table 10.3 Voice and data service link budgets

	AMR 12.2		AMR 5.15		16 kbit/s/1 TSL		32 kbit/s/1 TSL		48 kbit/s/1 TSL	
	Uplink	Downlink	Uplink	Downlink	Uplink	Downlink	Uplink	Downlink	Uplink	Downlink
Transmitter										
TX power[a]	30 dBm	45 dBm	30 dBm	45 dBm	26 dBm	43 dBm	26 dBm	43 dBm	26 dBm	43 dBm
Combiner losses	0	2.4 dB	0	2.4 dB	0	2.4 dB	0	2.4 dB	0	2.4 dB
Cable and connector losses	0	2 dB	0	2 dB	0	2 dB	0	2 dB	0	2 dB
Body loss	3 dB	0	3 dB	0	0		0	0	0	0
TX antenna gain	0 dBi	15 dBi	0 dBi	15 dBi	0 dBi	15 dBi	0 dBi	15 dBi	0 dBi	15 dBi
Receiver										
Antenna gain	15 dBi	0 dBi	15 dBi	0 dBi	15 dBi	0 dBi	15 dBi	0 dBi	15 dBi	0 dBi
Body losses[b]	0 dB	3 dB	0 dB	3 dB	0 dB	0 dB	0 dB	0 dB	0 dB	0 dB
Link										
SNR requirement	7 dB[c]	7 dB[c]	2.2 dB	2.2 dB	8.5 dB	8.5 dB	15.6 dB	15.6 dB	24 dB	24 dB
Receiver sensitivity[d]	− 109.4	− 104.4	− 114.4	− 109.2	− 107.9	− 102.9	− 100.8	− 95.8	− 92.4	− 87.4
Max. allowed path loss	151.4	157	156.2	161.8	148.9	156.5	141.8	149.4	133.4	141.0

[a] 8-PSK transmission is assumed to require 2 dB back-off in base station and 4 dB back-off in the mobile.

[b] Body loss is estimated at 3 dB for voice services due to the position of the mobile. The assumption is that for data services the terminal will be operated in position that does not introduce body loss.

[c] Estimated performance for 1% FER in TU3 multipath channel with frequency hopping.

[d] Base station receiver sensitivity is background noise level − 116.4 dB plus required signal to noise ratio in terms of energy per symbol to noise ratio (E_s/N_0). For the mobile assumption is 5 dB weaker implementation.

The path loss related availability of service is measured by location probability. The service is available over the cell area with a given probability. For example, one can require that there is 95% probability that the frame error rate of the voice service is less than 1% over the cell area. In order to guarantee this one additional slow fading margin is used in the link budget. The margin depends on the path loss factor and standard deviation of the fading. If the fading standard deviation is 7 dB and path loss factor 3.5 dB then the fading margin would be 7.3 dB.

For data services one can take a bit different approach. Depending on the quality of the service requirements, pretty big variations in the data rate can be tolerated. Non-real-time application users can cope with lower data rates if there happens to be a fade in the radio conditions, while during good channel conditions the throughput can go up with the better signal-to-noise ratio. Adaptation to the link conditions is implemented in GERAN by a number of modulation and coding schemes, link adaptation between these and retransmissions with the incremental redundancy. This results that even if voice service as such is pretty low bit rate, down to below 10 kbit/s, the same user can get significantly higher packet data throughput in terms of kbit/s at the same location. Also the uplink to downlink imbalance is not that big of a problem for data services as the assumption is that there is probably need for higher data rates in the downlink direction.

10.2 Overview of link performance enhancements

GERAN utilises many advanced techniques to carry bits over the radio connection in the most efficient and robust way. Improvements to the radio performance can be introduced by carefully studying the properties of these techniques. When operators are upgrading their equipment for the new generation of services there is a good opportunity to introduce some of these enhancements that improve the performance for both the new as well as the old services.

Frequency hopping

Frequency hopping is a standard feature of GSM and has been in use in the networks for several years. Together with channel coding and interleaving, power control and discontinuous transmission frequency hopping makes GSM a very efficient spread spectrum system that can distribute the load and utilise diversity of the total operator bandwidth. Frequency hopping mitigates frequency selective fading by changing transmission frequency burst by burst. In addition to the improvement on single link frequency hopping it also gives system level gains by randomising interference.

Table 10.4 Required performance for TCH/AFS 12.2

TCH/AFS 12.2	TU3 nFH	TU3 iFH	TU50 nFH	TU50 iFH
FER	22%	3.5%	6%	3.5%

Frequency hopping gives highest gains against fading for slow mobiles moving at speeds below 30 km/h on the 900 MHz band. Faster mobiles will see changing channel conditions through their mobility and as such do not need and will not gain against frequency selective fading, but they do still gain in terms of interference diversity. Frequency hopping performance depends on channel coherence bandwidth. In indoor conditions the channel is close to flat fading, which means that even a relatively large change in frequency does not bring different channel conditions. However, in outdoor conditions correlation bandwidth is smaller and, e.g. with the TU channel profile there is not much correlation left beyond the first adjacent channels.

For simulation and testing purposes two extreme conditions are used. In the non-hopping case (nFH) the frequency hopping is not on and consecutive bursts are transmitted on the same channel with time variation modelled. The other case is ideal frequency hopping (iFH), in which case a new multipath channel is created separately for each burst without any correlation between consecutive bursts. Table 10.4 shows the required performance for 12.2 kbps voice service. Figure 10.1 compares the simulated results of the TU3 cases to the required

Figure 10.1 Frequency hopping gain in TU3 environment, 900 MHz band. The type approval FER requirements are marked with horizontal lines.

performance from Table 10.4. Based on these results, the simulated receiver would pass the tests at around 6 dB SNR.

Uplink diversity reception

In order to compensate the imbalance between uplink and downlink path loss it has been a common practice to have diversity reception at the base station for the uplink direction. By using multiple antennas one can collect more energy and already this improves the signal-to-noise ratio. For additional diversity gain two or even four antennas are used in a configuration where they receive the signals with as small correlation as possible. This is achieved by using enough horizontal spatial separation between the antennas or antennas with different polarisations. Four-way diversity can be achieved by using two spatially separated cross-polarised antennas.

The signal received by diversity branches is combined at the receiver, preferably with maximum ratio combining (MRC). A MRC receiver weights the signals from each antenna based on the received signal-to-noise ratio before the combining operation, which provides best gain in signal-to-noise ratio.

Figure 10.2 shows the gains achievable by two- and four-way diversity reception. These results represent case where frequency hopping is used and the channel profile is typical urban at 50 km/h.

Exact gains for the diversity reception will depend on the environment and the service. In Figure 10.2 the weaker channel coding scheme MCS 8 gains more from diversity than the more robust MCS-5. As MCS-5 has pretty strong channel coding frequency hopping provides already diversity which can be utilised

Figure 10.2   Maximum ratio combining performance for two-way and four-way diversity reception.

Figure 10.3   Interference rejection combining performance for two- and four-way diversity reception.

through channel coding. Therefore there is about 2 dB gain on top of the theoretical 3 dB antenna gain increase that two receiver antennas provide. MCS-8 in its turn has weak channel coding so frequency hopping does not provide that much gain. Therefore the gain from multipath diversity contributes more than in the MCS-5 case to the performance.

Uplink interference rejection

When the limiting factor for system capacity is man-made interference, typically coming from co-channel and adjacent channel users, there is additional performance improvement that can be achieved by interference rejection methods. Each of the antennas receives not only the desired signal but also the interferers at different levels. In the case of interference rejection combining (IRC) the signals from multiple antennas are first decorrelated before the combining operation. Thus IRC provides in addition to diversity gain for the desired signal also damping of interference.

Figure 10.3 shows interference rejection performance for octagonal phase shift keying (8-PSK) modulation as function of carrier to interference ratio (C/I) and block error rate. The results are for TU3 channel with frequency hopping.

Interference rejection combining is very powerful method to reduce the harmful effects of interference and improve the system capacity. It would obviously be most advantageous if interference rejection could be used also in the downlink direction and thus improve the congested downlink. Joint detection schemes have been studied for this purpose for GMSK modulation [2]. For best performance in 8 PSK case this would require two antennas and receiver chains in the mobile, which would introduce additional complexity and cost.

10.3 Downlink enhancements

On the downlink direction delay diversity, phase hopping and antenna hopping are very attractive schemes for enhanced data rates for global evolution (EDGE) as they do not require any changes to the current mobile receivers designed for single antenna transmission. This not only simplifies the receiver design but also allows the legacy GSM mobiles to benefit from diversity gain. Another class of diversity techniques includes space-time block codes (STBC). STBC would be a promising candidate for GERAN systems as it could outperform delay diversity, but it requires changes to the GERAN system standard both at the base transceiver station (BTS) and mobile terminal side.

In most of the cases the GERAN networks are interference limited in the downlink direction. Any downlink performance enhancement is therefore directly translated into system capacity gains. Appendix C shows how link level gains are translated into system capacity gains under different network conditions. In this chapter transmitter configurations with different number of simultaneously transmitting transmitters are compared. In real implementation a dual transmitter will have double the transmission power and thus has about 3 dB link budget advantage over a single transmitter implementation. However, in the comparisons in this chapter the results are presented in a normalised way, which means that the total transmitted power is assumed to be the same regardless of the number of transmitter units.

Delay diversity

Delay diversity refers to a transmission scheme in which delayed copies of the modulated signal are transmitted over multiple antennas (Figure 10.4). By using sufficient antenna spacing or cross-polarised antennas at the base station, inde-

Figure 10.4 Transmitter with delay diversity.

pendent fading can be realised on the delayed signal components. The receiver constitutes an equaliser implemented using Maximum Likelihood Sequence Estimation (MLSE) or mobile application part (MAP) detection. Delay diversity is an especially attractive choice for a GSM/EDGE like system where an equaliser is already present in the receiver and it can potentially realise the desired diversity benefit. Most of the diversity gain can be reached using just two antennas.

Delay between the two antenna branches reduces the risk of total cancellation of received signal and the deep fades that this cancellation could cause. If delay diversity is used on a carrier that does not employ frequency hopping additional phase hopping can be introduced to randomise the fades. Phase hopping rotates the phase of the second antenna to induce time variations within the received signal. The phase shifting causes variations from burst to burst and introduces frequency hopping type of behaviour, which offers diversity gain in the form of interleaving gain when a strong channel code is employed.

Discrete phase hopping is well suited for GSM/EDGE burst structure. A constant phase offset is applied on the second antenna on a burst by burst basis. If the phase offset is kept constant for the whole burst span, the equivalent channel experienced by the traffic data and the training symbols is the same. The channel estimator estimates the composite channel and the equaliser operates as in the single antenna case. Hopping can be done periodically or randomly. In periodic hopping the coded data are interleaved over a number of bursts, say N, the phase on the second antenna can be changed in steps of $360/N$ deg so that we can assure 360 deg phase rotation during one radio block. Assigning full 360 deg phase rotation maximises the interleaving distance within the radio block. In the case of random hopping a random phase sample is applied to each burst. The phase value is selected from a Mary Phase Shift Keying (MPSK) constellation with uniform sampling. Since random hopping does not guarantee maximal interleaving depth for a given radio block it cannot outperform periodic hopping.

Antenna hopping

Antenna hopping is a scheme that can be used to improve link performance on carriers where frequency hopping is not in use or frequency hopping is not effective due to high correlation between the frequencies (Figure 10.5). This would be the case, e.g. in an indoor environment where delay spread is short and thus the channel is flat fading. In antenna hopping the signal transmission alternates between antennas burst by burst. As a result of these variations additional diversity gain can be realised with interleaving and channel coding. Since the received structure is unchanged relative to the single antenna case, the receiver designed for single antenna frequency hopping works also with antenna hopping. In general, phase hopping and antenna hopping are very similar in

Figure 10.5   Antenna hopping principle.

principle but phase hopping has an additional 3.0 dB power increase available through simultaneous transmission through multiple antennas.

The amount of gain that is achievable with antenna hopping depends on the correlation between the propagation paths. The main contributor is the correlation between the antennas. In environments with less multipath, such as rural outdoor, larger separation is needed between the antennas than, e.g. in an indoor environment where the angular spread of the signal is greater. Different polarisations give good, low correlation, but the signal level difference between the branches may in some environments decrease the gain. In order to get interference diversity frequency hopping can be combined with antenna hopping even in a flat fading environment.

As antenna hopping does not give additional gain in environments where frequency hopping already provides the diversity, the comparisons are feasible only in the non-hopping case. Figures 10.6 and 10.7 compare the gains that can

Figure 10.6   Downlink enhancement performance comparison for voice service (normalised power).

Figure 10.7 Downlink enhancement potential comparison for broadcast control channel (normalised power).

be seen over a single antenna non-hopping case for AMR 12.2 service and broadcast control channel performance.

Comparison of different services and different deployment cases would be very long. Table 10.5 summarises some results which compare downlink diversity techniques for voice services. The two AMR modes in this comparison are half-rate 7.4 kbit/s codec (AHS/7.40) and full-rate 12.2 kbit/s codec (AFS/12.2). Three different cases for frequency hopping are used. The non-hopping case (nFH) does not have frequency hoping at all. The reduced frequency hopping (6FH) case refers to the narrowband deployment scenario with only six consecutive frequencies to hop over. The ideal frequency hopping

Table 10.5 Downlink diversity gain for two AMR modes at 1% frame error

Gain (dB)		nFH	6FH	iFH
AHS/7.4 Kbps	One antenna	–	2.6	4.3
	AH	1.5	3.4	4.6
	DD + PH	3.0	5.4	5.9
AFS/12.2	One antenna	–	6.6	8.7
	AH	4.7	7.4	8.5
	DD + PH	4.3	8.2	9.2

(iFH) case corresponds to hopping over wider bandwidth with low probability of using the same frequency more than once per interleaving period. Results are given for the single antenna, antenna hopping and delay diversity case. Again, delay diversity results are normalised, so the 3 dB power gain that transmission from two antennas with equal power as transmission with a single antenna would give is not included in these values. Monte Carlo simulations have been run to get these estimated gains at 1% frame error rate (FER). Results should be seen as indicative as they are not accurate down to the last 0.1 dB. However, it can be seen from the results how full-rate speech service gains more from frequency hopping due to its larger amount of channel coding. On the other hand, half-rate speech service gains from the delay diversity as half-rate mode has relatively less channel coding and shorter interleaving depth.

10.4 Macrodiversity for GSM

Macrodiversity refers to a spatial diversity where the reception is performed simultaneously at two or more base station sites. The large physical separation of reception sites means that macrodiversity can combat slow fading, as the propagation paths can be largely uncorrelated for shadowing effects. In principle macrodiversity can also be used to combat fast fading, but local spatial or polarisation diversity implemented within one BTS site is usually sufficient for this purpose.

In GSM the first requirement for the implementation of macrodiversity is air interface synchronisation. Synchronisation is needed in order to achieve time-slot alignment that is required in order for the reception to be possible in multiple sites. Air interface synchronisation is presented with more detail in Chapter 8.

Uplink macrodiversity

Macrodiversity resembles soft handover in code division multiple access (CDMA) systems in a sense that when macrodiversity is applied the user is connected to multiple base stations simultaneously. It is used for users in marginal field strength or carrier/interference (CIR) condition where the slow fading effects can cause significant degradation of link quality. When such a condition is detected, the diversity reception is activated in one or more surrounding BTSs for which the field strength and CIR are found to be acceptable. The main received signal and the diversity signals are combined in the BSC, see Figure 10.8. The combining can be realised as a simple selection combining based on signal quality criteria or a more advanced scheme such as multiply detected macrodiversity (MDM) [3] or log-likelihood ratio (LLR)

Figure 10.8   Uplink macrodiversity.

[4] can be used. However, macrodiversity is not a replacement for a handover. A power budget handover to another BTS should happen normally when the network has observed that the path loss from another BTS is permanently less than the path loss from the serving BTS. More complete introduction to macrodiversity can be found in [5].

If macrodiversity is used, more receivers have to be installed to provide the extra hardware capacity. The required increase in the hardware capacity is directly proportional to the percentage of users for which macrodiversity is being applied at any one time and the average number of base stations involved in the reception of one user. Naturally, a transmission link to the base station controller (BSC) must also be provided for each extra receiver.

In order to effectively follow the shadow fading conditions the reaction time for the application of macrodiversity must be fast enough. Typically the radio link conditions must be evaluated at least every slow associated control channel (SACCH) period of 0.48 s but even more frequent evaluations are needed for optimum performance. This reaction time requirement doesn't leave room for the possibility of performing changes in the radio channel parameters such as frequency, timeslot or timing advance in conjunction with the application of macrodiversity. Thus, macrodiversity is transparent to the mobile station and changes in the radio channel configuration are not permitted when macrodiversity is applied. The fact that the timeslot cannot be changed leads to increased probability of macrodiversity blocking even if the BTS hardware capacity has been increased proportionally to the share of users applying macrodiversity.

Downlink macrodiversity

In principle macrodiversity could also be applied to the downlink direction. Whenever the user is subjected to marginal field strength or CIR condition, the downlink diversity transmission could be initiated in another BTS simultaneously with the uplink diversity reception. The user would then have two or more downlink signals available each having different propagation paths, which therefore increase the probability that the signal strength available to the user is sufficiently high despite of slow fading. Figure 10.9 represents schematically the downlink macrodiversity operation. Downlink macrodiversity can be based on either a simultaneous downlink transmission from multiple base stations or it is possible to use a switching principle where the downlink transmission is always performed from one BTS only. The BTS can be selected based on the path loss measurements reported by the mobile station.

Downlink macrodiversity has some problems and limitations that are not present in uplink macrodiversity. In the uplink direction the BTS receiver in a host BTS has time to adjust itself before the diversity reception is started whereas in downlink the mobile receiver has no prior warning of the application of downlink macrodiversity. From the mobile point of view the diversity transmission looks like additional multipath signal components. In order for them to be beneficial in reception, the diversity signal and its main multipath components must fit within the equalisation window of the mobile receiver. In GSM mobiles this window is 18.45 μs corresponding to about 5.5 km propagation distance. Therefore, it should be ensured that the propagation path distance differences should be clearly below 5 km. In the case of switching downlink macrodiversity the propagation distance difference introduces a

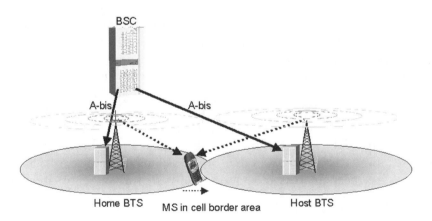

Figure 10.9 Downlink macrodiversity.

sudden change in downlink signal timing whenever a switch occurs. This is similarly likely to cause problems in the mobile receiver therefore limiting the maximum allowed propagation path distance difference. Another problem with the downlink macrodiversity may arise if the mobile station is moving. The movement introduces a Doppler shift in the received frequency that is corrected in the mobile receiver. It can happen that the speed vector is opposite relative to the host BTS than it is relative to the home BTS. This can be especially problematic if switching downlink macrodiversity is used as the mobile would see a sudden frequency change whenever the downlink transmission is switched to another BTS. As downlink macrodiversity involves fast and unpredictable changes in the points of downlink transmission while keeping the radio channel unchanged, the scheme is incompatible with DFCA that is presented in Chapter 8.

Macrodiversity simulations

The performance of a switching macrodiversity scheme has been evaluated by network level simulations. In these simulations the uplink received signal level was the trigger for a macrodiversity switch. The link was transferred to another base station whenever the uplink path loss was found to be lower in the other BTS. The uplink path loss evaluation was performed every 120 ms. If a switch was needed, it was performed without any additional delay. In this simulation both uplink and downlink were switched based on uplink path loss measurements. Power budget handovers were performed normally whenever the 6 dB path loss hysteresis was satisfied in a periodical check (Table 10.6).

Table 10.6 Simulation parameters

Cell layout	3-sectorised hexagonal
Site to site distance	1500 m
Frequency	900 MHz
Propagation model	COST-Hata
Standard deviation of slow fading	6 dB
Path correlation factor	0.15
Voice activity factor	50%
Power control UL and DL	RXLEV and RXQUAL
Power budget HO margin	6 dB
Power budget HO averaging	10 SACCH periods
Frequency plan	1/1 hopping
Speech codec	EFR
Macrodiversity evaluation and switching period	120 ms
Number of candidate BTSs for macrodiversity switching	6
FER sampling period	1 s

Figure 10.10   Uplink FER performance with 3 and 50 km/h MS speeds.

The simulation results in terms of FER performance are shown in Figures 10.10 and 10.11. The graphs show the share of FER samples that indicate higher than 8% FER during the past second. This share of 'bad' FER samples is referred to as an outage. This benchmark has been selected in order to capture even short periods of time when the user would have been expected to experience bad speech quality.

As can be expected, the results show that macrodiversity provides more gain with fast moving mobiles. This is due to the fact that fast moving mobiles move across cell border areas more frequently and they are therefore able to benefit more from macrodiversity. With fast moving mobiles the gain of downlink macrodiversity is clearly larger than that of uplink macrodiversity. With slow moving mobiles the uplink gain disappears in low load situations and in the downlink losses can be observed.

The drawback of downlink macrodiversity is that since the downlink transmission is switched away from the home BTS without changing the radio channel, it brakes the frequency plan and causes sudden changes in the interference situation. For example, in the 1/1 reuse case with random frequency hopping a switch of downlink transmission to a BTS in another site causes

Figure 10.11   Downlink FER performance with 3 and 50 km/h MS speeds.

intracell co-channel collisions because the switched connection is using a different hopping sequence number than the rest of the traffic in the cell. With slow moving mobiles and low loads this results in overall loss.

The impact of AMR speech codecs with more robust channel coding is expected to reduce macrodiversity gain. With AMR the effective frequency reuse distance can be extremely low leading to very high frequency load. With such high loads the intracell co-channel collisions are more frequent when the uplink reception or downlink transmission point is changed to another BTS using different hopping sequence number. This can cause a significant increase in the interference level limiting possibilities to apply macrodiversity reception.

The simulated macro-diversity gain is summarised in Table 10.7.

Table 10.7   Macro diversity gain summary (limiting link in parenthesis)

0.5% outage		1% outage		2% outage	
3 km/h	50 km/h	3 km/h	50 km/h	3 km/h	50 km/h
−30% (DL)	N/A	0%	29% (DL)	14% (DL)	32% (UL)

10.5 Conclusions

Macrodiversity sytem level gains are very dependent on mobile speed, becoming marginal if the users are slow moving. The assumption that all users are fast moving can't be maintained in most of capacity limited scenarios. On the contrary, most of the users tend to be slow moving and the macrodiversity gain turns to be quite limited.

As downlink macrodiversity involves transmitting the signal from two or more base stations either simultaneously or in a switched manner, there are some limitations related to timing and Doppler shift that can complicate the implementation. Further complications can be expected related to the shared GPRS channels and the transmission of the uplink state flag. The implementations of location services that rely on timing measurements performed by the mobile stations can also be problematic if the point downlink of trasmission is allowed to be changed during the location measurement. Furthermore, downlink macrodiversity cannot be used with DFCA that provides superior interference management and performance gains as presented in Chapter 8.

References

[1] 3GPP TS 45.005 Radio Transmission and Reception Specifications.
[2] Ranta, P.A.; Honkasalo, Z.-C.; Tapaninen, J, 'TDMA cellular network application of an

interference cancellation Technique', Vehicular Technology Conference, 1995 IEEE 45th , Volume: 1, 1995 Page(s): 296 -300 vol.1

[3] Haas Z. J., Li C.-P., 'The Multiply-detected Macrodiversity Scheme for Wireless Cellular Systems', IEEE Trans. Veh. Technol., Vol. 47, May 1998, pp. 506–530.

[4] Valenti M. C., Woerner B. D., 'Iterative Multiuser Detection, Macrodiversity Combining and Decoding for the TDMA Cellular Uplink', IEEE J. Selected Areas Commun., Vol. 19, No. 8, August 2001, pp. 1570–1583.

[5] Weiss U., 'Designing Macroscopic Diversity Cellular Systems', IEEE 49th Veh. Technol. Conf., Vol. 3, 1999, pp. 2054–2058.

11

Control Channels Performance and Dimensioning

Timo Kähkönen and Jorge Navarro

The performance analysis of GSM/GPRS control channels presented in this chapter supplements the performance analysis of preceding chapters where the focus has been on the traffic channels.

The first part covers possible synchronisation and signalling problems in demanding radio conditions. This can be an additional problem that degrades indirectly end user quality. This analysis is necessary when considering system coverage and capacity enhancements that use link adaptation techniques. On the traffic channels such as adaptive multi-rate (AMR) speech coding, which allows lower carrier-to-noise (C/N) and carrier-to-interference (C/I) ratios.

In the second part, the signalling capacity of different control channel configurations is evaluated. The main objective is to find the required configuration for cells with certain voice and data traffic loads. This analysis will provide guidelines for proper control channel dimensioning.

11.1 Introduction of control channels

Physical and logical channels

GSM allocates the physical resources of the air interface, i.e. frequency channels and timeslots, in order to control information by using two levels of channels:

physical and logical. Physical channels are described both in frequency and time domain in such a way that one channel occupies one fixed timeslot in every time division multiple access (TDMA) frame but may change the frequency channel according to its frequency hopping sequence (see Chapter 6) [1]. A physical channel can be allocated to several logical channels according to a multiframe structure consisting of 26, 51 or 52 TDMA frames. The logical channels divide into traffic and control channels that are used either in circuit or packet switched mode. Normal traffic channels (TCH) carry circuit switched speech or user data, and packet data traffic channels (PDTCH) packet switched user data. The control channels are used for synchronisation, radio link control and upper layer signalling procedures. In this chapter, the control channels that have been introduced for packet switched services are referred to as packet control channels, and those that were included in the original GSM specifications are called GSM control channels. The PDTCH and packet control channels are dynamically multiplexed on physical channels that are called packet data channels (PDCH) while the TCH and GSM control channels are mapped into fixed timeslots.

According to the use of radio resources the GSM control channels can be divided into the following three groups: broadcast channels, common control channels (CCCH) and dedicated control channels. The broadcast channels are the frequency correction channel (FCCH), synchronisation channel (SCH) and broadcast control channel (BCCH) on which base stations continuously broadcast frequency reference, time and system information. The common control channels are the random access channel (RACH), paging channel (PCH), and access grant channel (AGCH). They are used to establish a radio resource (RR) connection, after which the network can privately change signalling and radio link control messages with the mobile station on the dedicated control channels. The dedicated control channels are the standalone control channel (SDCCH), fast associated control channel (FACCH) and slow associated control channel (SACCH). The BCCH, RACH, PCH and AGCH have counterparts in packet control channels whose abbreviated names are PBCCH, PRACH, PPCH and PAGCH. Other packet control channels are the packet notification channel (PNCH), packet associated control channel (PACCH) and packet timing advance control channel uplink/downlink (PTCCH/U, PTCCH/D).

Transmission on the logical channels is based on four types of bursts. Frequency correction bursts are transmitted on the FCCH and synchronisation bursts on the SCH. On the RACH, PRACH and PTCCH/U only access bursts are used. On other channels normal bursts are used with two exceptions: on the FACCH and PACCH access bursts are used to transmit the first message in the uplink when the timing advance required for a transmission of a normal burst is unknown. On the control channels modulation is always Gaussian minimum shift keying (GMSK).

BCCH channel
Downlink   F=FCCH, S=SCH, I=idle

F	S	BCCH	CCCH	F	S	CCCH	CCCH	F	S	SDCCH 0	SDCCH 1	F	S	SDCCH 2	SDCCH 3	F	S	SACCH 2,0	SACCH 3,1	I

Uplink   R=RACH

SDCCH 3	R	R	SACCH 2,0	SACCH 3,1	R R R R R R R R R R R R R R R R R R R R R R R R	SDCCH 0	SDCCH 1	R	R	SDCCH 2

——————— 36 TDMA frames ———————————→←——— 15 TDMA frames ———→←

Figure 11.1   The combined configuration of GSM control channels.

Control channel configurations

The GSM control channels are mapped on to physical channels according to a 51-multiframe structure. The physical channel that carries the broadcast channels is called a BCCH channel. It always occupies timeslot 0 at a fixed carrier of the cell, and the broadcast channels use 14 frames in each of its 51-multiframe (a frame includes one timeslot in a TDMA frame). The remaining frames are allocated to the common and dedicated control channels, for which the operator can use either a combined or non-combined configuration depending on the number of carriers in the cell. With the combined configuration, the BCCH channel includes four SDCCH subchannels with their SACCHs and three blocks of four timeslots for the CCCH (Figure 11.1). In the downlink the PCH and AGCH share these CCCH blocks on a block by block basis. The operator can, however, reserve a fixed number of blocks for the AGCH that may not be used as the PCH. With the non-combined configuration all the blocks left from broadcast channels are reserved for the CCCH, and one or more additional physical channels must be allocated for the SDCCH usage (Figure 11.2).

BCCH channel
Downlink   F=FCCH, S=SCH, I=idle

F	S	BCCH	CCCH	F	S	CCCH	CCCH	F	S	CCCH	CCCH	F	S	CCCH	CCCH	F	S	CCCH	CCCH	I

Uplink   R=RACH

R R R R R R R R R R R R R R R R R R R R R R R R R R R R R R R R R R R R R R R R R R R R R R R R R R R

Any physical channel
Downlink   I=idle

SDCCH 0	SDCCH 1	SDCCH 2	SDCCH 3	SDCCH 4	SDCCH 5	SDCCH 6	SDCCH 7	SACCH 4,0	SACCH 5,1	SACCH 6,2	SACCH 7,3	I	I	I

Uplink   I=idle

SACCH 5,1	SACCH 6,2	SACCH 7,3	I	I	I	SDCCH 0	SDCCH 1	SDCCH 2	SDCCH 3	SDCCH 4	SDCCH 5	SDCCH 6	SDCCH 7	SACCH 4,0

——————— 36 TDMA frames ————————————→←——— 15 TDMA frames ———→←

Figure 11.2   The non-combined configuration of GSM control channels.

Additional physical channels can of course be allocated for the SDCCH also when the combined configuration is used. In the uplink, the RACH has 27 timeslots in the combined configuration, and all the 51 timeslots in the non-combined configuration.

An SDCCH subchannel transfers a complete data link layer (L2) frame in its block of four timeslots in every 51-multiframe, whereas its SACCH occupies only every other 51-multiframe. As indicated in Figs. 1 and 2, an SDCCH subchannel has always an offset between the downlink and uplink frames making it possible to receive the response for a transmitted L2 frame before the next transmission possibility and thereby minimise transmission delays. The network has a longer response time of 36 TDMA frame periods compared with 15 of mobile stations because L2 frames carry network layer (L3) messages, which have transmission delays between base transceiver station (BTS), base station controller (BSC) and mobile services switching centre (MSC).

(E)GPRS can use the existing BCCH and CCCH for broadcast and common signalling or accommodate packet common control channels (PCCCH) with the PBCCH on PDCHs. A PDCH is independently shared in the downlink and uplink directions according to a 52-multiframe structure, which is divided into 12 radio blocks of four consecutive timeslots. The first one to three blocks of one PDCH that contains the PCCCH are allocated to the PBCCH in the downlink. Other downlink blocks are allocated on a block by block basis for the PAGCH, PNCH, PPCH, PACCH or PDTCH, the actual usage being indicated by the message type. The network can, however, define a fixed number of blocks in addition to PBCCH blocks, which may not be used as the PPCH. On an uplink PDCH containing a PCCCH, all 12 blocks in a multiframe can be used as the PRACH, PACCH or PDTCH. The operator can optionally reserve a fixed number of blocks that can only be used as the PRACH.

The FACCH, SACCH and PACCH share the same physical channel with their traffic channels, except when one-block PACCH allocation is used in a physical channel containing the PCCCH during a PDTCH allocation. An L2 frame transmitted on the FACCH steals the bits of one 20-ms speech frame on a full-rate TCH and of two 20-ms speech frames on a half-rate TCH. The transmitter indicates this to the receiver by setting the stealing flag bits to one.

Usage of control channels

The purpose of the FCCH and SCH is to enable mobile stations to synchronise to the network both in idle and dedicated mode. All the bits that modulate a frequency correction burst on the FCCH are zero, and thus by the characteristics of Gaussian minimum shift keying (GMSK) modulation, transmitted bursts are periods of sine waves. By detecting these bursts a mobile station can find the BCCH channel of the cell and remove the oscillator frequency offset with respect to the base station. During or after adjustment it attempts to decode SCH information contained in a synchronisation burst. SCH information includes the current reduced TDMA frame number and the identity code of the base station. Once a mobile station has successfully decoded SCH information it starts to decode system information on the BCCH. System information includes the channel allocation in the cell, the BCCH frequencies of neighbour cells and a number of control parameters for the subsequent communication within the cell. Having this information the mobile station can decode the CCCH, transmit on the RACH and also synchronise to the neighbour cells for the cell reselection. In cells where PCCCH exists, packet data specific system information is transmitted on the PBCCH otherwise it is also transmitted on the BCCH.

Two basic types of signalling procedures using common and dedicated control channels are illustrated in Figure 11.3. In mobile originating procedures the mobile station starts a RR connection establishment by sending channel request on the RACH. This 8-bit message can be repeated on 3–50 access bursts to increase decoding success probability in bad radio conditions [2]. As several mobile stations may simultaneously attempt to transmit their channel requests, a Slotted Aloha protocol is applied on the RACH. Once the network has received channel request, it reserves an SDCCH subchannel for subsequent signalling and sends immediate assignment on the AGCH. If all the SDCCH subchannels are reserved network queues the request or rejects it by sending immediate assignment reject. In mobile terminating procedures, the network calls a mobile station by broadcasting paging request on the PCHs of the cells within the location area of the mobile station. By the receipt of paging request the called mobile station tries a random access as in mobile originating procedures. While using the SDCCH a mobile station, being in dedicated mode, regularly receives and transmits radio link control messages on the SACCH. The messages contain power control and timing advance commands in the downlink, and measurement reports and the current Tx-power level in the uplink.

The SDCCH usage depends on the signalling need, which can be a location update, a call establishment, or a short message service (SMS) message transfer.

Figure 11.3 Basic signalling procedures for circuit switched services.

Figure 11.4 Acknowledged mode signalling on the SDCCH and FACCH.

On the SDCCH, L2 frames are transmitted according to a send-and-wait automatic repeat request (ARQ) protocol as illustrated in Figure 11.4 [3]. Response frames carry acknowledgements, which can also piggyback on information frames when L3 messages are transmitted in consecutive blocks in the uplink and downlink. The first pair of frames establishes and the last pair releases an L2 acknowledged mode link.

Once a traffic channel has been allocated, the FACCH can be used for signalling. The principal purpose of the FACCH is to carry handover signalling, but it can also be employed in call establishments. As on the SDCCH, L2 frames on FACCH are transmitted in acknowledged mode. While a traffic channel is in use its SACCH carries radio link control messages every 0.48 s in both directions.

The (P)CCCH subchannels are used in (E)GPRS to establish a RR connection between a mobile station and the network for signalling, or a unidirectional temporary block flow (TBF) for data transfer. The establishment procedures are presented in Figure 11.5 [4]. While a mobile station has a TBF it is in a packet transfer mode; otherwise it is in a packet idle mode, see Chapter 1. The establishment procedures follow the principles described above for circuit switched

Figure 11.5 Basic signalling procedures for packet switched services.

services except that in a mobile terminated access the paging need depends on the mobility management (MM) state. A mobile station enters a MM READY state when it starts to transmit or receive data, i.e. when it changes to packet transfer mode. It stays in this state until it stops sending or receiving data, i.e. until it changes to packet idle mode, and a certain time elapses, after which it enters a MM STANDBY state. In the MM READY state, a mobile station performs cell updates, and no paging is needed. In the MM STANDBY state, the network knows only the routing area of the mobile station and it has to send packet paging request in all the cells within the routing area in order to contact to the mobile station. Once a TBF is allocated the PDTCH is used for data transfer and the PACCH for polling and acknowledgement messages.

Channel coding and interleaving

For the detection and correction of transmission errors, channel coding and interleaving are performed for the signalling messages at the physical layer as outlined in Figure 11.6. A message includes 8 bits on the RACH, either 8 or 11 bits on the PRACH, and 18 bits on the SCH. Other channels transfer 184 bits in a message, which is an L2 frame (link access protocol for the Dm channel (LAPDm) frame in GSM and radio link control (RLC) block in (E)GPRS) [5].

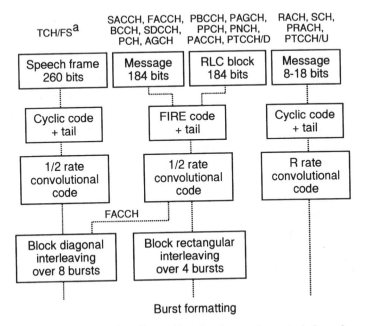

Figure 11.6 Channel coding and interleaving on the control channels.

[a]GERAN Rel'5 terminology is used. This corresponds to a full-rate TCH.

On all the control channels, a half-rate convolutional code is used for error correction. A small exception in regard to the code rate is the PRACH, on which 6 coded bits are not transmitted in the case of an 11-bit message. The RACH and PRACH use a cyclic code of 6 parity bits and the SCH uses a cyclic code of 10 parity bits for error detection. Other channels use a FIRE code with 40 parity bits, for enabling error correction besides error detection.

The BCCH, SDCCH, PCH, AGCH, and the packet control channels except PRACH and PTCCH/U, carry a message in four normal bursts in consecutive TDMA frames. As these channels have the same channel coding and interleaving, their physical layer reliabilities are equal. When the SACCH is associated with an SDCCH subchannel it is identical to the above channels, but when it is associated with a TCH/FS it has a longer interleaving over four bursts in consecutive 26-multiframes. As the RACH, SCH and PRACH carry messages in single bursts they do not have interleaving. The GSM full-rate speech channel (TCH/FS), which is used as a reference channel in Section 11.2, uses a cyclic code of 3 parity bits for the 50 most significant bits produced by the speech encoder during a period of 20 ms. For error correction, the bits are protected with a half-rate convolutional code, and the coded bits are interleaved over 8 normal bursts. In the case of full-rate speech, the FACCH has the same interleaving as the TCH. While in the case of half-rate speech the interleaving is over 4 bursts on the TCH and over 6 bursts on the FACCH.

11.2 Physical layer reliability

In this section, the physical layer reliabilities of control channels are compared on the basis of radio link simulations. The principal measure of physical layer reliability is the frame erasure rate (FER), which gives the ratio of detected, uncorrectable, erroneous frames on a logical channel in certain radio conditions. The upper layer protocols must take into account the possibility of undetected errors, but it is small enough to be discarded in this analysis. The focus is on coverage-limited conditions, where a radio link has a low C/N and experiences fading. In interference-limited conditions, where a radio link has a low C/I, also the unwanted signal varies in time and therefore error rates drop more slowly when the average signal quality improves. As this holds for all the logical channels, their respective performances are similar regardless of whether coverage- or interference-limited conditions are considered. Only GSM control channels are named in the analysis, but as one can see from Figure 11.6 results also apply to packet control channels.

Simulation model

The physical layer simulator that was used performs channel coding, interleaving, burst formatting and modulation according to specifications. Modulated bursts are transmitted over a typical urban (TU) channel model [6], which brings about multipath fading and time spread. In the simulations a carrier frequency of 900 MHz and mobile speeds of 3 and 50 km/h (TU3 and TU50 channels) were used. The results apply, for example, to the 1800 MHz band when the speeds are halved. With the same speeds reliabilities improve with the carrier frequency when interleaving is applied and frequency hopping is not used. When frequency hopping is used, it is assumed to be ideal so that the radio channel is uncorrelated between bursts. In the case of coverage-limited links the bursts are corrupted by independent Gaussian noise samples according to a given C/N per bit (E_b/N_0) value. In the case of interference-limited links a continuous co-channel signal is transmitted over a similar, uncorrelated radio channel and added to the wanted signal before reception. The simulated receiver follows a realistic implementation with floating-point arithmetics, adaptive channel estimation and 16-state log-MAP equaliser without antenna diversity. The FER was calculated over 20 000 frames on the SCH and RACH and over 5000 frames on the other channels.

Comparison of channels

The frame erasure rates in the TU3 channel are shown in Figure 11.7. In this channel, the mobile station moves half a wavelength in about 40 TDMA frames,

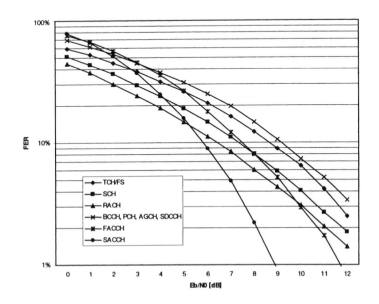

Figure 11.7 Frame erasure rates in the TU3 channel with noise (no FH).

and the signal power is almost unchanged during the shortest interleaving period of 4 TDMA frames. Thus, the signal power varies between frames on the channels like BCCH using this period, and the frame erasure rate decreases slowly as the signal quality (E_b/N_0) improves. The same applies also to the SCH and RACH, on which frames are not interleaved. Due to the longest interleaving period, the frame erasure rate drops most rapidly on the SACCH when E_b/N_0 improves. The TCH/FS and FACCH are intermediate cases that slightly benefit from the interleaving. Because the TCH/FS has a shorter frame length and poorer error detection than the FACCH, it has lower frame erasure rates at low values of E_b/N_0. On the other hand, the FACCH performs better at high values of E_b/N_0 due to its FIRE coding, which enables the receiver to correct errors that get through the convolutional code.

Figure 11.8 shows the results for the TU50 channel. Now half a wavelength takes only 2.4 TDMA frames and the number of bursts over which frames are interleaved has more significance than the interleaving period. Therefore, TCH/FS and FACCH have lowest error ratios above an E_b/N_0 of 4 dB.

By comparing Figure 11.7 with Figure 11.8 one can see that on all the channels, except the SACCH, SCH and RACH, physical layer reliability improves when the mobile speed increases. This is due to increased time diversity. The TU channel provides some frequency diversity within signal bandwidth due to its time spread, but more effective frequency diversity is achieved with frequency hopping, except the channels that are mapped on the BCCH channel. This can be seen from the frame erasure rates in the TU3 channel with ideal frequency hopping shown in Figure 11.9. To indicate the difference between E_b/N_0 and C/I

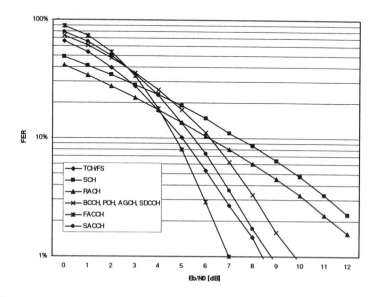

Figure 11.8   Frame erasure rates in the TU50 radio channel with noise (no FH).

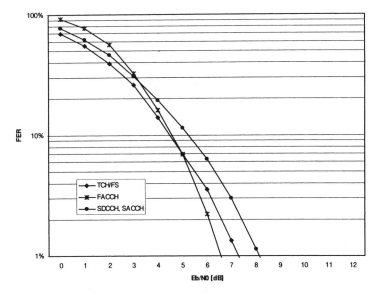

Figure 11.9 Frame erasure rates in the TU3 channel with noise (ideal FH).

results Figure 11.10 shows frame erasure rates in the TU3 channel with co-channel interference and ideal frequency hopping.

Because the FCCH is not decoded, it is not as simple a reliability measure as other channels. The algorithms that use the FCCH for finding the BCCH channel and correcting the frequency offset are implementation dependent, but usually they consist of burst detection, offset estimation and frequency correc-

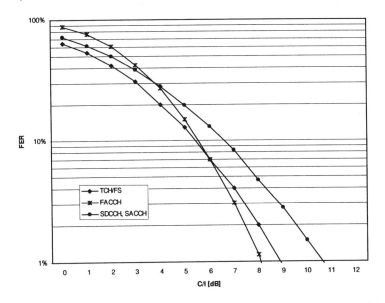

Figure 11.10 Frame erasure rates in the TU3 channel with co-channel interference (ideal FH).

tion. The convergence time of frequency correction depends on the initial frequency offset and the quality of received bursts. During initial synchronisation a mobile station receives over 20 frequency correction bursts per second, which is more than enough in practical algorithms even if the E_b/N_0 were 0 dB. When a mobile station needs to synchronise to a base station while it has a connection to another base station, it can receive the FCCH only every 0.6 s [1], on average. In that case the convergence time may be several seconds. The frequency offset that remains after convergence depends on the algorithm and C/N. Its effect on the detection of bursts increases with the burst length so that the loss in C/N is most considerable on the channels using the normal burst. The receiver can, however, employ a frequency correction algorithm also in the burst detection to reduce the effect.

11.3 Signalling reliability and delays

In this section transmission reliability and delays of control channels are analysed taking the data link layer into account besides the physical layer. For simplicity it is assumed that the radio channel is statistically stationary during signalling procedures that take only a few seconds at maximum (i.e. that the time average and correlation of C/N are constants). The message flows of the L3 procedures that are analysed have been obtained from signalling traces of a real network. The focus is on the signalling of circuit switched services, but the results that consider broadcast and common signalling also apply to (E)GPRS.

Probabilistic models

On the broadcast channels, L3 messages are transmitted repeatedly without acknowledgements at the data link layer. Assuming that the radio channel is uncorrelated between transmissions, the success probability after N-transmissions is

$$1 - p^N \tag{1}$$

where p is the frame error probability. By discarding undetected errors the frame erasure rate obtained from link simulations can be used as an estimate of p for a given average C/N. Thus, the success probability is a function of time and C/N. Transmissions can be assumed uncorrelated when the mobile stations move half a wavelength or more between them. Otherwise, the frame error probability depends on the success of previous transmissions, in which case the success probability can be estimated from the frame error sequences logged in physical layer simulations.

A successful RR connection establishment on the common control channels requires that paging request, channel request and immediate assignment are all decoded successfully. The network repeats paging request after a certain time without receiving paging response. Assuming the average C/N is constant in both directions and transmissions are uncorrelated, the probability of a successful establishment is

$$1 - (1 - (1 - p)(1 - (1 - P_k(1 - p))^{M+1}))^L \tag{2}$$

where p is the frame error probability on the PCH and AGCH, P_k is the decoding success probability of the channel request that is spread into k bursts (3–50), M is the maximum number of retransmissions for channel request (1, 2, 4 or 7) and L is the maximum number of paging attempts (not specified).

As described in Section 11.1, the data link layer increases transmission reliability with a send-and-wait ARQ protocol on the SDCCH and FACCH. That is, an information frame is retransmitted until both the frame itself and its acknowledgement are received correctly. Suppose that a signalling procedure consists of K information frame transmissions when all the frames are correctly received, and that a separate response frame carries the acknowledgement of each information frame. Then, the success probability of the procedure after N information frame transmissions ($N \geq K$) is

$$1 - \sum_{i=0}^{K-1} \binom{N}{i} ((1 - p_u)(1 - p_d))^i (1 - (1 - p_u)(1 - p_d))^{N-i} \tag{3}$$

when the frame error probabilities in the uplink and downlink are constants p_u and p_d. This requires that the average C/N is constant and all transmissions are uncorrelated. The expected number of information frame transmissions is then

$$\frac{K}{(1 - p_u)(1 - p_d)}. \tag{4}$$

Formulas (3) and (4) are inapplicable when acknowledgements piggyback on the information frames, which is possible if L3 messages are alternately transmitted in the uplink and downlink and they fit to single L2 frames. In this case probabilities and average times can be estimated with simulations. Nevertheless, (3) gives the lower limit of the success probability and (4) the upper limit of the expected number of transmissions.

The restrictions on the number of retransmissions that protocol specifications determine for noticing link failures can make signalling procedures fail. In an L2 acknowledged mode link establishment, the maximum number of retransmissions is five [3]. Since the establishment consists of one information frame and its response frame, the success probability is

$$1 - (1 - (1 - p_u)(1 - p_d))^6. \tag{5}$$

If the establishment fails, for example, when starting a call it blocks the call. After link establishment the number of retransmissions is limited to 23 on the SDCCH, and to 34 on the FACCH [3]. When these limits are reached a link failure is indicated to L3, which releases the link. At the other end the link is released after a timer expiry. For one information frame transmission, the probability of link failure is

$$(1 - (1 - p_{u})(1 - p_{d}))^{N} \tag{6}$$

where $N = 24$ for the SDCCH and $N = 35$ for the FACCH. When a signalling procedure comprises K information frame transmissions, and separate response frames carry acknowledgements, the probability of failure becomes

$$1 - (1 - (1 - (1 - p_{u})(1 - p_{d}))^{N})^{K}. \tag{7}$$

In other cases the link failure probability can be evaluated by simulation. For example, in the call establishment on the SDCCH link failure blocks the call, and in the handover signalling on the FACCH it may drop the call.

Formulas (2)–(7) take into account that a radio link can be in imbalance, in which case frame error probabilities are different in the downlink and uplink. When considering the limiting direction, a balanced link sets, the lower limit of C/N or C/I since improvement in one direction allows degradation in the other.

SCH information broadcast

A base station broadcasts synchronisation information on the SCH with an interval of 10 TDMA frames, or 46 ms. When the mobile speed is over 10 km/h, roughly, transmissions are uncorrelated and the decoding success probability is given by formula (1). In the TU3 channel, correlation between transmissions increases decoding delay as Figure 11.11 shows. The trade-off between the delay and E_{b}/N_{0} requirement is significant: repetitions increase the redundancy and time diversity, which lowers the E_{b}/N_{0} requirement, but the lower the E_{b}/N_{0} the longer the transmission delay.

As the transmission period is short in the initial synchronisation, E_{b}/N_{0} can be lower than 4 dB without significant decoding delay. In the pre-synchronisation for measurements and handovers, a mobile station can attempt to decode SCH information every 0.6 s, on average. Several decoding failures may thus cause a considerable delay though with this period transmissions are uncorrelated in the TU3 channel. For example, E_{b}/N_{0} needs to be 4 dB for that the success probability is close to unity and the decoding delay does not exceed 2.4 s (four transmissions). A delay of a couple of seconds may cause problems, for example, when a high-speed mobile station needs to synchronise to a new cell before a handover.

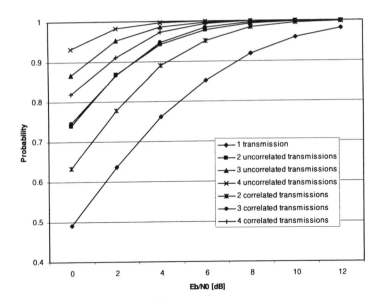

Figure 11.11 The success probability of decoding SCH information (TU3 channel, 900 MHz band).

System information broadcast

The system information that is broadcast on the BCCH is grouped into four message types. As the message types are alternately transmitted, the transmission period of a message type is four 51-multiframe periods (0.94 s), which is enough to make transmissions uncorrelated in the TU3 channel. Figure 11.12 shows the decoding success probability of a message type after 1, 2, 3, 4 and 5 transmissions. Since mobile stations decode the BCCH only in the idle mode, decoding delay is not a serious problem for circuit switched services. In (E)GPRS a lower delay in the BCCH decoding is more important since a mobile station performs cell re-selections during sessions. For example, with five attempts the success probability is close to one when E_b/N_0 is above 4 dB. This also means that the mobile station may be unable to decode the (P)PCH for 5 s when E_b/N_0 is 4 dB or lower in the new cell.

RR connection establishment

The success probability of a mobile terminated RR connection establishment, which is performed on the common control channels, can be evaluated with formula (2). Before that it must be examined how the number of bursts used to spread the channel request transmission impacts on the decoding success probability. Figure 11.13 shows the decoding success probabilities when using 3 uncorrelated bursts, and when using 3 and 10 consecutive bursts in the TU3 channel. It can be seen that in the case of three bursts in the TU3 channel, E_b/N_0

Figure 11.12   The success probability of decoding system information on the BCCH (TU3 channel, 900 MHz band).

must be over 10 dB in order that the success probability of the first transmission is close to one. By using 10 bursts the requirement drops to below 6 dB. The results show that the success probability substantially depends on the

Figure 11.13   The decoding success probability of a channel request message (TU3 channel, 900 MHz band).

correlation between transmissions. This suggests that, in order to maximise both the reliability and capacity of the RACH, using bursts for retransmissions is preferable to spreading single transmission. On the other hand, spreading messages into the maximum of 50 bursts minimises the delay if the capacity is unlimited.

To evaluate the success probability of the whole RR connection establishment it is assumed that channel request is spread into three bursts, the maximum number of retransmissions for channel request is four, and the collision probability on the RACH is negligible. Retransmissions can be assumed to be uncorrelated in the TU3 channel since the minimum retransmission period for channel request is 41 TDMA frames, or about 190 ms, and the minimum sensible period for paging attempts is about 1 s. It is furthermore reasonable to assume that the period between paging request and immediate assignment is of the order of 51-multiframe. Thus, all transmissions can be assumed uncorrelated and the success probabilities for different number of paging attempts are obtained from formula (2).

The success probabilities are shown as a function of E_b/N_0 for the TU3 channel in Figure 11.14. With a maximum of three paging attempts, the success probability is close to unity when E_b/N_0 is above 11 dB. With 5 paging attempts the minimum E_b/N_0 requirement drops to 5 dB, and with 10 paging attempts to 3 dB. The success probabilities are of course smaller when the RACH has a non-zero collision probability, which would decrease the value of P_k in formula (2). When channel request is spread only into three bursts, as assumed above, the collision probability can be, however, assumed negligible on the basis of Section 11.4.

Figure 11.14 The success probability of an RR connection establishment (TU3 channel, 900 MHz band).

Figure 11.15   The success probability of an L2 link establishment (TU3 channel, 900 MHz band).

L2 link establishment

The retransmission period of an L2 frame is 235 ms on the SDCCH and 120 ms on the FACCH. With the shorter period of 120 ms, the correlation between transmissions was found to have negligible impact on the signalling success probability in the TU3 channel. Therefore, transmissions can be assumed uncorrelated. Figure 11.15 shows the success probability of an L2 link establishment in the TU3 channel according to formula (4) with and without frequency hopping. Without frequency hopping establishment failures are possible when E_b/N_0 is below 8 dB, and with frequency hopping when E_b/N_0 is below 5 dB.

L2 link failure

Figure 11.16 shows the probability of an L2 link failure in the TU3 channel with and without frequency hopping for a call establishment and for a handover command transmission. During a call establishment on the SDCCH, an L2 link failure may occur when E_b/N_0 is below 3 dB, and during a handover command transmission on the FACCH when E_b/N_0 is below 2.5 dB. Frequency hopping decreases these limits by 0–1 dB, but on the other hand, it increases the probability of link failures when E_b/N_0 is below 1 dB.

Figure 11.16   The probability of an L2 link failure (TU3 channel, 900 MHz band).

Call establishment and location update

In a call establishment the network can choose between three different channel allocation strategies: (1) off air call set-up, (2) early assignment or (3) very early assignment. These strategies differ in the time when a traffic channel is assigned for the signalling on the FACCH. In strategy (1) the SDCCH is used until the called party answers the call. Because in this strategy the SDCCH is used for longest, it is considered in the following.

Since delays at L3 depend on the signalling needs between MS, BTS, BSC and MSC, the total time of an error-free procedure varies. The simulated case included a total of 13 information frame transmissions on the SDCCH. If these were transmitted only in one direction with the minimum time between transmissions, the total time would be 13×0.235 s $= 3$ s. The changes in the direction and unused frames increase the time to about 3.5 s. In the simulations, frame erasures were randomised according to the frame error ratios at each E_b/N_0 point and the link was assumed to be in balance. The average time of 10 000 simulated procedures with and without frequency hopping in the TU3 channel is shown in Figure 11.17. Link failures are taken into account, so that only successful procedures are included in the average. It can be seen, for example, that without frequency hopping the procedure takes 2.7 times longer at 3 dB, on average, than it would take without frame errors.

Figure 11.17   The average SDCCH usage time in call establishment and location update procedures (TU3 channel, 900 MHz band).

A location update is also signalled on the SDCCH after initial assignment on the RACH and AGCH. The simulated procedure included authentication signalling and temporary mobile subscriber identity (TMSI) reallocation. The average time of 10 000 procedure executions is shown in Figure 11.17. Because location updates cause significant load on the SDCCH, several retransmissions are undesirable. The excess load due to retransmissions is about 20% at 8 dB, 50% at 6 dB and 100% at 4 dB in the TU3 channel without frequency hopping.

Handover and channel transfer

When performing a handover to another cell, the network sends handover command to the mobile station on the FACCH using the old channel. The message assigns the new cell and the channel to which the mobile station shall change its transmission. In the case that is considered, the message fits into two L2 information frames, both of which are acknowledged by a response frame. Thus, formulas (3) and (4) apply with $K = 2$.

Figure 11.18 shows the success probability as the number of retransmitted frames increases from 0 to 15, or the transmission delay increases from 0.24 to 2.04 s. The combination of channel coding and ARQ protocol seems to give sufficient performance, i.e. high reliability and low delay, above 6 dB. According to Figure 11.16 a limit appears at 2 dB below which a link failure may occur.

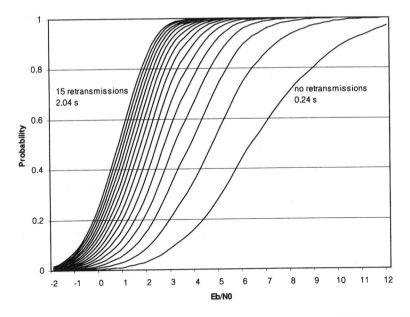

Figure 11.18   The success probability of a handover command transmission (TU3 channel, 900 MHz band, no FH).

However, if C/N drops rapidly, then any additional delay may cause a link failure. Besides transmission of handover command a successful handover procedure requires successes in measurement report reception (see Section entitled *Measurements and power control* in 11.3), pre-synchronisation (see Section entitled *SCH information broadcast* in 11.3) and in the access and link establishment on the new dedicated channel.

When the network needs to transfer a connection to another physical channel in the same cell, it sends assignment command to the mobile station on the FACCH using the old channel. This message fits also into two L2 frames and the success probability is the same as that of a handover command shown in Figure 11.18. The mobile station completes the channel transfer by establishing an L2 link and transmitting assignment complete on the FACCH using the new channel.

Measurements and power control

The most time critical messages carried by the SACCH include measurement reports in the uplink and power control commands in the downlink direction. Since a message is transmitted every 0.48 s, transmissions may be assumed uncorrelated in the TU3 channel. Decoding success probabilities according to formula (1) for 1–5 transmissions are shown in Figure 11.19. A higher frame erasure rate means longer measurement reporting delays in the uplink and longer power control command reception delays in the downlink. With one

Figure 11.19   The success probability of decoding a message on the SACCH (TU3 channel, 900 MHz band, no FH).

decoding attempt the success probability is close to unity when E_b/N_0 is 11 dB. With two, three, four and five attempts the corresponding E_b/N_0 values are 7, 5, 4 and 3.5 dB.

To be able to release the physical channel when the signal is irrevocably lost, the mobile station updates a radio link counter based on the frame erasures on the SACCH [7]. Every correct frame increases the counter by one and every lost frame by two. Therefore, depending on the initial value of the counter, a link failure indication means that the FER has been greater than 66% for several seconds, or on the basis of Section 11.2 that the E_b/N_0 has been lower than 1 dB. The maximum value of the counter is a network parameter, which can take values from 4 to 64 corresponding to a radio link timeout from 2 to 30 s when all frames are lost.

Conclusions

The preceding paragraphs considered probabilities of completing signalling procedures as a function of time and C/N. It was assumed that the short-time average of C/N is independent of time. On the basis of results, it can be concluded that an L2 link establishment on the SDCCH sets the highest E_b/N_0 requirement of 8 dB in the TU3 channel without frequency hopping. The FER with this E_b/N_0 is 13.5%. L2 link establishments are most vulnerable to frame erasures because the number of retransmissions is limited to five and errors may occur in both directions. With ideal frequency hopping the requirement is decreased to 5 dB, which is also met by the other channels if delays of a few seconds are accepted. The GSM full-rate speech channel requires an E_b/N_0 of 7 dB or higher for that the FER is smaller than 1% with ideal frequency hopping in the TU3 channel. When frequency hopping is non-ideal the requirement is higher; without hopping the required E_b/N_0 for a 1% FER is 14.5 dB. Since control channels allow lower signal qualities without signalling failures, using more robust channel coding on the traffic channels can decrease the E_b/N_0 requirement.

11.4 Control channels vs. AMR TCH

As described in Chapter 6, adaptive multi-rate (AMR) speech coding can be used to increase the capacity and coverage of GSM networks thanks to its improved robustness against channel errors in full-rate mode and better voice quality in half-rate mode compared with old GSM speech codecs. The improved error resilience of AMR full-rate speech codecs lowers the C/N and C/I requirements of traffic channels, but the performance of control channels remains unchanged. In this section the performance of AMR full-rate speech traffic channels (TCH/AFS) is compared with the performance of dedicated control channels in order to find if signalling failures or delays at low C/I levels limit the maximum system capacity.

Physical layer comparison

Figure 11.20 shows the physical layer performances of the SACCH, SDCCH and FACCH signalling channels and of the TCH/AFS of different data bit rates. Results are for the TU3 channel with co-channel interference and ideal frequency hopping. According to the figure the TCH/AFS12.2 allows 3 dB lower C/I than the SDCCH and SACCH when the frame erasure rate is 1%. This is mainly due to a half longer interleaving of the full-rate TCH. Due to lower code rates besides longer interleaving the AFS7.4–4.75 are 5–8 dB better than the SDCCH and SACCH. The FACCH, which has the same interleaving as the full-rate TCH, is slightly better than the AFS12.2, but less robust than the AFS of lower bit rates.

   With the same C/I the FER is considerably higher on the dedicated control channels than on the low bit rate AFS traffic channels. But whereas frame erasures decrease the speech quality on the TCH, they increase the signalling delay or failure probability on the control channels that use retransmissions to increase reliability. In order to analyse the lowest acceptable C/I, the question of how much higher frame erasure rate is allowed on the control channels than on the traffic channels has to be resolved.

System level comparison

It was concluded in Section 11.3 that the SDCCH has the highest E_b/N_0 requirement of control channels when the link is in balance and statistically stationary during signalling procedures and it is required that the probability of signalling failures is close to zero (< 0.1%). The FER on the SDCCH corresponding to the E_b/N_0 requirement was 13.5%. According to Figure 11.20 this corresponds to a C/I requirement of 6 dB with ideal frequency hopping in the TU3 channel. As opposed to E_b/N_0, C/I is however uncorrelated between the downlink and uplink, and in the limiting link establishment signalling the C/I requirement of the worse direction is smaller if the FER in the other direction is smaller. For example, if the FER is zero in one direction, the C/I requirement of the other direction drops from 6 dB of the balanced link to 4.5 dB. Thus, the C/I requirement of the SDCCH is 4.5–6 dB depending on the link balance whereas the C/I requirement is independent of the link balance on the TCH.

To evaluate the system level reliability, the part of SDCCH signalling that is successful when the C/I is below 6 dB has also to be taken into account. Thus, the signalling reliability in an interference-limited network needs to be estimated by weighting the success probabilities of a link establishment of different C/I points with the C/I distribution of the network. The cumulative C/I distribution of the downlink shown in Figure 11.21 has been obtained from the network simulation that gave the highest system capacity with frequency hopping and AMR when the FER on the TCH was used as the quality criterion. The quality

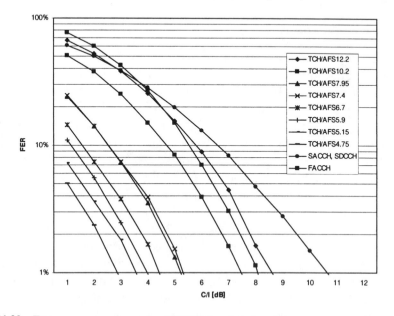

Figure 11.20 Frame erasure rates on the TCH/AFS and dedicated control channels (TU3 channel, ideal FH).

Figure 11.21 The distribution of the C/I average of 0.48 s in an interference-limited network with frequency hopping and the success probability of an L2 link establishment on the SDCCH.

criterion was such that frame erasure rates were calculated over 2 s, the samples with a FER greater than 4.2% were classified as bad, and it was required that no more than 2% of the samples are bad (the same quality criteria was used in Section 6.5). Figure 11.21 also shows the success probability of an L2 link establishment in the TU3 channel with ideal frequency hopping.

The C/I samples are averages over a measurement reporting period of 0.48 s. Using their distribution, the C/I-weighted probability of a link establishment failure was found to be 2.9% when the FER on the uplink and downlink was assumed to be always equal, and 1.7% when one direction was assumed to be always error free. These figures are comparable with the frequency of bad TCH FER samples, which was 2%. Because the C/I samples of network simulation imperfectly correspond to link simulation samples, failure probabilities may differ when frame erasures on the SDCCH are determined during network simulation as has been done for the traffic channels. In addition a link establishment attempt can take 0.94 s (four 51-multiframes) at maximum, i.e. longer than a measurement period, and with a longer averaging period the samples with low C/I become slightly less frequent. Nevertheless, as the impact of these inaccuracies is relatively small, the results clearly show that the SDCCH is not robust enough to be allocated on to frequency hopping channels without limiting the service quality when using AMR to maximise system capacity.

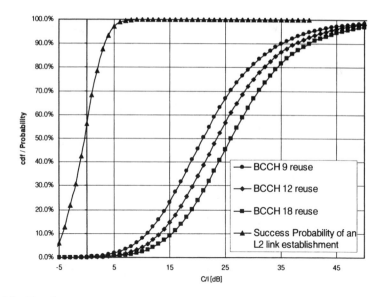

Figure 11.22 The distribution of the C/I average of 0.48 s in interference-limited networks without frequency hopping and the success probability of an L2 establishment on the SDCCH.

To compare the performance of AMR traffic channels and control channels at the BCCH frequencies, the C/I distribution of the downlink of a non-hopping network and frame erasure rates of a non-hopping link need to be used. Figure 11.22 shows the C/I distributions obtained from network simulations with different reuses and the success probability of an L2 link establishment in the TU3 channel without frequency hopping. The AMR TCH qualities of these simulations are presented in Chapter 6.

The calculated failure probability of an L2 link establishment is shown for different reuses in Table 11.1. The upper limit has been obtained by assuming all the links balanced and the lower limit by assuming that the uplink is error free. Because a mobile station moves in network simulations at 3 km/h only about one wavelength in 0.48 s, and therefore multipath fading, which is included in link simulations, slightly widens the C/I distribution, the results are a bit pessi-

Table 11.1 The upper and lower limit for the failure probability of an L2 link establishment in networks with reuse 9, 12 and 18 without frequency hopping

Reuse	Upper limit	Lower limit
9	0.5%	0.1%
12	0.2%	0.0%
18	0.1%	0.0%

mistic. Thus, it can safely be concluded that signalling reliability is acceptable with reuse 9 and failures are almost absent with higher reuses. As the AMR TCH required reuse 12 or higher in order that the same quality criterion was met as on frequency hopping channels, control channels are less restrictive on the BCCH frequencies.

In statistically stationary radio conditions the C/I needs to be below 2 dB (FER = 60%) in order that lost frames on the FACCH can cause a link failure during transmissions of handover and channel assignment commands. If a link failure occurs during the link establishment on the new channel the mobile station shall try to continue on the old channel. Because at a C/I of 2 dB the FER on the AFS/TCH4.75 is moderate (2.4%), handovers and channel transfers must be performed before the speech quality is unsatisfactory. The criterion for stationary conditions is, however, insufficient as lost frames affect power control and handover processes whose one purpose is to maintain acceptable signal quality. Figure 11.23 illustrates the delays of these procedures in error-free conditions. The impact of frame erasures on these delays must be considered to find out possible limitations of the SACCH and FACCH.

Frame erasures on the SACCH mean lost measurement reports in the uplink and lost power control commands in the downlink direction. Thus, they increase the delay of power control in reacting on the changes in the radio signal quality. The Rx-quality thresholds of power control found appropriate in network simulations with AMR, correspond approximately to a target window from 8 to 16 dB for C/I. Therefore, it is reasonable to consider delays at a C/I level a little below 8 dB. For example, at 5 dB the FER on the SACCH is 27% in the TU3 channel with frequency hopping. According to formula (1), with this FER the probability of

Figure 11.23   Delays in power control and handover processes.

loosing more than two consecutive frames is 2% and more than four consecutive frames 0.1%. This means that with a probability of 98% the delay is below 1.5 s and with a probability of 99.9% below 2.5 s. The same reasoning also applies to the power control commands in the downlink. As a mobile station must be able to change its Tx-power 2 dB in every 60 ms after receiving a command, these delays may not considerably decrease the TCH quality.

Lost measurement reports also cause additional delay in the handover process, which is based on the measurements performed by the mobile station as indicated in Figure 11.23. As in the case of power control their impact is small if the lower threshold of Rx-quality is high enough in handover algorithms.

The impact of low signal quality on the transmission delay of the FACCH was considered in Section 11.3. It was noted that the signal quality requirement decreases significantly when short additional delays due to retransmissions are accepted. Since the retransmission period is only 120 ms, the delay due to retransmissions is moderate compared with the whole handover process including the downlink measurements and the averaging of measurement reports in the BSC before the handover decision. For example, with an E_b/N_0 of 4 dB the maximum delay in the transmission of handover command is 2 s in the TU3 channel (see Figure 11.18). This corresponds to a maximum FER of 36% and to a C/I of 3.5 dB with ideal frequency hopping. For example, if handover decision is made when C/I is 8 dB, the handover command transmission is successful unless C/I drops more than 4.5 dB in 2 s, which plausibly is a rare event. One should note that in handover decision the BSC can only rely on the Rx-level of neighbour cells in the downlink. Therefore, in interference-limited networks handover is typically performed if the Rx-quality goes below the lower threshold on the old channel.

Conclusions

L2 link establishment failures on the SDCCH are experienced when the C/I of the limiting direction is below 6 dB in the TU3 channel with ideal frequency hopping. In an interference-limited network, with acceptable quality for 98% of quality samples using the full-rate AMR codecs channels, the link set-up failure probability was estimated to be 1.7–2.9%. Therefore, link establishment failures may limit the service quality more than traffic channels, if the SDCCH is allocated to the hopping channels. This limitation can be avoided by allocating the SDCCH to the non-hopping BCCH frequencies where traffic channels are more limiting than control channels. The frame erasures on the SACCH and FACCH increase delays in power control and handover algorithms. The impacts of these delays on the TCH quality and link failures depend on the time correlation of C/I and on the lower Rx-quality threshold. More accurate evaluation of the impacts requires modelling delays in network simulations.

11.5 Signalling capacity

The way current networks are dimensioned accounts for a certain blocking probability to avoid costly hardware overdimensioning. Thus, the resources are dimensioned to allow a certain blocking probability, which is a design criteria, and the 'hard-blocking' capacity of these resources is defined as the traffic that causes this blocking probability. This section presents a study of signalling channel capacity of typical GSM cell configurations, and includes guidelines about how to optimally configure signalling channels according to the traffic carried by the network. In this section signalling channels include those control channels which load varies in time and are allocated independently from traffic channels, i.e. the broadcast and associated control channels are not considered in this section.

The next subsections describe the criteria used to dimension the signalling channels, as well as the signalling procedures and the methodology used to calculate the capacity of these channels. This capacity is shown for GSM voice calls and (E)GPRS data sessions, and finally some conclusions about channel dimensioning are drawn.

Signalling capacity criterion

The most common way to dimension the traffic channels for voice calls in GSM is dimension the network for a 2% blocking probability, which can be calculated using the Erlang B formulas. However, calls can also be blocked due to lack of signalling resources. Assuming independent blocking probabilities for traffic and signalling channels, the overall blocking experienced by the end users is the sum of both blocking probabilities. Therefore, a criterion is needed in order to obtain the capacity of the channels for signalling purposes. There are some approaches to solve this problem. The first intuitive idea is to use a low blocking probability for the signalling procedures, so all the blocking is caused due to lack of resources for traffic channels. In this way, a blocking probability 10 times less can be used (0.2%). The second criterion is to allow more blocking in order not to overdimension the resources for signalling purposes. This makes it reasonable to check the capacity of signalling channels for a blocking probability of 1 and 2%.

Consequently, the capacity of signalling channels has been obtained for a blocking probability of 0.2, 1 and 2%. As the signalling load depends on the call arrival rate but not on the duration of calls, the capacity of signalling channels is determined as the number of calls per hour that can be supported with a given blocking probability, instead of traffic load.

Signalling capacity for GSM voice

This section describes the methodology to estimate the capacity of GSM signalling channels, as well as the traffic assumptions taken into account. Then, the capacity obtained is exposed and finally some conclusions are presented.

The channels used for call establishments in voice services are the CCCH (common control channel) and the SDCCH (stand-alone dedicated control channel) channels. The number of location updates and the SMS traffic also affect the load and therefore the call establishment capacity of these channels.

The configurations of these signalling channels and the procedures they involved were described in Section 11.1. The procedures can be modelled using queue systems, since if there are no free SDCCH subchannels channel request is queued to await subchannel release. Figure 11.24 shows the model for the RR connection establishment procedure.

Since these queue systems are complex including multiple queues, they have no simple analytic solution. Also, the use of timers and reattempts increase complexity. Therefore, simulation analysis is required in order to study the capacity of these channels taking into account all the queues, collisions in the random access, the maximum number of retransmissions and expiring timers.

Methodology
The procedure used to find out the capacity of the signalling channel is different for SDCCH and CCCH. The procedure for SDCCH consists of oversizing CCCH, so only the load of the SDCCH influence the blocking probability. For the CCCH channel, the capacity of each type of subchannel (RACH, AGCH and PCH) is calculated for a blocking probability of 0.1, 0.5 and 1%

Figure 11.24   Example of signalling procedure modelled with queue systems.

(half than allowed blocking), assuming that the other CCCH subchannels and SDCCH are overdimensioned. This makes it possible to separately compute the capacity of each subchannel. By using half-smaller blocking probabilities than the blocking criteria it can be ensured that the overall blocking probability does not exceed 0.2, 1 or 2%. This is also applied for the PCCCH in Section entitled *Signalling capacity for (E)GPRS* in 11.5.

GSM traffic assumptions

The signalling load in GSM consists of call establishments, SMS messages and location updates (periodic and forced).

The arrival of the voice calls is assumed to follow a Poisson distribution and the duration an exponential distribution with a mean of 100 s. The proportions of mobile originating and mobile terminating calls are assumed to be equal.

In every call the mobile station attempts to access on the RACH by transmitting channel requests until it gets a response from the network or until it reaches the maximum number of retransmissions (1, 2, 4 or 7). The number of slots between transmissions is a random variable with a uniform distribution in a set $\{S, S + 1, ..., S + T-1\}$ where T is the number of bursts used to spread a transmission and S is a parameter (from 41 to 217), which depends on T and channel configuration [4]. In mobile terminated calls a paging message is previously sent on the PCH to the mobile station. Then, if there is a free SDCCH subchannel, immediate assignment is sent on the AGCH. The SDCCH is used 2.8 s in call establishments, 3.5 s in location updates and 3.5 s in SMS transfers.

The SMS message transmissions and periodic and forced location updates are also supposed to be Poisson processes. In these procedures traffic channels are not used, but all the information is carried on the SDCCH. Based on real networks, the following traffic assumptions are taken: 0.3 SMS messages are sent per hour per call, 0.3 forced and 0.25 periodic location updates are executed per hour.

SDCCH capacity

Simulation results show that the blocking probability due to SDCCH subchannels follows a behaviour similar to the Erlang-B curves. However, there are some differences in the region of interest (from 0 to 2%). Figure 11.25 compares the evolution of the blocking probability for some configurations (from 4 to 20 SDCCH subchannels), which was obtained from simulation results, with the Erlang-B curves.

Figure 11.26 exposes the number of calls per hour that the configurations support with a blocking probability of 0.2, 1 and 2%.

Figure 11.25   Evolution of the blocking probability due to SDCCH channel.

SDCCH capacity for different SMS traffic load   This section studies the impact of different SMS loads on the call establishment capacity of the SDCCH, since different operators may have a substantially different SMS load.

The arrivals of calls and SMS messages follow a Poisson distribution, and they use an SDCCH subchannel 2.8 and 3.5 s, respectively. As the SDCCH reservation time is similar for both services, and they follow the same statistical distribution, the capacity of SDCCH depends on its usage time. Assuming that there are x calls per hour, and a SMS messages per call, the SDCCH usage time per hour can be expressed as:

$$T_{\text{SDCCHusage}} = x \cdot 2.8 + a \cdot x \cdot 3.5 \text{ [seconds]}.$$

Figure 11.26   SDCCH capacity for different blocking probabilities.

If the SMS load increases by a factor b, the call rate has to decrease by a factor c to keep the blocking probability. Thus, the SDCCH usage time per hour would be

$$T_{\text{SDCCHusage}} = \frac{1}{c} \cdot x \cdot 2.8 + a \cdot b \cdot x \cdot 3.5 \text{ [seconds]}.$$

Assuming that the blocking probability remains equal for the same SDCCH usage time, this factor c can be calculated by making both expressions equal. The expression obtained is

$$\frac{1}{c} = 1 + \frac{3.5}{2.8} \cdot a \cdot (1 - b).$$

This expression has been verified by means of simulations. Figure 11.27 shows the relationship between the increase of SMS load and the decrease of call rate needed to maintain the same blocking probability.

The following example illustrates how this figure and the previous formulas shall be used. Assuming a blocking probability of 0.2% and 12 SDCCH subchannels, the call arrival rate is around 5000 calls (using Figure 11.26) per hour and 0.3 SMS messages are sent per hour per call ($a = 0.3$). If the SMS load was doubled ($b = 2$) then the reduction factor of the call rate would be $1/c = 0.625$. This means that the voice load shall be reduced with 37.5% in order to maintain the same blocking probability (0.2%) when the SMS load increases 100%. The figure makes this calculation easier. The increased factor of the SMS load is represented in the X-axis, and the curve associates this value to the corresponding reduction factor of the call rate. Hence, the value 2 in the X-axis (b) is associated with the value 0.625 in the Y-axis ($1/c$).

Figure 11.27 Relationship between SMS messages load and calls for a SDCCH capacity.

CCCH capacity

In this section, the capacity of the CCCH subchannels (RACH, AGCH and PCH) is described.

RACH capacity   The RACH channel is used according to a Slotted Aloha protocol for the random access. However, the theoretical blocking formula cannot be used because it doesn't take into account GSM particularities: timers (i.e. time between retransmissions if there is no response from the network), retransmissions (the maximum number of retransmissions is a network parameter, typically 4) and the number of slots used for a message. A typical number of slots is 10, but the results exposed in Section 11.3 show that repetition on consecutive slots is not an efficient way of using resources in slow fading radio channels. Therefore, the chosen value for this parameter was 3.

Figure 11.28 displays the evolution of the blocking probability when the load increases, for combined and non-combined configuration. The capacity with different blocking probabilities are shown in Figure 11.33.

Figure 11.29 exposes how the number of consecutive timeslots used for the transmission of channel request (all timeslots contain the same information) affects to the capacity of the RACH in non-combined configuration, showing the evolution of the blocking probability.

AGCH capacity   With the immediate assignment extended feature, the access grant channel can support two assignments per block, doubling its effective capacity. It is assumed that this feature is used, and the network puts the assignment messages into a queue. In a combined configuration, a

Figure 11.28   Evolution of the blocking probability due to RACH channel.

Figure 11.29 Dependence of the capacity of RACH channel with number of consecutive TSLs.

51-multiframe contains three and in a non-combined configuration nine blocks shared between the AGCH and PCH. The operator can reserve 0–2 blocks in the former and 0–8 blocks in the latter case for the AGCH, which are not used for paging, to ensure a mobile station satisfactory access to the system [3]. Thus, the capacity of the AGCH is computed for all the possible number of blocks reserved for the AGCH. Figure 11.30 presents the evolution of the blocking probability due to the AGCH channel. The capacity depending on the number of blocks assigned to the AGCH channel and the allowed blocking probability is exposed in Figure 11.32.

PCH capacities   There are two paging methods, depending on the type of mobile identification used (IMSI or TMSI), which can provide different number of paging messages per paging block. When using IMSI, only two MS can be paged in a paging message. Instead, up to four MS could be paged in a paging message using the TMSI method. In this section, the average number of paging messages per block is assumed to be 3.

If the network does not receive paging response in a predefined time, it repeats the paging request. The maximum number of retransmissions is a network dependent parameter. The value selected in this study is 3 (see the reliability results in Section entitled *RR connection establishment* in 11.3).

It is assumed that paging requests are produced according to a Poisson distribution, and they are stored in a buffer. The evolution of the blocking probability on the PCH is shown in Figure 11.31, and its capacity is exposed in Figure 11.32.

Figure 11.30 Evolution of the blocking probability due to AGCH channel.

Summary The capacity of the CCCH subchannels is summarised in Figures 11.32 and 11.33 by showing the number of calls per hour that the channels support. All the possibilities for the number of radio blocks assigned to each subchannel are represented in the *X*-axis. The values have been obtained using immediate assignment extended (two assignments per AGCH block) and three paged mobiles per PCH block.

To maximise the capacity of the CCCH, the recommended number of blocks reserved for the AGCH is two in the combined configuration (one block avail-

Figure 11.31 Evolution of the blocking probability due to PCH channel.

Figure 11.32   Capacity of the CCCH subchannels for different blocking probabilities.

able for the PCH), five in the non-combined configuration (four blocks available for the PCH). These distributions have approximately the same capacity for PCH and AGCH channels. In other cases one of these channels would limit the capacity while the other would be overdimensioned. In combined configuration, the RACH channel is overdimensioned with respect to AGCH and PCH channels, but, however, in non-combined configuration the RACH channel has less capacity than AGCH and PCH channels.

Number of TRXs supported
To determine the number of transceivers (TRXs) signalling channels support

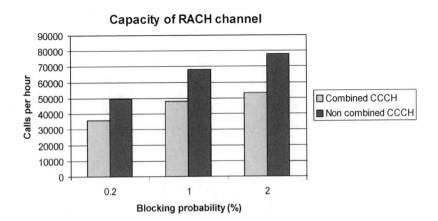

Figure 11.33   Capacity of RACH channel for different configurations.

(with a blocking probability of 0.2%), call arrival rates are translated into traffic loads in Erlangs:

$$\text{Traffic load (Erlangs)} = \frac{\text{Capacity (calls/hour)} \times 100 \text{ (seconds/call)}}{3600 \text{ (seconds/hour)}}.$$

Then, the Erlang-B formula is used to obtain the number of channels needed for a blocking probability of 0.2%. Taking into account that one TRX carries eight physical channels, the number of supported TRXs can be easily calculated by means of the following formula:

$$\text{No. of supported TRXs} =$$

$$\frac{\text{No. of supported traffic channels} + \text{No. of timeslots for signalling channels}}{8 \text{ (channels/TRX)}}.$$

The obtained results are shown in Figure 11.34. They clearly show that the number of physical channels allocated to the SDCCH determines the signalling capacity in GSM networks as the CCCH is overdimensioned.

Signalling capacity for (E)GPRS

As explained in Section entitled *Control channel configurations* 11.1, there are two possibilities for initiating a packet data transfer in (E)GPRS: using the existing CCCH channel or the new packet common control channel (PCCCH). In this section, the signalling capacities of these two options are evaluated.

(E)GPRS traffic assumptions

In order to obtain the blocking due to (E)GPRS data sessions, the following types of communications are considered: email downloading, web browsing

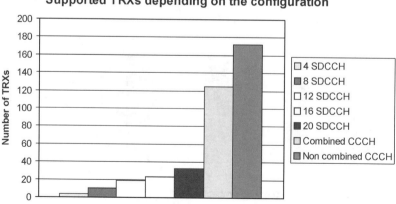

Figure 11.34   Supported TRXs depending on the configuration.

and wireless application protocol (WAP). The traffic of data sessions includes these types with the same probability of occurrence (33.3% for each type of session). The traffic sources send data by using a unidirectional TBF connection between the mobile station and network. The signalling procedures for TBF establishments, which depend on the data source (the mobile or the network) and on the mobility management state of the mobile, explained in Section 11.1.

Also, the measurements report messages sent for cell reselection are considered, because they represent a high percentage of the load on signalling channels. They are transmitted on (P)CCCH when the mobile is in the MM READY state and in packet idle mode, that is, when the mobile is not transmitting data and the READY timer has not expired. The reporting periods are determined by network parameters (NC_REPORTING_PERIOD_T for packet transfer mode and NC_REPORTING_PERIOD_I for packet idle mode).

Other messages like attach, detach, routing area update, and cell update messages are not taken into account because they represent a negligible percentage of the load on (P)CCCH compared with the previous ones. The signalling on the associated channels (PACCH) and the data traffic are not taken into account although they can be transmitted in the (P)CCCH timeslots, because they have lower priority than other (P)CCCH signalling and thus they have no impact on the blocking.

GSM and (E)GPRS sharing CCCH

This section analyses the remaining capacity of the CCCH that can be used for signalling in GPRS without increasing the blocking of GSM calls. This free capacity is obtained for the maximum GSM traffic load of a large cell with 12 TRXs. Using the Erlang-B formula, this load, calculated for a blocking probability of 2% due to traffic channels congestion, is 83.2 Erlangs. Assuming that the duration of a typical call is 100 s, this value can be translated to calls per hour:

$$\frac{83.2\ \text{Erlangs} \times 3600\ \text{seconds/hour}}{100\ \text{seconds/call}} = 2995.2\ \text{calls/hour}.$$

Both the combined configuration and non-combined configuration are considered. On the basis of previous simulation analysis, the signalling capacity for (E)GPRS sharing the CCCH does not depend on the GSM load, as the signalling load on the CCCH due to voice services is very low.

Combined configuration   Figure 11.35 presents the evolution of the blocking probability when the number of (E)GPRS sessions increases. The capacity that can be used for the data sessions with this configuration is shown in Figure 11.37.

Figure 11.35   Blocking probability for (E)GPRS sharing CCCH.

Non-combined configuration   The blocking probability for the combined configuration is shown in Figure 11.36 and the values of the capacity with this configuration are shown in Figure 11.37.

Summary   Figure 11.37 exposes the remaining capacity that can be used by (E)GPRS communications with the previous two configurations.

Figure 11.36   Blocking probability for (E)GPRS sharing CCCH.

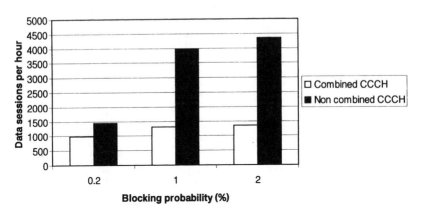

Figure 11.37   Capacity of CCCH for (E)GPRS communications.

*Number of TRXs supported*As it was commented above, the number of TRXs that the CCCH supports for (E)GPRS services is nearly independent of the TRXs used for voice services as their impact on the CCCH load is low.

With the traffic models used, the average amount of data per session is 445 kbits. In order to get a first approximation of the number of additional TRXs that can be supported for (E)GPRS sessions, it is assumed that the TRXs are transmitting/receiving all the time and using CS-2 (with a maximum throughput equal to 12 kbps). With this assumption, one TRX can transmit 290 Mbits per hour. Therefore, one TRX can carry 650 sessions per hour, and so one CCCH supports 1–2 TRXs with the combined configuration and 2–7 TRXs with the non-combined configuration. The exact value depends on the allowed blocking probability.

(E)GPRS on PCCCH

As described in Section 11.1 (E)GPRS capable cells can accommodate a PCCCH on one or more packet data channels with a maximum of 12 available blocks in a 52-multiframe. This section evaluates the capacity of this option.

PRACH capacity   The capacity of the PRACH was calculated for all possible numbers of blocks (1–12). As for the RACH, the maximum number of retransmissions used is 4. Figure 11.38 presents the evolution of the blocking probability for different numbers of blocks reserved for the PRACH.

PAGCH capacity   As above, all possible values of blocks assigned to the PAGCH channel are taken into account. As opposed to GSM, immediate

Figure 11.38 Evolution of the blocking probability due to PRACH channel.

assignment extended messages cannot be used in (E)GPRS, and thus only one message can be carried per block. Figure 11.39 exposes the blocking probability of the PAGCH as a function of the number of data sessions per hour.

PPCH capacity In (E)GPRS mobile stations are identified by either IMSI, TMSI or P-TMSI and depending on the identifier from one to four mobile stations can be addressed in a paging message. As with the PCH, the average number of paged mobiles per paging message is assumed to be 3. The evolution of the blocking probability is shown in Figure 11.40.

Summary The capacity of the PCCCH subchannels for different blocking probabilities is shown in Figure 11.41.

Number of TRXs supported The recommended distribution of blocks between the PAGCH and PPCH channels is such that it makes the capacity of these channels equal. Thus, according to Figure 11.41, seven blocks should be reserved for the PAGCH in which case the remaining four blocks are available for the PPCH channel. Since it was approximated above that one TRX can carry about 650 sessions per hour, with this block allocation, one PCCCH supports from six to nine additional TRXs dedicated entirely for (E)GPRS.

Figure 11.39 Evolution of the blocking probability due to PAGCH channel.

Comparison between CCCH and PCCCH

Figure 11.42 compares the capacity of the PCCCH and CCCH. It can be seen that the PCCCH has more capacity than the CCCH. However, using non-combined configuration and a blocking probability of 2%, the capacity of these two channels is similar. Figure 11.42 shows the capacity of both options for (E)GPRS using the recommended configurations.

Conclusions

For GSM networks the SDCCH channel is the limiting signalling channel since the CCCH subchannels are overdimensioned for standard TRX configurations.

Figure 11.40 Evolution of the blocking probability due to PPCH channel.

Figure 11.41   Capacity of PCCCH subchannels (data sessions per hour).

For a blocking probability of 0.2%, a combined configuration, with 2 blocks for AGCH and 1 for PCH, supports 125 TRXs, a non-combined configuration, with 5 blocks for AGCH and 4 blocks for PCH, supports 172 TRXs. With the same blocking probability 4 SDCCH subchannels support 4 TRXs, 8 SDCCH subchannels 10 TRXs, 12 SDCCH subchannels 19 TRXs, 16 SDCCH subchannels 23 TRXs, and 20 SDCCH subchannels 33 TRXs. A typical high loaded cell, which must support high GSM load, has 12 TRXs, requires 8 SDCCH subchannels (i.e. one physical channel dedicated to SDCCH). The numbers of blocks for AGCH and PCH and the number of SDCCH subchannels depend of course on the traffic assumptions used.

The dimensioning of the location area has a high impact on the number of paging and forced location update messages. With a larger location area, the

Figure 11.42   Comparison between the capacity of CCCH and PCCCH.

number of forced location update messages decreases, but the number of paging messages increases because there are more subscribers in the location area. Hence, the appropriate proportion of blocks for AGCH and PCH channels depends on the dimensioning of a location area, which is a compromise between the signalling load due to paging messages and the signalling load due to location updates.

(E)GPRS signalling procedures can use CCCH or PCCCH channels. Assuming cells with 12 TRXs dedicated for GSM speech, the CCCH with combined configuration supports one TRX for (E)GPRS with a blocking probability of 0.2. The number of supported TRXs increases to two when using the CCCH with non-combined configuration. Since the PCCCH allocated on one PDCH supports more than six TRXs, the required number of PDCHs for the PCCCH can be approximated by dividing the number of TRXs dedicated to (E)GPRS by six.

References

[1] 3GPP TS 05.02 V8.10.0 (2001-08), 3rd Generation Partnership Project; Technical Specification Group GSM/EGGE Radio Access Network; Digital Cellular Telecommunications System (Phase 2 +); Multiplexing and Multiple Access on the Radio Path (Release 1999).

[2] 3GPP TS 04.08 V7.10.0 (2000-12), 3rd Generation Partnership Project; Technical Specification Group Core Network; Digital Cellular Telecommunications System (Phase 2 +); Mobile Radio Interface Layer 3 Specification (Release 1998).

[3] 3GPP TS 04.06 V8.1.1 (2001-01), 3rd Generation Partnership Project; Technical Specification Group Core Network; Digital Cellular Telecommunications System (Phase 2 +); MS–BSS Interface; Data Link (DL) Layer Specification (Release 1999).

[4] GSM 04.60 V8.1.0 (1999-11), European Standard (Telecommunications Series), Digital Cellular Telecommunications System (Phase 2 +); General Packet Radio Service (GPRS); Mobile Station (MS)–Base Station System (BSS) Interface; Radio Link Control/Medium Access Control (RLC/MAC) Protocol (GSM 04.60 version 8.1.0 Release 1999).

[5] 3GPP TS 05.03 V8.6.1 (2001-08), 3rd Generation Partnership Project; Technical Specification Group GSM/EDGE Radio Access Network; Digital Cellular Telecommunications System (Phase 2 +); Channel Coding (Release 1999).

[6] GSM 05.05 V.8.3.0 (1999-12), European Standard (Telecommunications Series), Digital Cellular Telecommunications System (Phase 2 +); Radio Transmission and Reception (Release 1999).

[7] 3GPP TS 05.03 V8.11.0 (2001-08), 3rd Generation Partnership Project; Technical Specification Group GSM/EDGE Radio Access Network; Digital Cellular Telecommunications System (Phase 2 +); Radio Subsystem Link Control (Release 1999).

Part 3

3G Evolution Paths

The International Telecommunications Union (ITU) launched the International Mobile Telecommunications-2000 (IMT-2000) programme, which, together with the main industry and standardisation bodies worldwide, targets to provide the framework for the definition of the third-generation (3G) mobile systems.

Several radio access technologies have been accepted by ITU as part of the IMT-2000 framework. From the existing IMT-2000 accepted standards, two main differentiated 3G evolution paths are available for current cellular operators. Such paths are the *UMTS 3G multi-radio* and the *cdma2000* evolution paths.

The UMTS Forum, an association with over 250 member organisations, and endorsed by the GSM Association and the Universal Wireless Communications Consortium (UWCC), is focused in the successful commercial realisation of universal mobile telecommunication services (UMTS) globally. On the other hand, the CDMA Development Group (CDG) is the equivalent organisation supporting the successful evolution and commercialisation of cdma2000.

The third and last part of this book will introduce the main aspects related to different 3G radio access technologies and evolution paths. Chapter 13 introduces the main code division multiple access (CDMA) based 3G technologies, analysing their associated performance, both in terms of coverage and capacity. It presents as well a technical comparison between different 3G radio technologies (global system for mobile communications (GSM)/enhanced data rates for global evolution (EDGE), wideband CDMA (WCDMA) and cdma2000). Finally, it introduces the UMTS 3G multi-radio integration concept with its possibilities and performance benefits analysed. Finally Chapter 14 presents a global market analysis to assist defining the technology evolution strategy within the existing 3G evolution paths in the global frame that characterises the cellular business.

12

Automation and Optimisation

Volker Wille, Sagar Patel, Raquel Barco, Antti Kuurne, Salvador Pedraza, Matias Toril and Martti Partanen

Due to the complexity that the third-generation (3G) evolution will bring, planning, optimising and operation of radio networks is a tough challenge which operators have to tackle at a minimum cost. The following chapter presents new concepts based on automation and control engineering, designed to deal with all these issues, with functionality capable of fully automating the planning, optimisation and operation of radio networks, which not only decrease the operational expenditure (OPEX) of radio networks but maximise their performance as well.

12.1 Introduction to radio network optimisation

The coming years will bring profound change to the mobile telecommunications industry. This change is due to the changing nature of services offered by mobile operators, the introduction of new air interface technologies as well as the increased level of competition. Most importantly, the business of operators is changing fundamentally as 'mobile phones' will not only be the preferred method of voice communication but also the preferred choice for connecting to the Internet. This change is driven by the desire of an end user to have Internet access at any time regardless of their location.

To enable fast and reliable Internet access via the mobile, cellular systems will undergo several technical enhancements. In the past mobile networks were designed solely for circuit switched speech traffic, as speech telephony was virtually the only service offered. In the future the majority of the traffic will be packet switched data generated by mobiles that are connected to the Internet. This means that mobile networks will have to be constantly re-designed as the ratio of packet switched to circuit switched traffic is constantly growing. Furthermore, a growing amount of speech traffic and the additional packet traffic will have to be carried in the network within the same frequency band. Hence, additional requirements on the network will emerge in terms of providing further network capacity and being more spectrally efficient.

To enable the convergence of mobile systems with the Internet, new functionality will have to be integrated into existing cellular systems. For operators using global system for mobile communications (GSM) as the second-generation cellular telephony system, this meant adding general packet radio system (GPRS) related hardware and software to existing networks. Similarly the core/transmission network will have to be upgraded to facilitate data throughput for the new services. The addition of GPRS will bring data speeds of up to 160 kbps while with enhanced GPRS (EGPRS) 473.6 kbps will theoretically be possible, see Section entitled *(E)GPRS* in 7.1. In practice however, these data rates will not be available to all end users. In order to enable new services that require even higher data rates, operators will be deploying wideband code division multiple access (WCDMA) (see more information in Chapter 13). This technology is capable of transmission speeds of up to 2 Mbit/s, thus providing the foundation for intranet and Internet services that are technically more advanced than many of those offered today via fixed-line access.

As WCDMA has been designed with packet switched traffic in mind, quality of service (QoS) considerations such as packet scheduling and prioritisation have been embedded in the system. As mentioned in Chapter 3, the same QoS considerations are now being standardised for GSM EDGE radio access network (GERAN). An additional advantage of WCDMA is the fact that it comes with designated spectrum. This effectively increases the potential network capacity significantly. However, operators are facing the task of seamlessly integrating these technologies into one system. This integration will require significant efforts in terms of dimensioning, planning and operating the new system, some of which are outlined in Chapter 13.

The addition of new technology increases the complexity of operating the network. When speech telephony was the only service offered, indicators such as dropped call rate, speech quality and blocking of call attempts were suitable indicators to give operators an understanding about the performance of the network. With the addition of packet switched service, completely new performance indicators are required. Here data transfer speed, delay and availability are some of the most important indicators that describe the performance of the data service. For operators integrating WCDMA and/or wireless local area network (WLAN) into the network a multitude of other indicators will be required for assessing the performance of the network. The increased complexity of the network makes it far more difficult to analyse and tune its performance.

The huge fees that operators had to pay for WCDMA licences have increased their debt burden significantly. In addition further investment is needed to purchase the WCDMA equipment and finance its installation. Consequently operators are under increasing financial pressure to produce returns on their investments. At the same time new 3G entrants have been admitted into several markets hence further increasing competition in the mobile industry.

These changes of the network go along with a continuing network roll-out even in relatively mature cellular markets. The deployment of additional sites is mainly required to provide increased levels of in-building coverage as mobile users expect to be offered service in all geographical locations. But continuing network growth is also needed to enhance system capacity to cater for the rapid increase in the level of packet switched traffic.

In the past operators managed to cope with rapid technological change and growth of the network by increasing their workforce. Table 12.1 gives some indication on the growth of employee numbers experienced by D2Vodafone in recent years. However, due to mounting financial pressures this is not a viable strategy anymore. Therefore, the only feasible option to maintain network quality with the existing workforce whilst also integrating new technology into the network is to increase the level of automation with a view to enhance the productivity of the employees. This will free resources, which can tackle upcoming challenges brought about by the continuing evolution of the network.

Operational efficiency

Reducing the cost of transferring information with a given grade of service is a clear aim of network operators and equipment vendors alike. The lower the cost of providing an 'x kbps pipe' for a subscriber, the more competitive such a service can be offered and ultimately the higher the financial success of the operator. For a cellular operator using radio waves as a means of transport, spectral efficiency of the radio access technology is a very important factor to achieve this aim. The higher the spectral efficiency, the more information can be transferred within the same spectrum. Given the high cost of radio spectrum –

Table 12.1 Growth of operator workforce
(*Source*: D2Vodafone annual reports).

1996	1997	1998	1999
4800	5800	7000	7900

operators paid about 420 million euro per MHz during the spectrum auction for WCDMA in Germany – increasing the spectral efficiency of the radio access system is of paramount importance. Several ways to increase the spectral efficiency of GERAN system have been outlined in previous chapters.

Spectral efficiency is mainly a characteristic of the radio technology and its implementation by the equipment vendor. In addition to spectral efficiency another crucial factor in providing cost-efficient data transfer is operational efficiency. Operational efficiency is the work-effort that is required to provide a given spectral efficiency at a given grade of service. While it is relatively easy to formulate spectral efficiency by means of mathematical equation, the formulation of operational efficiency is less straightforward in mathematical terms as diverse aspects such as operator process and the salary of employees have to be considered. However, the validity of the concept of operational efficiency in providing cost-efficient data transfer can be shown with a simple example. If a given feature enables an increase of spectral efficiency by 10%, but the related work-effort has to be increased by 50% to enable this improvement, then it is doubtful if the overall data transfer cost can be reduced by this feature.

The relationship between data transfer cost, spectral efficiency and operational efficiency is given by

$$\text{Data transfer cost} \approx \frac{1}{\text{Spectral efficiency} \times \text{Operational efficiency}} \tag{1}$$

It can be seen in equation (1) that the higher the spectral efficiency and the operational efficiency, the lower the data transfer cost. On the other it can also be noticed that an increase in spectral efficiency that brings about a reduction in the operation efficiency might lead to higher cost of transferring data. Only if the product of spectral efficiency and operational efficiency is increased, will the data transfer cost actually be reduced.

To further explain the concept of operational efficiency it is useful to draw on the similarities between a telecommunications network that transports information and a vehicular transport system such as a bus. In this analogy spectral efficiency of the cellular network is similar to fuel efficiency of the motor vehicle. Spectral efficiency like fuel efficiency is a very important factor in determining the cost of providing the transport service, but it is not the only one. Apart from the purchasing cost of the system it is the cost of running the system that determines the overall cost of offering a service. For example, if a bus requires a conductor and a driver then its operational efficiency is ultimately lower compared with one where only a driver is required. The same holds true for a cellular network, if the same (grade of) service can be provided with only half the number of staff, then this is as important a factor as providing high spectral efficiency. This means that in order to reduce the cost of providing an x kbps pipe it is extremely important to increase the operational efficiency of running the network. The only feasible way to increase operational efficiency is to increase the level of automation in the planning as well as in the operations and maintenance phase of running the network.

Characteristics of automation

There are two main areas in which automation can be deployed to enhance the operational efficiency of the workforce.

Automation of existing tasks to reduce the workload

When running the network a number of tasks have to be repeated at regular intervals. Amongst these are frequency and adjacency planning. With the addition of each new cell, it is required to determine which are the most suitable handover (HO) candidates. On all new cells several cells have to be defined as HO candidates to ensure the satisfactory HO to and from the new cell. On all the existing cells it is required to remove those HO definitions that will be obsolete through the addition of a new cell. In addition, the new cell also requires a set of appropriate frequencies to enable proper operation. When deciding on these frequencies, it is necessary to consider the frequencies on all existing cells in its vicinity. Depending on the situation it might be required to re-plan a larger number of cells around the new cell or to pick the least interfered frequencies available. Providing the means to automate the process of integrating new cells seamlessly into the existing network is just one example of routine tasks that lend themselves to the application of automation. Here the purpose of automation is to provide the means to increase operational efficiency by reducing the workload associated with existing tasks.

One positive side effect of increasing the level of automation for existing tasks is that staff is freed from mundane and tedious tasks and can apply their skills in new areas which bring additional value to the operator. In this sense automation ensures that staff is working on challenging tasks which in turn helps to motivate and retain staff.

Automation of new tasks to improve the performance of the network

Certain tasks cannot be carried out without the application of automation. For example determining the optimum parameter values that are unique for each cell is impossible in a large network without the use of automation. Areas that require automation to offer a cost-effective implementation are those involving heavy calculation and/or the evaluation of several input parameters. In contrast to humans, computers can easily carry out analysis of complicated input data. Furthermore, computers are able to repeat this method many times and thus enable the application of a method to all cells in the network. In the case of troubleshooting, for example, it is required to establish the most likely reasons for a given network problem. In this scenario the application of automation enables determining the most probable fault by combining a large number of factual pieces of information. This information can be in the form of hardware alarms, short-term and long-term performance data as well as configuration data such as parameter settings on all surrounding cells. The benefit of automation to these kind of tasks is an increase in network performance as without the application of automation these tasks could not be carried out on a wide scale, i.e. a large number of cells.

The following sections highlight areas that lend themselves to the application of automation with a view to increase operational efficiency while also increasing network performance. Thus it is possible to obtain optimum performance from the existing network at the lowest cost. The technical focus in this chapter is on GERAN but the trend of increasing automation in cellular systems is also applicable to UTRAN networks.

Areas of automation

Figure 12.1 displays several technical factors contributing to the performance of a GERAN network and their interdependence. Non-technical issues, which also have an influence on network performance and operational efficiency, such as equipment deployment strategy, use of tools and operator processes are not being listed here.

The foundation of a well performing network is the basic radio platform. Issues such as the location of sites as well as positioning of antennas have a major impact on the distribution of the radio signal and thus on call quality. For example the closer sites are located near a traffic hot spot, the easier it is to provide good call quality with low signal levels. This will reduce interference to surrounding cells.

Figure 12.1 Dependency of network planning tasks.

In addition, the clearance of faults is also very important in order to ensure that the network operates exactly as designed. If a cell is temporarily non-operational, the performance of all cells in the vicinity is impaired. Ensuring that this cell is speedily brought back into operation is a crucial task to ensure optimum network performance. Methods aimed at improving and automating the fault clearance will be outlined in Section 12.5.

Furthermore, it is the adjacency plan that is the major contributor to network quality. Only with the definition of correct adjacencies will HO to most suitable cells be possible. Without the definition of the optimum adjacencies, HOs occur to cells which are sub-optimal in terms of their radio link performance and subsequently dropping a call might not be avoided. Automating the process of updating adjacency definitions by deleting under-used ones and creating missing ones is outlined in Section 12.3. The automation of this task will enhance operational efficiency and depending on the quality of the existing adjacency management process, also increase network performance.

Similarly, a good frequency plan is of paramount importance for network performance. The better the frequency plan the less co- and adjacent channel interference is experienced by subscribers. This means better call quality and subsequently smaller dropped call rate. Thus frequency planning is a crucial task in building a well performing network. The application of automation to frequency planning is described in Section 12.2. Here the emphasis is on improving the quality of the frequency plan and reducing the work-effort associated with its creation, i.e. enhancing operational efficiency.

Another important factor that determines network quality is parameterisation of network features. Traditionally standard parameters have been used network wide, but this is not optimal when trying to enhance network performance. Each cell operates under unique conditions and requires customisation of parameter values to provide optimum performance. Insight into this aspect of automation is provided in Section 12.4.

Only once all the above-mentioned tasks have been taken care of in the optimum manner will sophisticated network features such as the ones explained in the previous chapter.

Automation can be applied to all the tasks shown in Figure 12.1 with a view to enhance network performance and at the same time reduce the man-effort required to carry out the task. This means that through the application of automation to these tasks the data transfer cost (equation (1)) can be reduced.

On a conceptual level, automation of the above-mentioned tasks is achieved by:

- collecting information from the network
- analysing the obtained information and
- implementing required changes in the network

In effect a feedback loop is being created which is common to any control system. This concept is shown in Figure 12.2.

It is worth noting that an additional task can be integrated into the feedback loop between steps 2 (suggesting changes) and 3 (implementing changes). This step 'verification of the suggested changes' is optional and is most likely to be used in the initial stages of using an automation feature. Once confidence on the validity of the suggested changes is gained, this task will not be used anymore.

For operators a very important aspect of this feedback loop is the fact that information to tune the network is provided for free. The majority of information that is being collected is made up of measurement reports that ordinary subscribers are providing free of charge when in call mode. This means that it is not required to build a separate network model, in which to simulate the behaviour of the network. For example, in the case of frequency planning it is common practice to maintain a hugely complex image of the network to determine the best frequency plan. By using information from the network to plan the network the creation and up-keep of such a model would not be required anymore.

In all subsequent sections this basic form of a feedback loop is being applied to the above-mentioned tasks. The control unit used in the feedback loop depends on the kind of automation that is to be carried out. In the case of adjacency deletion, it might be a simple threshold for the number of HOs between cells at which an adjacency definition should be deleted. In the case of troubleshooting, it is a specialised technology that combines observations, their probabilities and logical reasoning in order to determine the most likely

Figure 12.2 Automation feed-back loop.

reason for a fault and the most cost-efficient action to take.

The decision whether to close the feedback loop and allow the control system to implement changes rests with the operator. Alternatively the feedback loop might also be open and the control system only suggests changes, the verification of these and their implementation will still be carried out by operator personal.

12.2 Mobile measurement based frequency planning

Outline of the problem

Over the last few years cellular networks have experienced a remarkable growth. The large number of subscribers in the network and the increasing usage of existing mobile telephony services together with upcoming new services such as video and audio streaming, will force operators to significantly increase capacity offered by the network. Capacity requirements exist particularly in a city centre environment where the greatest subscriber density is experienced. Sectorisation and cell splitting, two standard methods to increase capacity from macro cells, have generally been employed in GERAN networks. Often operators also utilise advanced network features such as IFH to reduce the frequency reuse factor and thus further increase network capacity of macro cells. Due to the limited spectrum and because of hardware constraints regarding channel separation required on a cell or a site basis, providing additional capacity from existing macro cells is often impossible without degrading call quality. Due to the fact that in a city centre environment the distance between macrosites of an operator is typically as low as 300 m and because several operators are competing for new sites, suitable locations to establish additional macro cells are very difficult to obtain. Therefore, network operators also have to use alternative cell types such as micro cells and in-building cells to meet capacity requirements. Distances between micro cells can be of the order of a few tens of metres when deployed on a large scale. Dedicated in-building cells provide tailored solutions in office buildings for corporate customers and in public places such as shopping centres and underground stations. One advantage of the utilisation of small cells is that additional capacity can be provided at 'traffic hot spots' and it is thus possible to target areas with high capacity requirements directly.

The reduced cell sizes also create new challenges. The main challenge is frequency planning. Traditional frequency planning methods rely on the accurate prediction of radio wave propagation. Accurate prediction of propagation is relatively straightforward in non-undulated, rural environments where the signal is transmitted from tall, isolated radio masts towering over all other surrounding buildings. In that scenario empirical models, such as those developed by Okumura–Hata [1], are able to predict signal level with a good degree of accuracy. However, only a minority of cells is deployed under this kind of conditions. The majority of cells are located in city centres with antennas fixed to ordinary buildings in the midst of taller buildings.

For the prediction of radio wave propagation in micro cell environments, other empirical models such as Walfish–Ikegami [2] have been developed. However, even using these bespoke models signal level prediction accuracy in dense urban environments is relatively low. The low prediction accuracy is due to several reasons. Firstly, reduced cell sizes require more accurate data to describe topology and morphography of the propagation environment. With macro cell seize in the order of around 150 m from the location of the base transceiver station (BTS) it is required to utilise digital map resolution in the order of a few tens of metres. For micro cells with cell sizes of about 50 m even higher resolution is required, typically of the order of a few metres (1–5 m). For cost reasons and to speed up computation times, this kind of high-resolution map data is normally not being used.

In addition, propagation models normally determine signal levels about 1.5 m above ground level as this is close to the typical subscriber height. However, a significant share of the subscribers is at elevated locations such as in tall office buildings. The vertical displacement between the location of the signal level predictions and the actual location of subscribers has an adverse effect on prediction accuracy. Moreover, the fact that these elevated locations are normally indoors further increases prediction inaccuracy. The prediction of signal levels from external cells to indoor environments is particularly difficult, as detailed information on the propagation properties such as penetration loss into the building and propagation loss within the building is generally not available. With the increase of dedicated in-building cells it is also required to frequency plan these accurately in coexistence with outdoor cells. However, commercially available propagation prediction tools do not normally cover these two aspects. Due to these difficulties in frequency planning indoor and outdoor cells operators tend to dedicate spectrum for indoor cells when deploying indoor cells on a large scale. In this way both cell types are separated in the frequency domain. While this approach is easing the planning requirements it is not optimum from the spectral efficiency point of view.

Deterministic techniques based on uniform theory of diffraction (UTD) have been studied extensively in the academic arena [3,4] and it is shown that these models can deliver fairly accurate propagation predictions. A prerequisite for obtaining accurate coverage predictions is a high-resolution clutter database. Clutter resolution of the order of 1–2 m is normally required, describing the outline of each building as a polygon. Depending on the use of 2D or 3D prediction methods, the polygon has to describe the buildings in the desired dimension. Obtaining such detailed clutter information can be difficult for some cities and, generally, the cost associated with a high-resolution database is relatively high. In addition prediction times rise exponentially with clutter resolution and huge memory as well as computational requirements exist for the calculation platform. Therefore, UTD based radio planning tools are currently not widely used by operators.

Due to these reasons it is required that a new approach is taken to simplify frequency planning by easing requirements and processes while also improving the quality of the frequency plan. In this novel approach signal level measurements provided by ordinary mobile stations while in dedicated mode are being used. The reported measurements are sent in regular intervals (usually every 480 ms) and contain the signal level of the serving cell and up to six neighbouring cells. Based on this information which is provided to the operator free of charge, it is possible to create a better frequency plan than is currently possible using standard tools and techniques.

The ability to create a high quality frequency plan even for complex propagation environments such as city centres and in-building cells enables operators to be more spectrally efficient as less channels have to be used to provide a certain grade of service. These 'saved' frequencies can then be used to increase the capacity of the network. Alternatively the increased interference prediction accuracy can also be utilised to provide higher network quality.

One additional benefit of using mobile measurement information is the fact that the subscriber distribution within a cell is explicitly included in the information. For example, more signal level measurements will be reported from locations with a high density of mobile stations as compared with locations where a low mobile density is experienced. In this way it is possible that the signal levels from neighbouring cells are weighted depending on the amount of cell traffic that is 'interfered'.

In systems using dynamic frequency and channel allocation (DFCA), see Chapter 8, the frequency of the BCCH TRXs can not be dynamically controlled. These TRXs have to be frequency planned using normal, non-dynamic methods. The benefits of the method outlined above can be applied to these TRXs. This means that the reuse of the BCCH can be reduced by the use of this method. Hence DFCA and mobile measurement based frequency planning work in harmony to increase spectral efficiency in all layers of the network.

Traditional frequency planning

This section provides further detail on current frequency planning by describing the process and methods to carry out frequency planning based on propagation predictions. The new method of mobile measurement based frequency planning is then outlined in the subsequent section.

Frequency planning principles

In cellular time division multiple access (TDMA) systems, like GERAN, each TRX in the network requires at least one frequency to operate. As the number of frequencies that an operator has been allocated is far smaller than the number of TRXs in the network, it is required to repeat the use of a frequency several times throughout the network. For example, in a network consisting of 20 000 TRXs using 50 channels, each channel has to be reused about 400 times.

The term reuse factor is normally used to denote the 'distance' at which the use of frequencies is repeated. For example a reuse factor of 16 implies that in case frequency f_1 is used on a cell A, only on the cell that is 17 cells 'away' from cell A will f_1 be used again. It is the aim of frequency planning to ensure that all cells that are 'closer' to A in propagation terms, i.e. cells 2–16, will have frequencies different than f_1.

Operators aim to obtain maximum performance from the existing network with the given number of frequencies. This can only be achieved if the available frequencies are reused in an optimum manner. Frequency assignment is the process of providing each TRX in the network with the required number of frequencies while ensuring that interference between cells is minimised, i.e. the network quality is maximised. Frequency assignment is a very crucial task as it has a direct impact on network performance. In order to be able to assign frequencies correctly, it is required to accurately determine interaction between cells within the network in interference terms.

The interaction between cells is normally determined by network planning tools that use propagation prediction methods such as those outlined in the previous section. In planning tools the geographical area is described by regular squares, called bins. A commonly used bin size is 50 m × 50 m. The size of these bins corresponds to the resolution of the map component of the planning tool.

Figure 12.3 displays two pictures that show the same geographical area. The picture on the left-hand side of Figure 12.3 uses a bin size of 2 m whereas the picture on the right-hand side uses a bin size of 50 m. It can be seen in Figure 12.3 that with bins of 2 m it is possible to visually recognise many features of the actual city, such as roads, buildings and open areas. With the larger bin size, on the other hand, such identification is not easily achieved. A bin size of 50 m also means that structural features that are smaller than 50 m disappear or are increased in size to a minimum of 50 m. For that reason the street width is set to 50 m or the street is not shown at all. Figure 12.3 thus visualises the importance of an accurate description of the propagation environment by a small bin size, as the bin size determines how the propagation prediction tool 'sees' the propagation environment.

Each bin has also a certain attribute that describes its propagation characteristics. In Figure 12.3 this attribute is denoted by difference in colour. Typical morphographic attributes include water, open, road, forest, rural, suburban,

Figure 12.3   Effect of map bin size.

urban and dense urban. The more attributes are being used to describe the propagation environment, the more information can be provided about its morphographic make-up. This means that bin size and number of morphographic attributes of the bin have a significant impact on the propagation accuracy of the planning tool. In general, the smaller the bin size and the larger the number of morphographic attributes, the better the propagation accuracy. However, this increased level of accuracy can only be provided with increased cost for the map component of the planning tool and increased levels of computational power to execute propagation predictions.

Based on propagation predictions the signal strength from each cell is determined within each of the relevant bins. Within each of these bins signals from several cells are received. Thus it is possible to determine the interaction between cells within a bin and consequently within a larger area made up of a large number of bins. This interaction between cells is expressed as the carrier to interference (C/I) ratio and indicates to what extent cells interfere if they were going to use the same or an adjacent frequency. Analysis is carried out to determine the number of bins in which the C/I between a pair of cells is below a certain threshold such as 9 dB. A threshold of 9 dB is normally chosen since below this threshold poor speech quality would be obtained whereas above this value good quality can be provided. By carrying out this analysis for all surrounding cells it is possible to determine a ranking of the interference obtained from these cells.

The exact formulation of the level of interference between cells is known as the interference matrix (IM) or inter-cell dependency matrix [5]. Element x_{ij} in the IM is an indicator of the potential interference produced by cell j on mobiles attached to cell i, assuming that both use the same frequency.

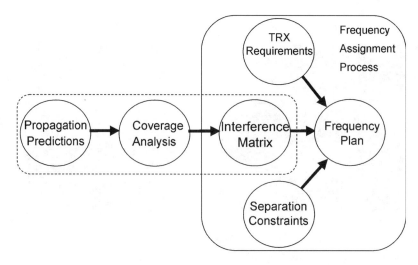

Figure 12.4   The traditional frequency planning process.

Conceptual view of frequency planning process

Figure 12.4 displays the frequency planning process on a conceptual level. It can be seen that the creation of a frequency plan is a sequential process where the final frequency plan is based on the results of the previous steps. Propagation predictions are the starting point to create a frequency plan. Based on these the coverage is analysed to establish the level of interference between cells. This is formulated in the form of the IM. Based on the IM together with TRX requirement and separation constraints on a cell-by-cell basis the actual frequency plan is created.

Commercially available network planing tools normally include all this functionality. But it is also becoming more widespread that the tasks associated with creating the IM (shown in dashed box) are located within one tool, whereas the actual frequency assignment is carried out in a dedicated frequency assignment tool (denoted by solid line).

The difficulty with this sequential process is that the quality of the last stage of the process is still dependent on the quality of the first stage of the process. This means that all deficiencies of propagation predictions are still reflected in the final frequency plan. Hence even if an ideal frequency assignment process is being assumed, this process is only as good as the input into the process, i.e. the propagation predictions.

Importance of planning data consistency

For propagation predictions to be accurate it is required that the actual network configuration is correctly expressed in the planning tool. For each BTS in the network a large number of parameters have an influence on the signal level that mobiles receive. These factors are BTS output power, feeder and combiner loss, as well as the gain, height, tilt and direction of the antenna. These parameters have to be correct in the planning tool to be able to model the network correctly. Ensuring that changes in the actual network are also reflected in the planning tool is a tedious and time-consuming task which does not only encompass the addition of new cells into the tool but also any alteration on existing cells. However, any discrepancies between the tool and the actual network will lead to wrong propagation predictions and consequently to a sub-optimal frequency plan. This means that both the database of the planning tool and the network have to be constantly maintained to be able to provide a good frequency plan. When creating a new frequency plan it is the maintenance of this database that consumes the largest amount of time. Building of the IM and the actual allocation of frequencies is a comparatively fast process as this can be carried out by the respective tools without much human intervention.

The need to maintain two databases is undesirable as it reduces operational efficiency significantly with each task having to be carried out twice. Any method that can reduce the need to maintain two databases will greatly reduce the work-effort associated with creating a frequency plan.

Support of different frequency hopping strategies
Over the past few years a high ratio of operators have started to use frequency hopping. Frequency hopping provides two advantages over the non-frequency hopping case. These are frequency diversity and interference diversity (see Chapter 5). Frequency diversity provides higher link resilience as the effect of deep fades (temporal reduction in signal level) is normally only experienced on one of the frequencies and hence the others provide the required link stability. Due to the effect of interference diversity it is possible that high levels of interference on one frequency can be offset by low levels of interference experienced on other frequencies. As non-hopping networks lack these benefits operators do generally apply frequency hopping wherever possible.

There are two types of frequency hopping, baseband (BB) and synthesised/radio frequency (RF). In BB hopping frequencies are planned as in the non-hopping case, but during the course of a call the frequency is changed burst by burst, cycling through the frequencies available on the cell, i.e. a call hops from TRX (frequency) to TRX (frequency). In terms of planning BB hopping, there are no differences to the case of non-hopping as each TRX in the network requires one frequency. This means that the above-mentioned process of assigning frequencies is still applicable.

In the case of RF hopping, only the BCCH TRX is non-hopping and thus has to be planned as normal. The other TRXs in the cell are not provided with one frequency, but a set of frequencies through which an active call will be hopping. A possible reuse pattern in RF hoping is 1/1, see Section entitled *Frequency hopping* in 6.1. This means that each non-BCCH frequency is available to each non-BCCH TRX in the network. Frequency collision (interference) between neighbouring cells is prevented by allowing only a relatively low level of loading of the cell. Due to the fact that all non-BCCH frequencies are used in each TRX, frequency planning is not required. However, for RF hopping schemes different than 1/1, accurate knowledge of the interaction between cells in the networks is still useful as it is possible to customise the frequency sets that calls are using in RF hopping. Through customisation of these sets performance benefits can be obtained.

One advantage of BB over RF hopping is the fact that remote-tune (cavity) combiners can be used, which have a lower insertion loss than the wideband combiners that are required for RF hopping. Due to the lower loss of the combiner higher output powers are achieved with BB and thus larger cell sizes and higher in-building penetration is provided.

RF hopping is the simplest approach to frequency planning as it minimises the planning effort for the traffic channel (TCH) TRX, requiring frequency planning only for the BCCH TRX. RF hopping with regular reuses like 2/2 and 3/3 area also easy to plan and do not require too much planning effort in the non-BCCH layer, see Chapter 6. However, from the spectral efficiency point of view, this is not the optimum approach. For scenarios where building an accurate IM is not possible, this hopping strategy is certainly beneficial. However, the more accurate the ability to determine interference between cells and the more accurate the IM, the more useful it is to create a tailored frequency plan that is making use of the knowledge of interference levels between cells. The knowledge that two cells interfere strongly (or not at all) can be used by the frequency assignment tool to improve the quality of the frequency plan.

The new frequency planning concept

The traditional frequency planning process was described in the previous section and illustrated in Figure 12.4. The new concept described here is called mobile measurement based frequency planning (MMFP). The idea was proposed about 5 years ago; but in 2002 not many operators really use MMFP. This is due to the fact that without support of the equipment vendor access to the required mobile measurement data demands huge effort if these data are to be collected on a large number of cells. The main part of the process, shown in Figure 12.4 still applies to MMFP, as it is only the data source that is different. Instead of using radio wave propagation and small-scale *ad hoc* measurements, each call made by the subscribers in the area to be planned contributes to the interference matrix. The idea is useful from an optimisation point of view, as there is no longer a need to predict the radio path or the subscriber distribution, both of which are extremely complex to estimate. The subscriber distribution being implicitly present in mobile measurement reports allows really optimising the frequency plan based on the actual traffic hot spots. The IM reflects exactly *where* and *when* subscribers make calls.

Storing the mobile measurement reports replaces the two phases in Figure 12.4 "Propagation predictions" and "Coverage analysis". There are many different methods to construct the IM from RXLEV values in Mobile Measurement Reports (MMR). These are discussed in Section entitled *Review of interference matrix types* in 12.2. After the IM has been built the MMFP process continues just like normal frequency planning process. The advantage is that the IM is more accurate than what is obtainable by conventional methods.

Increased accuracy of the measurement based IM over the prediction based arises from two distinct sources. Firstly with conventional method, carrier strength and potential interferer strength are estimated in each map bin as explained in Section entitled *Traditional frequency planning* in 12.2. Each bin contributes one sample, and collecting all samples from each cell service area determines the interference relations for that particular cell. The inevitable error made here is the fact that not each map bin is equally often used for a call by subscribers. This error is avoided when measurement reports are used to construct an IM. This is because each sample now corresponds to a fixed length portion of an actual call. Therefore those locations where calls are made more often get to be measured more often and thus contribute more to each cell's interference relations. Secondly, in addition to the fact that number of samples affecting each cell's interference relations is distributed according to subscriber call locations, each sample is also more accurately estimated in measurement based method. Each sample is based on measured values of a carrier and a potential interfering cell. Even though there are errors in these measurements, they are of smaller magnitude than errors in propagation predictions.

Signal level reporting in GERAN

Radio link measurements done by mobiles were defined for use in HO and RF power control processes. For this purpose, mobiles report not only the serving cell level, but also that of the neighbouring cells, which are defined as HO candidates. It is important to analyse the characteristics of these measurements in order to use them for MMFP.

The signal level from the neighbouring cells can be measured because each BTS continuously transmits a *beacon frequency* (BCCH carrier) at a constant power level containing a 6 bit BSIC. In the specifications [6] it is stated that mobiles should measure the power level of as many beacon carriers as possible, and decode their BSIC in order to determine the cell emitting the signal. After basic processing, mobiles send on the slow associated control channel (SACCH) messages (mobile measurement reports (MMR)) to the BTS containing the measured values.

The following process applies: within a period of 120 ms there are 25 intervals in which the mobiles can measure the power of the BCCH carriers indicated in the BA list, normally performing one measurement per interval [6,7]. That means that the number of samples per SACCH measured on each BCCH carrier will depend on the number of carriers defined in the BA list. For example, if the number of carriers in the BA list is 16 and the mobile station (MS) can measure only one carrier during each interval, the number of samples per carrier in a SACCH multiframe (480 ms) will be 6 or 7 (= 100 intervals/16 measurements).

The average of these samples is taken in dBm over one SACCH multiframe and mapped to an RXLEV value between 0 and 63, as follows:

$$\text{RXLEV} = 0 \Leftrightarrow \text{level} \leq -110 \text{ dBm}$$

$$\text{RXLEV} = n \Leftrightarrow -110+n-1 \text{ dBm} < \text{level} < -109+n-1 \text{ dBm} \qquad n = 1....62$$

$$(2)$$

$$\text{RXLEV} = 63 \Leftrightarrow \text{level} > -48 \text{ dBm}.$$

Then, the MS transmits a measurement report to the BTS, which contains the RXLEV for the BCCH carriers of up to the six cells with the highest RXLEV among those with known and allowed BSIC.

In conclusion, ideal mobile measurements are subject to restrictions:

- quantisation and truncation (equation (2));
- the fact that only the six strongest cells are reported;
- the uncertainty in the BSIC decoding procedure.

It is important to analyse the effect of these MMR limitations on MMFP [8]. Figure 12.5 shows the ARP (see Section entitled *Review of interference matrix types* in 12.2) received in one cell from the surrounding cells (normalised by the total number of MMRs). The figure compares the interference for ideal measurements (Ref) with those of real MMRs including their limitations: firstly, quantisation and truncation (RefQ), secondly only six strongest reported (N6) and finally all limitations considered (N6QBSIC).

It can be observed that quantisation does not have a significant effect when the measured power is inside the margins. Only for cells producing low level interference (less than −110 dBm), the value is overestimated, due to lower truncation. Upper truncation can be a problem in dense urban areas, where values greater than −47 dBm may be very common.

Another problem observed in the figure is due to the fact that mobiles only provide information about the six strongest interfering cells and therefore other potential strong interferers are not reported. This effect is also clearer in the interferers with low level.

Finally, the BSIC decoding process also has an influence in the interference matrices because the MS has to identify the main interfering cells amongst the ones sharing the same frequency. Decoding can fail if the signal level is too low or there is not a dominant interfering cell on that frequency. Two opposite effects can be pointed out. First, if BSIC is decoded, a power level higher than

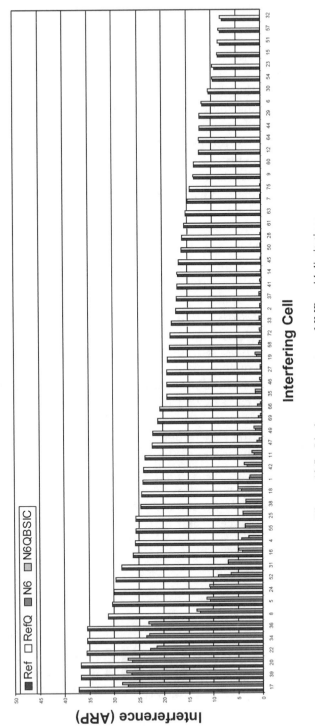

Figure 12.5 Ideal measurements *vs.* MMRs with limitations.

the real one is reported. This is due to the fact that the reported power is not only from the cell whose BSIC is decoded, but the sum of the signals from other cells sharing the same BCCH. This can be observed in the figure on cells producing a high-level interference. On the other hand, if BSIC is not decoded, zero power level is assumed, which is less than the real one. This effect is more important on cells creating low-level interference because the interference level received from cells sharing the same BCCH is usually very similar.

It can also be noticed that even if only the six strongest interfering cells are reported, after a short time several tens of cells are seen. This will allow to build accurate IMs (see Section entitled *Review of interference matrix types* in 12.2).

In more recent versions of the GSM specifications [6], other types of MMRs, apart from the ones described above, are specified. These are extended measurement reports (ExMR) and enhanced measurement reports (EnMR).

Before sending an ExMR, the MS should receive an extended measurement order (EMO) message requesting the MS to measure the signal level of those frequencies specified in the message. Then, the MS should send one ExMR containing the RXLEV for the carriers specified, considering that BSIC decoding is not required for those frequencies.

It has been found that ExMRs are not very useful for MMR based frequency planning because measurements are associated with frequencies but, not with cells, due to the fact that BSIC is not decoded. In order to use these measurements for MMR based frequency planning each frequency should be associated with a cell. The first drawback is that only information for a restricted set of cells is available (one for each BCCH frequency). The most logical procedure would be to associate each frequency with the cell with the highest interfering level among the ones sharing that frequency.

Figure 12.6 shows a comparison of the FER_{nw} (see Section entitled *Review of interference matrix types* in 12.2) obtained from ideal, real and extended measurement reports for a macro cell scenario. It can be observed that extended measurement reports overestimate the ideal value of interference, because they include interference from all cells sharing the same frequency.

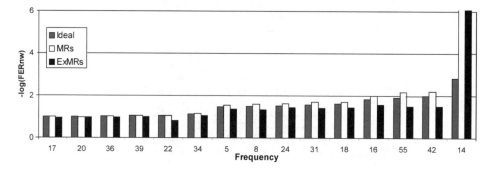

Figure 12.6 Comparison of FER_{nw} for ideal, real and extended MMRs.

The network can also request the MS to send EnMRs. They provide a better estimation of the received signal level from neighbouring cells than conventional MMRs because the maximum number of reported cells in one MMR is not limited to six and also the mapping function from measured power to RXLEV is changed to:

$$\text{RXLEV} = 0 \Leftrightarrow \text{level} \leq -110 \text{ dBm} + \text{SCALE}$$

$$\text{RXLEV} = n \Leftrightarrow -110 + n - 1 \text{ dBm} + \text{SCALE} < \text{level} < -109 + n$$

$$- 1 \text{ dBm} + \text{SCALE} \quad n = 1....62 \tag{3}$$

$$\text{RXLEV} = 63 \Leftrightarrow \text{level} > -48 \text{ dBm} + \text{SCALE}$$

where SCALE is an offset indicated in a measurement information message.

As stated before, in microcells it is common for the signal level to be higher than -47 dBm and therefore the level would be underestimated when using the traditional mapping function. But if the SCALE parameter is incorporated the maximum reported level can be increased by up to 10 dB.

Another issue to be considered is that, in some scenarios, for EnMRs or multiband mobiles, the number of neighbours reported from several serving cells in different frequency bands could be different to six. Figure 12.7 shows the error between the ideal case (all neighbours reported) and only the 3, 5, 6 or 12 highest neighbours reported for FER_{nw} (see Section entitled *Review of interference matrix types* in 12.2). It can be observed that, the higher the number of neighbours reported the closer the results are to the ideal values. But even when the number of neighbours reported is low the results are still quite accurate. For cells creating small levels of interference (right-hand side of graph), the reduced number of reported cells does not have a significant effect.

Review of interference matrix types

An intercell dependency matrix (IM) is a matrix that describes the interaction between cells. This interaction can be described in several ways, although it normally means radio interference. That is, the element x_{ij} in the matrix is an indicator of the potential interference produced by cell j on mobiles attached to cell i, supposing that both use the same frequency channel. There are several classifications of IMs depending on the aspects considered: radio wave propagation prediction based or measurement based, uplink or downlink IMs, expected values or percentile IMs, etc.

Generally GSM networks are downlink interference limited. Furthermore, in the future, services such as Internet access will be very commonly used, where mobiles will send short requests in the uplink, while receiving large amounts of

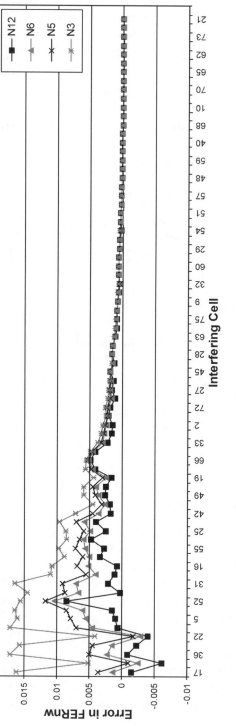

Figure 12.7 Error in FER$_{nw}$ for different maximum number of reported neighbours.

data in the downlink. These are the main reasons to concentrate on downlink MMR based IMs. Several types of IMs can be built based on the processing made on the raw MMRs.

Several types of IMs can be built based on the processing made on the raw MMRs. Besides, each of these types can be divided into two groups: traffic-weighted or non-traffic-weighted measurements.

ARP_w (average received power) represents potential interference from neighbouring cells within the analysed serving area. The non-weighted ARP, ARP_{nw}, is ARP_w normalised by the total number of MMRs received in the cell.

$$ARP_{w_{ij}} = \sum_{k=1}^{N_{ij}} p_{ij}(k) \tag{4}$$

$$ARP_{nw_{ij}} = \frac{1}{N} ARP_{w_{ij}} \tag{5}$$

where

N_{ij} = number of measurements reported of interfering cell j to serving cell i

$p_{ij}(k)$ = k interfering power level (dBm) from cell j reported to serving cell i

N = total number of measurement reports received by serving cell

Another type of IM is the FER (frame erasure rate) matrix [9]. FER is used because it is a very good indicator of speech quality. It represents the percentage of frames being dropped due to an inability to correct bit errors in the vital part (category 1A) of the speech frame, see Section entitled *Transmittring and receiving chain* in 5.1. In order to obtain FER, CIR values are calculated for each MMR and then mapped to FEP (frame erasure probability). FEP represents the probability of frame erasure associated with the current MMR. FER is obtained by averaging FEP corresponding to multiple MMRs.

$$FEP_{ij}(k) = f_{FEP}(CIR_{ij}(k)) \tag{6}$$

$$FER_{w_{ij}} = \sum_{k=1}^{N_{ij}} FEP_{ij}(k) \tag{7}$$

$$FER_{nw_{ij}} = \frac{1}{N} \sum_{k=1}^{N_{ij}} FEP_{ij}(k) \tag{8}$$

where f_{FEP} is a mapping function from CIR to FEP [10].

The last analysed IM is the SFER (step FER) matrix [5]. SFER is an estimation of the probability of CIR being lower than a threshold (9 dB). It can be seen as a rough approximation to a FEP–CIR mapping function.

$$f_{\text{SFER}}(x) = \begin{cases} 0 & x \geq 9 \text{ dB} \\ 1 & x < 9 \text{ dB} \end{cases} \tag{9}$$

$$\text{SFER}_{w_{ij}} = \sum_{k=1}^{N_{ij}} f_{\text{SFER}}(\text{CIR}_{ij}(k)) \tag{10}$$

$$\text{SFER}_{n_{ij}} = \frac{1}{N} \sum_{k=1}^{N_{ij}} f_{\text{SFER}}(\text{CIR}_{ij}(k)). \tag{11}$$

It is important to point out that FER and SFER provide a better approach to the quality perceived by users than ARP.

Hardware architecture for MMFP

The scope of this section is to provide a framework for the architecture required for MMFP and a general overview of the component functionalities. The application requirements of MMFP include planning from a single cell to large geographical areas either by a single user or by multiple users. To satisfy such diverse scenarios it is necessary to have a framework behind the planning interface that seamlessly caters for the need of the planner(s) and provides all the necessary data to support those needs.

MMFP process overview

To understand the architectural requirements for MMFP it is necessary to have a brief overview of the course of events involved. A planner would need to select a set of cells to plan, allocate them for measurement acquisition, define various planning parameters and constraints, perform the allocation, verify the plan, and finally download such a plan to the network. This brief outline of the events is illustrated in Figure 12.8.

Normal practice with current planning tools is to have separate planning tools to plan the required areas. These tools are offline to the actual network and always require network data to be accurately and painstakingly synchronised before any planning can be undertaken. In the case of MMFP, the architectural components shown in Figure 12.8 have constant interaction with the network and are therefore always reflective of the real network situation. Hence ongoing network changes and network evolution aspects are captured by the various components and can be considered during any planning tasks. This presents a big advantage as the most current network elements, configurations, parameters and measurements are utilised to produce the IM for frequency allocation.

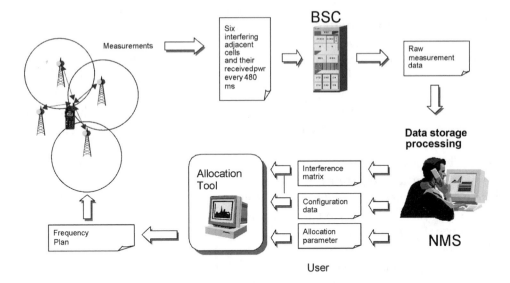

Figure 12.8   MMFP process outline.

Planning scope
The use case of planning using MS measurements is applicable to a single cell, a selected area or even the whole network. The technique described in the previous section is generic for the scenarios mentioned. However, as the area selected for planning gets larger, it becomes necessary to consider the planning in relation to the neighbouring buffer area.

As such it is necessary to understand the concept of global and regional areas in the cellular network domain. These differ from the geographical areas used in planning that can encompass several network management subsystem (NMS)/ BSC areas and cross over the network element borders too. Regional network management systems handle only a certain number of BSCs each and hence NMS/BSC areas differ from the geographical planning based areas. To be able to plan any area independently of NMS/BSC borders requires that all necessary measurement data be captured and be available with the necessary network configuration data. A global management functionality is therefore necessary for effective measurement based planning.

MMFP architecture requirements
Mobile measurement based planning requires easy accessibility to the measurement data. The data should be available over short as well as long-term periods for comparison as well as historical use. Data stored should be accessible at a global level to allow network element border crossover so that the planning scope is not compromised.

Data storage and availability can therefore be seen to be an important and integral architectural requirement. Obtaining the necessary measurement data requires changing the BA lists so that mobiles in dedicated mode are made to measure all BCCHs. The architecture should support obtaining the measurements automatically and the process be somewhat invisible to the planner.

With different operation subsystem (OSS) components being utilised by MMFP, it is important that there is seamless operability between the separate components. If file transfers are necessary they should appear transparent to the user, if objects are selected for planning the related measurement availability is checked for automatically. Support for the feature usage by multiple users is another important requirement. Another vital requirement for the architecture is to provide all necessary data to the automated frequency planning (AFP). This includes data like separation parameters, constraints, allocation parameters, etc. Normally such data are available from the planning tool and not in the OSS. These important data must therefore be provided for and it becomes another vital requirement for the MMFP feature. Again it can be seen that seamless operability is a requirement between components and open interfacing is necessary. Integrating an AFP to the IM building blocks is an option that helps to overcome such obstacles.

Further important requirements include having workflow management for the MMFP feature. Catering for multiple users performing separate tasks to either the same or different parts of the network requires efficient tracking and logging of those tasks. This allows for better visualisation of the planning tasks and introduces a higher level of planning control. Some degree of automation can also be handled by such a workflow management system and approved network changes can be scheduled for downloading to the network.

Planning strategy plays an important role within the MMFP concept and an efficient planning solution should be determined from factors like planning needs, network status, network growth and NMS/BSS coverage areas. These factors influence the use of MMFP for either regional or global application. An interpretation of the MMFP architecture catering for global and regional planning is shown in the Figure 12.9.

MMFP trial results

MMFP was tested in a city with a population of 1.5 million (Network 1). Retune area was located in the centre of the city and had 82 cells, all on 900 MHz band. The spectrum used in the trial area and surroundings was about 8 MHz wide. The network was BB hopping with an average of 2.5 TRXs per cell. During weekdays the average traffic served by trial cells was 3500 Erlang hours per day.

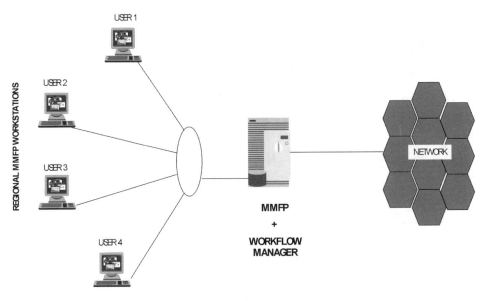

Figure 12.9 MMFP architecture.

Prior to trial the average dropped call rate (DCR) on the trial area was 1.29%. When only drops due to radio reason are counted, RRDCR (radio reason dropped call rate) was 1.16%. The corresponding number of drops per Erlang hour were 0.51 and 0.46. Downlink (DL) and uplink (UL) bit error rate (BER) were 0.41% and 0.32%. The values are based on 8 days of data, from Monday to Monday.

Figure 12.10 shows quality indicators before and after the retune. The solid vertical line marks the retune. Carried traffic is displayed on secondary Y-axis as Erlang hours, while the quality indicators have percentage values.

After retuning the trial cells based on mobile measurements, the quality indicators were checked again to compare the quality of different frequency plans. DCR was 1.14% and RRDCR was 1.01%, corresponding to improvement of 12% and 13%. Drops and radio drops per Erlang hour were 0.46 and 0.41, indicating a gain of 10% on both. Bit error rates had decreased to 0.34% for DL and 0.29% for UL, therefore improving by 18% and 10%. Greater improvement in DL was expected, since the mobile measurements only provide DL information. It can be seen however, that using only DL measurements also improves quality in the UL direction. Total number of HOs per carried traffic also decreased after the retune by 2%. This is due to smaller number of triggered quality HOs.

After the retune the reason for remaining dropped calls was investigated . It was found out that virtually all of these were due to low signal level. Therefore frequency planning could not improve DCR in this network any further. The

Figure 12.10   Quality in the network before and after MMFP.

magnitude of the gain obtainable from this method is strongly dependent on the quality of the existing frequency plan. The poorer the performance of the propagation based plan, the larger the gains that can be obtained from the MMR based approach.

The trial has proven that despite the measurement limitations inherent to GSM, which were listed in section entitled *Signal level reporting in GERAN* in 12.2 it is possible to utilise these measurements for frequency planning and significantly improve network performance.

12.3 Automated measurement based adjacency planning

Subscribers can move freely in a cellular network whilst still maintaining traffic connections. This is made possible by the use of HOs. The basic concept involves the subscriber moving from the coverage area of one cell to the coverage of another hence establishing a new connection with the target cell and releasing the connection with the old cell. Such cell change operations are referred to as HOs and occur while the MSs are in *dedicated mode*, i.e. whilst a call is ongoing.

To perform such HOs the target or adjacent cells need to be evaluated using adjacent cell measurements. These measurements provide the serving BSC with information about the received power levels from the target cells that are to be considered for HO. The basic reasons for triggering a HO are poor call quality or received signal strength falling below required thresholds and also when the traffic capacity of the serving cell is approaching its limits. Adjacent cell measurements are therefore carried out for all the required target cells and a HO is initiated to the best target cell. Without defining correctly defined cells, call quality and performance are clearly compromised.

To obtain the adjacent cell measurements the mobile stations must be informed of the frequencies that are used on all surrounding cells, which need to be measured. These frequencies have to be defined in the BCCH allocation list (BAL) of the cell to be analysed. Two types of BAL can be defined; IDLE which is used by MSs in idle mode for cell re-selection and ACTIVE, which is used by mobile stations during dedicated mode for performing handovers. The ACTIVE BAL therefore usually consists of the list of frequencies of the adjacent cell BCCH carrier frequencies normally defined in a neighbour definition list. A maximum of 32 such frequencies or adjacent cells can be defined in a neighbour list in the BSC as per GSM specifications 05.08 in Section 7.1.

The goal of adjacency management is therefore to have the right BCCH carrier frequencies in the neighbour definition list so as to always provide the best target cells to perform HOs. Since any group of adjacent cells, neighbour cells, or target cells for a cell A is a set of cells to which HOs from cell A are permitted it becomes important that this group of cells is correctly defined. The responsibility of defining these HO target cells is currently a function of network planning tools that operate by simulating the cellular network operations using various definable parameters and constraints. Operators also perform drive tests and through trial and error generate adjacency lists for key areas of the network. Some advanced operators also utilise Abis data to capture adjacency performance metrics and then optimise their neighbour definition lists.

The techniques mentioned are however laborious and the complexity of managing adjacencies increases as the network evolves and it's behaviour alters. An automated and accurate adjacency management method would be able to react to evolution changes quickly and hence greatly benefit operators, reducing operational efforts and increasing the efficiency of the network.

The following sections provide a description of how capable a measurement based adjacency management technique is in performing the task of always providing the best possible candidates as handover targets.

Maintaining adjacencies

Network performance and quality depend on various operational factors but one of the first stages in a network optimisation loop involves frequency and adjacency cell planning. To maintain and manage the adjacency relations involves the definition of the correct adjacencies for handovers and eliminating the rogue or under performing adjacencies. Both these steps aim to reduce drop calls and congestion, improve call quality, support the traffic density patterns and hence provide better network performance. However, this is generally an extremely tedious task. In a network of 30,000 cells where each cell may have up to 20 adjacencies it requires handling 600,000 adjacency relations. Assuming that on average it takes 5 minutes to manage one adjacency, the total time required for 600,000 adjacencies is around 5 man years. This provides a synoptic view of the resource and time required within the daily operational tasks of running a cellular network. The present technique most frequently employed by operators is to accurately try and model their network using offline-planning tools utilising propagation algorithms. Although this can prove to be somewhat successful, modelling of the actual network is a painstaking, laborious and resource consuming task. The constantly changing network means plans generated may quickly become out dated and efficiency is lost due to the operational speed of such tools.

Solution description

In light of the problems and failures of the existing methods and solutions, an automated mobile measurement based adjacency management technique has been developed. A major advantage of using measurements comes from the fact that they intrinsically take into account all propagation effects, subscriber density and movement conditions. Planning tools tend to be inaccurate espe-

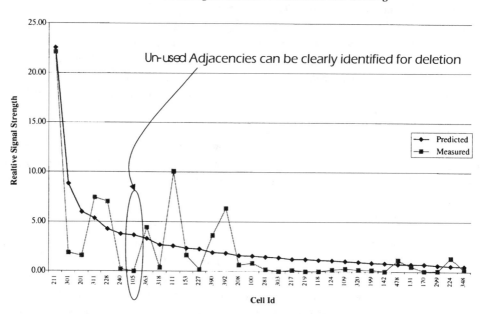

Figure 12.11 Measured vs. predicted cell ranking.

cially in complex radio propagation environments, such as city centres, where the radio wave propagation is difficult to predict. A measurement based technique can therefore be seen to offer advantages in terms of more accurate adjacency definitions than possible with a theoretical based approach.

Tests using measurement data show the differences between a prediction based approach and a measurement based one. The data shown in Figure 12.11 comes from real network tests conducted in a busy city environment where mobile measurement data from several days were analysed. The results are best analysed by displaying the 32 strongest predicted adjacencies ranked against the measured signal strengths. It can be seen that the predicted ranking is quite different from that obtained using measurements and that some of the cells predicted to be good target cells were not even measured by mobiles in dedicated mode. The results show that for example the sixth strongest predicted adjacency was not even measured and is therefore incorrectly defined as an adjacency. This adjacency should therefore be recommended for deletion to reduce the chance of drop calls occuring.

The various performance metrics of adjacencies, such as HO attempts, HO successes, etc., are normally stored in the network management database. Such data can easily be utilised for deletion purposes and is readily available without having to facilitate for any kind of special data collection. However, for the more difficult task of identifying missing adjacencies, it becomes necessary to measure all the surrounding cells and then deduce the best candidates for HOs.

Any method to generate IM (interference matrix) values described in the Section entitled *Review of interference matrix types* in 12.2, can also be used to decide whether a cell should be added to the neighbour definition list of another cell. If the IM value corresponding to cells A and B is high, it indicates a high degree of coverage overlap between the two cells. Therefore, the likelihood is high that HOs will be made between the cells.

A description of the adjacency management process

This process has been identified as having two parts; deleting obsolete adjacencies and creating missing ones with both processes exploiting MS measurements. The entire process is repeated on a regular basis to maintain optimal adjacency performance in the network.

Deletion of obsolete adjacencies

It is important when looking at the deletion process to first identify and benchmark the performance criteria that are required. Factors like network growth and strategy are important and contribute towards the benchmarking of the performance requirement. Any existing adjacencies that do not then meet the required performance criteria are deleted from the network. An example

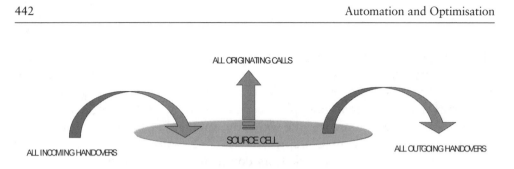

Figure 12.12   Handover traffic percentage.

performance metric applied to the HO performance data is shown in Figure 12.12, where data from existing adjacencies are compared with a value known as HO traffic percentage. This is evaluated as being the ratio of the calls set up and handed into cell A which are then later directed successfully to cell B to the total successful handovers of all defined adjacencies of cell A. This together with the handover success percentage to a target cell metric provides an effective benchmark to allow detection of badly and under-performing adjacencies. The overall deletion process can be outlined as follows and is shown in Figure 12.13.

The majority of the steps highlighted in the deletion procedure are automated. The benchmarking and verification steps however need to be carefully assessed and implemented in the procedure.

Figure 12.13   Adjacency deletion process.

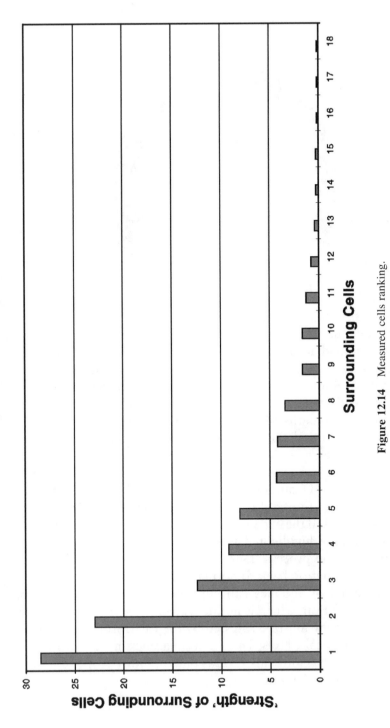

Figure 12.14 Measured cells ranking.

Figure 12.15   Adjacency creation process.

Creation of missing adjacencies

The other part of the adjacency management procedure concerns the aspect of identifying missing adjacencies. This process is a little more complex than the deletion of obsolete adjacencies and requires measurement data from all surrounding cells. Ranking of the measured cells as shown in Figure 12.14 is a clear way of identifying any of the measured target cells. Such a ranking shows the relative signal strengths of the measured cells or the extent of coverage overlap between cells and the most overlapping cells can be selected as the new target adjacencies.

As with the deletion process, a benchmark must be applied that allows the selection of the right HO candidates from such a ranking. An example benchmark for target cell selection could be determined from the product of the average signal strength value and the average number of samples of a measured cell. The creation process can be highlighted as is shown in Figure 12.15.

Most of the creation process steps can be automated and this also includes restoring the BAL situation back to normal. Always having 32 BCCH frequencies to be measured by mobiles can contribute to some degree of network performance degradation, which is not desirable. Once the newly created adjacencies are created they must be assessed by their performance in the network and must be seen to be take HOs. The whole adjacency management process has to be repeated at regular intervals such as on a monthly basis, to keep up with network evolution and changes.

Network test results

To test the functionality of this new approach a trial was designed and implemented during the early part of 2001 using a network management system as a central feature control system. An entire area managed by a BSC in a busy city environment was the subject of the trial where the trial objectives were to:

- Determine and delete unused adjacencies for a BSC.
- Determine and create missing/required adjacencies for the same BSC.
- Verify the network performance improvements related to the new adjacency plans.

The trial was also focused into looking at the 900 layer part of the dual band network and an important secondary objective of the trial became the mapping of 900 layer adjacencies to 1800 layer cells in the network to determine 1800 layer adjacencies. The possibility of using a mapping technique successfully was based on the fact that all the 1800 layer cells were co-sited to the trial subject 900-layer cells and were all in a high cell density area.

The deletion process was performed first and allowed the selection of obsolete adjacencies. A performance target was determined and any adjacencies failing to meet the criteria were deleted.

The creation process required measurements from all neighbouring cells and these were obtained using all 32 of the operators' 900 MHz BCCH frequencies. The measurements yielded the BCCH, BSIC, number of samples and average

CASE 1:

Mobile Measurements

SOURCE-BTS	SUM_DL_SIG_LV	SAMPLES	T-BCCH	T-BSIC	T-CI
073	2835662	67630	113	53	134
073	2034602	60600	530	53	434
073	674451	24639	103	51	234
073	373939	15562	106	50	334

Performance Statistics

BTS ID	Att_HO=>	<=HO_Att	TargetBTS
073	1550	378	434
073	1165	2256	134
073	657	317	889
073	384	74	737
073	324	647	293
073	231	378	393
073	92	139	192
073	54	110	221
073	50	8	612
073	48	73	334
073	25	95	234
073	17	0	534
073	15	42	312
073	2	0	634
073	1	0	534

Figure 12.16 ADCE trial results – case1 (source microcell).

measured signal strength and example data shown in Figure 12.16. A threshold was generated based on performances of the existing adjacencies to select suitable candidate adjacencies. Results obtained this way were verified before being downloaded to the network and performance data studied to determine the success of the newly added adjacencies.

Figure 12.16 also shows the results and performances of the newly created adjacencies in case of a measured microcell.

The data shown contains the target cells identified by their BCCH and BSIC, their measured coverage overlap in terms of the sum of downlink signal strength and the number of times the measured target cell was seen. Once these measured cells were benchmarked and created as new adjacencies, their performance was monitored. The data presented shows the newly created adjacencies are clearly handling a large percentage of the overall handovers in terms of successful attempts made in both directions. The newly created adjacencies are highlighted so their performance can be easily compared to the existing adjacencies and are seen to perform better than the best existing adjacencies.

The HO utilisation of the added adjacencies clearly show that using a measurement based technique is valuable, providing a viable solution, which can also be easily automated.

Outlook for 2G–3G planning

With the deployment of 3G cells, HOs from 2G to 3G (2G3G HO) become important. 2G3G HOs can be coverage, load or service based. The scenario, especially in the early days of 3G, will be that the network environment evolves with rapid changes. As a result it will become tedious to manually handle neighbour relations from the 2G to the 3G system (2G3G neighbours). The adjacency management system described earlier is not applicable directly to 2G3G adjacency management. This is because the system is designed for 2G2G neighbour definition lists that contain BCCH frequencies and not cells. Using these BCCH frequency lists, the mobile measures a cell that happens to be the strongest one and reports this to the BSC. But the 2G3G neighbour list contains cells, which means that any cell no matter how strong it is, it is not reported if it is not listed. Therefore, unlike the 2G2G neighbour list, it is not possible to add new previously unlisted neighbours based on mobile measurements to the 2G3G adjacency list simply because measurements can only be obtained from cells that are already listed.

One possible approach is to keep one or more slots in the 2G3G neighbour list as temporary slots which are used for neighbour candidate cycling. These candidates are taken from the neighbour candidate list which is much longer than the actual neighbour list. A candidate list contains all cells within a certain geographical area from the serving cell. Each cell from the candidate list is moved to the neighbour list for a certain time period. After the overall measurement period, the worst performing adjacencies from the 2G3G neighbour list are deleted. These worst performing adjacencies are chosen by the same method as in the case of a 2G2G neighbour list, namely by the number of successful HOs performed to the cell. New candidates are moved from candidate list to real neighbour list for a time and the process is continued until the neighbour candidate list has been exhausted. At that point all cells have been trialled as real neighbours in turn, and only best performers have been kept as real neighbours.

One drawback of this method is that if the neighbour candidate list has many entries, it can take a long time before each candidate has been on the real neighbour list. If the candidate list is shortened, there is a risk that an important cell is left out meaning that it cannot end up on the real neighbour list. One way to overcome this is to allow more temporary slots for trialling cells with the obvious side effect of potentially reducing the performance of the network whilst the measurements are being taken.

12.4 Automated parameter optimisation

Outline of problem

Current procedures for network design estimate the radio network performance based on empirical propagation models. The accuracy of the models is largely dependent on the precision with which the actual environment has been modelled in the tool. Adjustments, such as base locations, antenna tilt and orientation, BTS output power or frequency plans, are normally carried out during the design phase, following a trial-and-error approach.

During the deployment phase, some limited further field-based tuning of the previously mentioned parameters is performed based on the feedback data obtained from drive tests or key performance statistics. Closing the optimisation loop with actual performance indicators ensure that the network is running at the desired operating point.

Finally, during the operation phase, homogeneous parameter setting is normally considered, as the large set of parameters makes the detailed parameter planning process on a cell-by-cell basis a time-consuming task. Therefore, operators fix parameters to common sets of default values shared between cells, even if not optimum performance in terms of quality/capacity is achieved. This homogeneity hypothesis may be far from reality, as propagation, interference and traffic distribution vary both in space and time throughout the network. This uneven distribution may result in further degradation of system performance.

In cases where existing specific features are extremely sensitive to network peculiarities (e.g. traffic balancing in complex urban scenarios), tailoring of parameters is sometimes performed by means of field trials carried out on a few cells which define the default values to be applied across the network.

In order to take into account non-homogeneity in the space domain, few advanced operators extend parameter optimisation to different scenarios such as rural, urban, tunnel, indoors or layer/band like Macro900/1800, Micro900/1800, Pico1800, Motorway900. Retuning is only carried out on those cells where performance problems exist. Cell-by-cell based parameterisation is rarely considered due to time and cost constraints. Therefore, full potential of parameters that can be adjusted on a cell-by-cell basis is not fully exploited.

Even if optimum parameter values were reached through trials for a certain area, network behaviour is still changing with time. Temporal variation results from a combination of both long-term trends (i.e. user growth) and periodical changes (i.e. season, day and hour periods). Whilst slow trend variation can be easily coped with by means of network re-planning, no automatic instantaneously reactive process is currently in use, still remaining as a challenging optimisation task.

From this analysis, it can be concluded that the flexibility emerging from the existing large set of parameters included in various network algorithms cannot be currently fully utilised because of the complexity in current networks and a lack of tools to fully support this activity. Obviously, a great potential gain can be obtained from an *automatic parameter tuning and optimising process* that takes advantage of cell and adjacency parameters. Such strategy would help operators during the deployment and operational stages, providing benefit from cost savings and improved performance despite changes of the environment both in space and time.

Automatic optimisation has clear and direct benefits for the operators, since it aims to maximise radio performance for a given network infrastructure through a minimum manual effort. Keeping the network in a continuous optimum point, the operator can minimise the over-dimensioning factor, thus reducing the infrastructure costs to have a certain grade and QoS. At the same time, network radio performance is improved, increasing revenue and user satisfaction.

The increased complexity involved in parameter consistency checks would be the main disadvantage derived from this cell-basis optimisation approach. Hence, only a subset of radio parameters must be subject to optimisation, while the rest are set to appropriate default values. This drawback may be relaxed if traditional value-oriented checking is substituted by valid interval-oriented checking.

The next section gives a brief introduction to general control engineering concepts and techniques. A control system architecture for telecommunication networks optimisation is then presented. Later on, some examples of optimisation algorithms are described. Finally, the last section provides some thoughts on future principles of automation.

Introduction to control engineering

Automatic control has played a vital role in the advance of engineering and science [12], providing the means for attaining optimal performance of dynamic systems, improving productivity and relieving the drudgery of many routine manual operations.

The first significant work in automatic control was James Watt's centrifugal governor for the speed control of a steam engine in the eighteenth century. During the decade of the 1940s, frequency-response methods made it possible for engineers to design linear close-loop control systems that satisfied strict performance requirements. Since the late 1950s, the emphasis in control design has been shifted to the design of optimal system in some meaningful sense.

During the years 1960–1980, optimal control of both deterministic and stochastic systems, as well as adaptive and learning control of complex systems, were fully investigated. In recent years the artificial intelligence approaches have been used extensively in complex control system design.

Now that computers have become cheaper and readily available, they are used as integral parts of control systems. Recent applications of modern control theory include such non-engineering systems as biological, biomedical, economics, and socio-economic systems [13].

Control of complex man-made systems
The world of artificial systems is reaching complexity levels that escape human understanding. The control of surface traffic, electricity distribution, airplanes, mobile communications, etc., are examples of problems that are beyond classical scientific or engineering knowledge.

As the complexity of man-made systems grew it became apparent that there were many important problems that could not be related to a specific engineering discipline but required an holistic view. It became essential to consider the interaction of parts that form a system instead of the parts themselves. This led to emergence of new systems oriented disciplines such as automatic control [14].

Although many control approaches and solutions have been developed so far that solve linear and a few non-linear control problems, no general method for control system design has been found for all complex systems. Control systems have a large complexity and it is formed of smaller parts integrated in common control architecture.

Control principles
System engineering theory is the language used to discuss control problems. It has its roots in the mathematical theory of dynamic systems. Automatic control provided a new element in the theory because it was necessary to explicitly account for the fact that the system could be influenced externally.

The foundation of system theory consists of ordinary and partial differential equations, but many other mathematical tools such as operator theory, algebra and differential geometry have also had a strong impact.

Feedback is a fundamental system principle which is essential in control system design. Feedback control, or close-loop control, refers to an operation that, in the presence of disturbances, aims to reduce the difference between the output of a system and some reference input and that does so on the basis of this difference, the error. Here only unpredictable disturbances are so specified, since predictable or known disturbances can always be compensated for within the system (see Figure 12.17).

The principle is simple – if it is possible to measure something in a process and also do something that affects that process in a predictable way then keep checking it and carry out whatever can be done to improve it. If the impacts of an action on the process are known, it may be possible to calculate what actions to take to improve the performance.

The simplest linear control system that can be used is a constant gain. Thus, the control action is proportional to error signal. Generally, the farther the system is from the stable point, the more actuation is feed into the system. This controller is easy and can be applied for poor performance specifications. The most frequently used controller in industrial applications is the proportional, integral and derivative (PID) controller. In this case, the control action is a sum of a proportion to the error, a proportion of the integrated error, and a proportion to error derivative. Basically, the integral term helps in correcting steady-state error and the derivative term permits fast system response.

Open-loop control systems refer to those systems in which the output has no effect on the control action. In other words, in an open-loop control system the output is neither measured nor fed back for comparison with the input. This type of control can only be used when the behaviour of the system under control is very well known and no major disturbances affect it. In this case the control system is a static mapping function from target space into control action space.

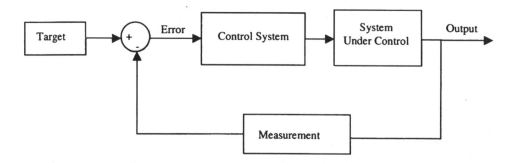

Figure 12.17 Feedback control system diagram.

Generally speaking, in control system design the trial-and-error repetition method plays an important role. After a control system is designed, the designer checks to establish if the designed system satisfies all the performance specifications. If it does not, then the design process is repeated by adjusting the parameter setting or by changing the system configuration until the given specifications are met. Although the design is based on a trial-and-error procedure, the ingenuity and know-how of the designer will play an important role in a successful design. The knowledge on the system under control is also a critical issue in control system design.

Since there are not precise mathematical models of mobile network systems, analytical design methods are not feasible. Artificial intelligence techniques, especially expert systems, and other heuristic approaches will play an important role in control system design for this type of systems.

Control system for telecommunication system performance optimisation

Once general control techniques have been briefly introduced, their applicability to the mobile telecommunications field must be analysed. From a system engineering point of view, the radio access system (RAN) in a telecommunication network is a high dimension and non-linear complex system. One of the most complex systems created. It has millions of inputs, hundreds of parameters in thousands of cells that define the overall system behaviour and performance. The complexity of the system is not only due to the huge number of parameters but also because the complexity of the non-linear dynamics involved.

The RAN performance optimisation problem can be formulated as a control problem in which the inputs of the system are the radio parameters (a) and the outputs are statistic counters (b) (see Figure 12.18). Although the system under control is composed of RAN and an external system, the customers using the resources (c), only parameters in RAN are available as control input.

The inputs to the optimisation system (d) are those operator definable parameters which permit determining the performance: costs of different events in the systems like a block call or a drop call, gains derived for system utilisation as revenue of speech call and packet data calls, and restriction in key performance indicators (KPIs).

The outputs of the system (e), which measure the overall performance, are the revenue and the actual KPI values, which are obtained from raw counters. The optimisation system defines the optimum set of parameter values performing a translation from operator inputs and with feedback control from counters.

Control system architecture
The great size and high number of different dynamics involved in the system make it impossible to develop a unique control system which tackles optimisa-

Figure 12.18   Control of telecommunication radio access system.

tion of all parameters. Instead, a modular analysis of the system is required. First, a control hierarchical architecture (see Figure 12.19) and multiple control system are designed, which will be later mapped into network elements.

The definition process of the controller architecture is as follows:

- Firstly, the RAN is split into smaller subsystems that can be analysed and controlled separately. Those subsystems should be as small as possible and at the same time as separate as possible to simplify the control system design. Some examples of those subsystems could be, as shown in Figure 12.19, traffic management, power control, GPRS dimensioning, etc.
- Then, a control subsystem is designed for each radio subsystem (first layer in control architecture). The control approach selected in this phase depends on the subsystem characteristics, linearity, dynamic involved, etc.

Figure 12.19   Control system architecture.

Finally, the algorithm is integrated in the overall control architecture. Splitting the system into subsystems, dynamic independence has been assumed. Actually, this is not true and soft interaction between subsystems exists. This interaction is considered in higher layers of control architecture. Synchronisation and communication between algorithms is also done in those levels. The main characteristics of this hierarchical control structure are:

- *Scalability.* The control system can be built in different phases with minimum impact to other parts of the system. Whenever a new control system is developed it can be integrated relatively easily into the existing system.
- *Maintainability.* As the control system is split into smaller subsystems, any change in one system has no effect on other systems and thus the maintenance effort is minimised.
- *Reusability.* Control subsystems developed for a RAN subsystem can easily be adapted to similar problems.
- *Capacity to distribute.* The control system can be distributed through different elements along the network (BTS, BSC, NMS, etc.) in order to minimise the information flow hence the control system can be sparse.
- *Multisystem.* Such a control system architecture deployed for RAN of telecommunication networks can be applied for any radio access technology (GSM, EDGE, WCDMA, WLAN, etc.) and even for other network subsystems (core, IP network, etc.).

Final implementation of this hierarchical structure requires a mapping process of control subsystems to network elements. In this process, control algorithm requirements, both in time response and amount of statistical information, must be considered. Agile algorithms (i.e. seconds time scale) will sit on elements close to the radio resource management algorithm (i.e. BSC in GERAN or RNC in WCDMA networks). Those algorithms that require area or network level statistics will be placed on elements with wider and longer visibility (i.e. OSS).

The next section will intend to give the reader an understanding about how previously discussed control techniques may be applied to increase network performance in cellular networks. Several design examples of control elements for different GERAN radio subsystems will be proposed, which would be finally integrated into the outlined control architecture.

Applications for radio network parameter optimisation

The starting point of any controller design is the definition of the parameters to be controlled. As a first approach, any parameters that could be tuned to take into account temporal or spatial non-uniformity in the network is candidate for adjustment. Nonetheless, a good understanding of the main reasons for space and time inhomogeneities in the system is extremely helpful to identify the main areas, that may benefit from improved performance provided by parameter optimisation. Once these areas are determined, subsequent studies will aim at analysing the impact of specific parameter tuning processes on network performance.

The first cause of cell differences is the *radio propagation* environment. Several basic cellular structures (e.g. macro cell, micro cell, overlay, underlay) are deployed in current systems to accommodate high traffic density while still keeping high QoS. Signal propagation peculiarities such as slow fading due to shadowing and fast fading due to multipath effects prevent network algorithms from performing well in all cells with common parameter values. In addition, due to frequency planning and propagation related issues, the interference (and subsequent quality) is not uniform throughout the network. Hence, adaptation to local interference and propagation conditions is desirable.

Traffic is another main reason for network performance changes, translating into hard (i.e. hardware) or soft (i.e. interference) blocking as its main negative consequences. Due to user mobility, small areas with instantaneous high user density tend to be rather usual but not spatially fixed (e.g. hot spot, events). In order to cope with these large peaks of capacity demands, networks may be over dimensioned in the design phase, leading to an inefficient usage of resources. To solve this problem, several strategies that take advantage of the overlapping area among neighbouring cells have been proposed in the literature under the name of load sharing. In this direction, both off-line and real-time adjustment of handover parameters proves to be instrumental in relieving persistent and temporary congestion problems.

It can be concluded that many fields in radio access networks can be optimised automatically by applying control and optimisation algorithms. In the following, several examples of control based optimisation algorithms are presented. It is important to remember that this set is not meant to be comprehensive but give a rough idea of their possibilities.

Basic handover parameters

In cellular systems, one of the most important tasks is to maintain a call while a mobile station moves between two base stations, which is accomplished by the handover (HO) process. Thus, optimisation of the parameters associated to this mechanism is crucial.

Any HO parameter tuning strategy will trade-off outage probability and handoff delay against number of HOs (that is to say, dropping probability and co-channel interference against switching load and connection quality). The final performance metrics to be used for evaluation of the optimisation process will largely depend on the purpose of the HO algorithm to be optimised (e.g. whether emergency or power budget HO) and the overlapping coverage situation among considered cells. [22–25]

Parameters related to HO algorithm may be grouped into three main categories: measurement averaging, triggering condition and target cell evaluation.

The *averaging interval* is the first HO parameter considered for optimisation. Long averaging windows tend to reduce the number of unnecessary HOs due to slow fading deviations (as fast fading is normally averaged out by long measurement interval). The main drawback of a long averaging window is a delay increase, which may jeopardise call quality in scenarios where sudden changes in signal or interference occur (corner effects in microcells). Hence, a tradeoff must be achieved depending on the local propagation conditions.

Subsequently the *handover evaluation* process is triggered either by a threshold crossing event or a periodically enabled timer. Monitoring of absolute quality, field strength or distance measurements have been common practice for urgent HO triggering. The adjustment of these parameters is somewhat limited because its likelihood to produce outage problems.

Finally, the *target cell evaluation* consists of the assessment of cell attractiveness from a candidate list, based on received signal levels. A HO to an adjacent cell is proposed whenever the level difference (i.e. power budget) exceeds a predefined margin, which may be different depending on the triggering cause. Consequently, margins are the key parameters to ensure hysteresis so that unnecessary HOs are avoided. In combination with the averaging parameters, margins will smooth out the random nature of the level crossing process due to shadow fading. Additionally, guard timers are added as a timing constraint to reduce ping-pong HOs.

The optimisation of this evaluation process is first focused on the requirement for a cell to be included in the candidate list, which is built based on minimum signal level constraints. These level thresholds may be individually tuned to adapt to local interference and propagation cell conditions. Hence, an estimation of the minimum signal level to ensure acceptable call quality on a cell basis is required, a process that will be discussed in detail later in this chapter.

Likewise margins have shown to be a powerful mechanism to easily control the operational cell area. By means of changing margins for both directions of the adjacency, while still maintaining the overlapping region for stability purposes, the crossing point where HO occurs between both cells is displaced. Cell shrinking (or cell breathing as it is also known in the literature) may be instrumental in balancing cell performance and relieving time/space localised problems across the network, as proposed in [15]. This topic will be also explained later in this chapter.

Cell breathing

In GSM networks, candidate cell comparison takes place whenever an urgent or better-cell HO is triggered. Connection to the cell with a higher priority is assured by the power budget evaluation of neighbouring cells. Within this

Figure 12.20   Cell breathing as a consequence of margin parameter tuning.

process, the figure of merit of a cell is proportional to the received downlink level received by the user from the beacon channel. In equation (12), a simplified power budget equation for handover is shown:

$$PBGT_{serving \to adjacent} = RxLev_{adjacent} - RxLev_{serving} \geq Margin_{serving \to adjacent} \qquad (12)$$

Handover margins play an important role in the definition of the operational area of cells in a system, as they may be used to bias the level budget evaluation to request a HO, apart from its use for hysteresis purposes. Margins modifications will impact the instant where HO takes place between two adjacent cells and this will produce changes in the service area covered by the base station. This effect is known as cell breathing.

As adjacency parameters, margins may change both size and shape of the serviced area of a base station, as depicted in Figure 12.20. This approach proved to be useful when relative (i.e. not absolute) comparisons of performance indicators are used for cell prioritising and a certain degree of cell overlapping is present in the network.

Applications for cell resizing will aim at levelling (or balancing) some cell problem indicator throughout the network. Traffic, quality or interference balancing strategies have been proposed in the literature as a potential means for adaptation to spatial distribution of radio traffic (or interference).

Dynamic handover margins for load sharing
In a network where load is unevenly distributed, hard blocking probability may increase above acceptable limits. Various congestion relief techniques are currently in use, such as directed retry and traffic reason HO.

In the first strategy, a *directed retry* HO is performed from a source cell to an adjacent cell whenever a new call attempt is raised in a congested cell. Minimum level thresholds are required to assure sufficient quality when redirecting a call to another cell. These thresholds may be optimised by the tuning procedure proposed in the cell breathing section that makes use of measurement reports in the candidate cell. Although useful for hard blocking relief, this HO is not optimal from the interference point of view as a call close to BTS A may be directed to a distant BTS B.

Traffic reason handover partially solves the problem by temporarily modifying the power budget margins from congested to empty cells, while ensuring a constant hysteresis region. Thus, optimisation of adjacency margins can help to cope with uneven user load distribution problems while users are mainly connected to the closest base stations. Consequently, interference in the network is reduced compared with the directed retry feature. Equations (13) and (14) show the basic principle for traffic balance parameter optimisation

$$PBGT_{serv \to adj} = RxLev_{adj} - RxLev_{serv} \geq Margin_{serv \to adj} = f(Load_{serv}, Load_{adj}) \quad (13)$$

$$Margin_{serv \to adj} \cong K^* [f(Congestion_{adj}) - f(Congestion_{serv})] = - Margin_{adj \to serv} \quad (14)$$

where

$PBGT_{serv \to adj}$ = power budget between serving and adjacent cell
$RxLev_x$ = received field level from the beacon channel of cell x
$Margin_{serv \to adj}$ = value to be surpassed to trigger a HO for this reason
$Congestion_x$ = any suitable congestion indicator in cell x
$f(.)$ = mapping function used to derive the final problem metric
K = gain factor to control the sensitivity to cell differences

A practical implementation of this tuning process is shown in Figure 12.21. An incremental controller procedure is launched for every adjacency, which proposes changes in current power budget margin values. Direction and magnitude of these changes are based on traffic indicators in analysed cells.

In the first block, congestion performance indicators are mapped into a meaningful metric related to problems, namely 'cost' in the optimisation terminology. The congestion-to-cost mapping process will take the form of a non-linear function, based on problem threshold detection, controlled by the operator strategy.

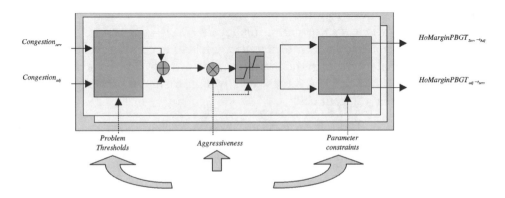

Figure 12.21   Feedback control system diagram.

Comparison of the cost metric will give information about the deviation from the equilibrium situation, where costs are equal in both cells and so balance is reached. The control signal is proportional to the deviation, from the equilibrium making use of some scaling factor, namely the aggressiveness. Speed and stability issues will be accounted for through the application of control theory principles over this parameter.

It is worth noting that proposed changes in margins for both directions (source > target and target > source) are calculated in one go, resulting in the same magnitude and opposite direction. Hysteresis region is so maintained in order to avoid unnecessary HOs due to ping-pong effect.

Operator supervision will govern both the optional activation of the optimisation on cell/adjacency basis and parameter valid intervals by the constraint control block.

The basic principle of traffic balancing through parameter tuning can also be applied to other HO algorithms intended for traffic management between layers in a hierarchical structure. Velocity thresholds in mobile speed sensitive HO [16] or level thresholds in umbrella HO are such examples.

Other balancing strategies

The *power budget* concept is known to effectively reduce interference in homogeneous networks. With this assumption, the level crossing point between two base stations match up with the quality crossing and path loss crossing point. Either because of differences in transmission power or interference/propagation conditions, deviations between ideal crossing points are encountered.

Similar to traffic balancing, quality balancing also prioritises cells based on the estimated offered QoS. The hysteresis and offset term would be optimised separately to give a final pair of margins for any symmetric adjacencies. Thus, the crossing point between cells would be the quality crossing (instead of level crossing) situation. A simplified method to include interference and signal propagation differences between adjacent cells based on the quality-to-level mapping values is later proposed.

Gathering of statistical information related to traffic congestion, connection quality problems and produced interference, will help to combine cell individual and network global interests in the margin tuning process. The basic implementation, shown in equation (15), represents a weighted sum of problem indicator differences between adjacent cells

$$\text{Margin}_{\text{serv}\rightarrow\text{adj}} = \sum K_i^*[f_i(\text{Problem indicator}_{\text{adj}}) - f_i(\text{Problem indicator}_{\text{serv}})]$$

$$= K_1^*[f_1(\text{Cong}_{\text{adj}}) - f(\text{Cong}_{\text{serv}})] + K_2^*[f_2(\text{Qual}_{\text{adj}}) - f_2(\text{Qual}_{\text{serv}})] + K_3^*[f_3(\text{Interf}_{\text{adj}})$$

$$- f_3(\text{Interf}_{\text{serv}})]$$

$$(15)$$

where

$\text{Margin}_{serv \to adj}$ = power budget margin from serving cell to adjacent cell
K_i = relative weights for problem prioritisation
Problem indicator = any metric related to congestion (Cong_x), quality (Qual_x) or interference (Interf_x)
$f_i(.)$ = mapping function to derive problem severity from performance indicators.

Quality-to-level statistics for level-related parameter tuning
From the previous analysis, it can be deducted that *margins* are a means for smooth cell area optimisation while *minimum level thresholds* help in adding constraints to the optimisation procedure.

Access to a cell that is unable to provide an adequate connection quality, should be prevented both in call set-up or HO. Minimum signal level thresholds ensure of that responsibility, so tuning of these parameters is a key factor in the optimisation for clamping and HO processes.

Information about the exact signal level that will ensure acceptable call quality is reflected in cell statistics related to connection measurement reports. In these measurement reports, connection quality and received level are continuously monitored. By means of this piece of information, the relationship between received signal level and perceived signal quality may be derived in a cell level, thus reflecting signal propagation and interference peculiarities combined. Doing so, traditional HO signal level thresholds are easily upgraded into signal quality thresholds.

In these collected quality-level statistics, practical issues related to memory size in network elements limit the number of counters involved. Thus, the full range of reported values is segmented into intervals. Figure 12.22 represents an

Intervals (dbm)	q0(%)	q1(%)	q2(%)	q3(%)	q4(%)	q5(%)	q6(%)	q7(%)	Share in range
(-100, -110]	0.07	0.03	0.02	0.05	0.09	0.11	0.007	0.02	0.46
(-95, -100]	0.24	0.07	0.11	0.12	0.08	0.03	0.02	0.02	0.67
(-90, -95]	1.36	0.12	0.11	0.06	0.05	0.03	0.01	0.00	1.74
(-80, -90]	5.67	0.41	0.36	0.34	0.31	0.24	0.11	0.05	7.49
(-70, -80]	25.36	1.66	1.54	1.26	1.10	0.68	0.20	0.08	31.88
(-48, -70]	50.39	2.04	1.89	1.48	1.19	0.59	0.15	0.01	57.75
Share in RxQual range	83.08	4.33	4.04	3.31	2.82	1.68	0.57	0.18	100.00

Figure 12.22 Quality-level distribution in a tabular form.

example of quality-level distribution in a tabular form. RxQual reported values extending from q0 (best quality) to q7 (worst quality) represents the perceived quality, while the received signal level measurements are grouped into six intervals. On this table, it can be observed that the relationship from level to quality is not deterministic but probabilistic (i.e. there could be two measurement reports with the same level but different quality). Factors such as user position or instantaneous interference levels are reflected within the stochastic nature of this information.

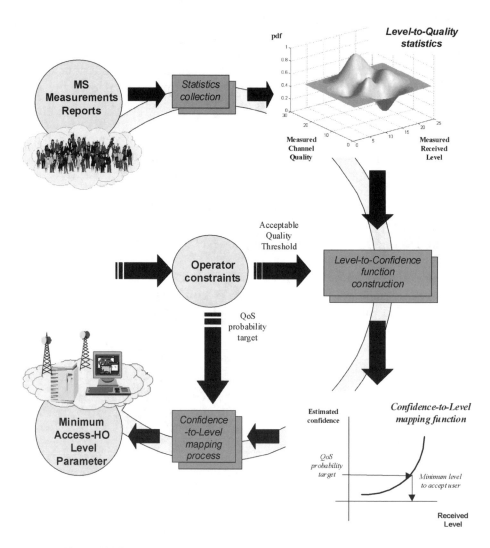

Figure 12.23   Basic structure of tuning process for minimum level thresholds.

From these statistics, the mapping function between signal level and quality must be built. Figure 12.23 summarises the basic ideas of the reckoning process. The quality-level statistical distribution is built based on the number of reports in a certain (level, quality) band, which basically forms a jointly-distributed probability density function. Later customisation includes accounting for operator specific quality and certainty constraints.

Figure 12.24 depicts in detail the construction of the quality-to-level function. Figure 12.24 (left) represents the cumulative quality distribution (i.e. probability of quality better than certain value) against level ranges. Obviously, higher signal level values lead to better quality, which translates into a higher probability of obtaining acceptable quality. The reckoning process translates the quality probability distribution in every level range into a single quality indicator, so that a deterministic level-to-quality (or quality-to-level) relationship is established. This ascription process is controlled by the operator through the selection of the percentage of quality reports to be exceeded by the quality value that is finally assigned to a certain level range, which clearly resembles to a confidence factor. As an example, a confidence equal to 50% has been chosen, thus assigning the median quality value to every level band.

Figure 12.24 (right) represents how the confidence value controlled by the operator determines the mapping curve from quality to level values. The exact curve is the intersection between the quality-level distribution 3D surface and the 2D operator selected confidence plane selected by the operator.

Once the *quality-to-level* mapping function has been determined, operator definition of quality level to be assured translates into a certain minimum level value. By means of this relationship, usage of level thresholds as quality thresholds for cell candidate list construction in call set-up or HO algorithms is rather straightforward.

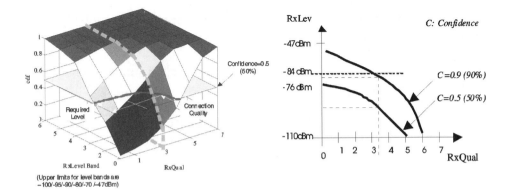

Figure 12.24 Cumulative quality–level distribution (left) and quality-to-level mapping functions for different confidence values (right).

Future principles of automation

The growing complexity of *current cellular networks* makes network optimisation a very challenging task. Possible interactions between active features further complicate the situation. Hence, understanding the relationship between individual strategies is key to reach the global optimum performance.

Within this scenario, *future wireless systems* (comprising both MSs and networks) shall provide a wide range of services in many different situations (even different technologies). An acceptable user QoS must be provided in an efficient way for the operator. New radio technologies and services will appear and will be combined in a non-homogeneous complex system. This evolution will end up with a great degree of freedom that will increase the complexity of configuration and optimisation tasks further. Hence, a huge manual effort is to be performed by a human operator in order to keep the system in an optimum operational point.

On the other hand, the combination of the rapid progress in computer devices, with their associated increase of computation power at lower prices, and the evolution of intelligent and complex control techniques may be the perfect mixture to develop intelligent devices with capacity for *auto-configuration* and *auto-optimisation*.

In order to achieve *self-configured* networks, it will be necessary to develop algorithms for automatic set-up, auto-configuration and auto-calibration, based on internal and external information. Compared with a centralised control structure, a distributed control strategy, which is spread between several system elements, will allow operators to increase the performance and efficiency of their networks, based on its time- and space-adaptive nature.

Likewise, if continuously *optimised networks* are targeted, it will be necessary to develop automatic optimisation and control systems so as to reach their optimum operating point, regardless of the surrounding conditions such as load, traffic or interference. Self-learning approaches will be instrumental in reconfiguring and adapting control systems in such a dynamic environment.

Both types of future automatic systems will be based on similar and common approaches. Development in several common research topics will greatly contribute to deal successfully with these problems, areas such as identification system, system engineering and automation, supervision, fuzzy control, neural networks, evolutionary systems, stochastic simulation, agents and self-learning systems.

12.5 Automated troubleshooting of cellular network based on Bayesian networks

Introduction

Troubleshooting (TS) in cellular systems – the identification and correction of problems in the network – is an integral part of the ongoing network operation and optimisation process. As a cellular system consists of several thousand elements, even with highly reliable hardware not all of these elements are always performing in an optimum manner. Furthermore, in a growing network sometimes it is difficult to control that all elements interact as they are expected or that the configuration is the right one. To ensure that the network is providing maximum performance continuous TS should be applied.

The TS of a cellular network has three different aspects: first, identification of cells which are facing problems in terms of performance, second, establishing the reasons why a particular cell is not performing well, and finally, implementing a solution to the problems. Of these three tasks establishing the cause of the fault is by far the most difficult and time-consuming one since a large number of causes could result in detrimental performance. Being able to identify the cause of poor performance in a cell can be one of the most difficult issues, especially if there are several problems masking the real cause.

Currently, TS is largely a manual process, in which the person looking into the reasons for the fault has to carry out a series of checks in order to establish the causes of the problem. This process is characterised by eliminating likely problem causes in order to single out the actual one. In the process of TS several applications and databases have to be queried to analyse performance data, cell configurations and hardware alarms.

The speed of detecting problems is dependent on the level of expertise of the troubleshooter, the level of information available and the quality of the tools displaying the relevant pieces of information. This means that, in addition to a good understanding of the possible causes of the problem, a very good understanding of the tools available to access the sources of information is also required. Due to the complexity of the system it is almost impossible for newcomers to carry out this task in a proficient manner. Therefore, some of the most experienced members of staff are involved in rectifying problems in the network rather than looking after its further development. Hence, several operators consider TS as one of the most time- and resource-consuming tasks associated with the operation of the network.

The growing size of cellular networks, together with the increasing complexity of the network elements, creates an apparent need for an automated troubleshooting tool (TST). With the help of a TST the time required to locate the reason for a fault that is causing a problem will be greatly reduced. This means that network performance will be enhanced as the downtime and the time with reduced QoS will be limited significantly. In addition, by automating the TS process fewer personnel and thus fewer operational costs are necessary to maintain a network of a given size. By creating an application that is based on the knowledge of the expert troubleshooter, the process of fault rectification will be de-skilled as the majority of problems can be rectified with the help of the automated TS system. The knowledge of this highly experienced staff, which is released from the TS work, can be then utilised for other aspects of the network optimisation, thus further increasing network performance. Thereby, the automated TS system will ensure that operators will increase returns on their investment.

The aim of such an automated tool is then to de-skill and speed up the TS process by automatically gathering the required information from the data sources and making reasoning with the information related to the faulty cell. The TST can guide the less experienced engineer in the direction of the problem or directly indicate the reason. It is important to point out that the efficiency of the TST largely depends on the skill level of the TS expert who 'taught' the system because the reasoning is based on his knowledge. The higher the level of expertise of the 'teacher', the higher the proficiency of the TST.

One additional benefit is that the knowledge from different experts can be stored in the tool, therefore not being dependent on the staff working for the company at any time. Furthermore, with the TST the work is more systematic than when it was a manual procedure where the knowledge was only in the head of the experts.

The gains of the automated TST for an operator are significant as fewer TS personnel with a lower skill level can resolve more network problems in less time. For an average size network the gain that can be obtained through increased network performance, which is brought about by automated TS, is in the order of several million euro per year.

In summary, automatic TS of problems in cellular networks has an important role in network operations to provide maximum QoS with minimum downtime. That means that TS will contribute to an increase in the network performance, whereas cost will be highly reduced, therefore increasing operational efficiency. On the other hand, once the number of active faults in the network is greatly reduced by TS, the performance benefits of adaptive parameterisation of the network will be significantly increased.

Troubleshooting process

Automatic TS consists of three steps:

- Fault detection: automatic detection of bad performing cells based on performance indicators, alarms, etc.

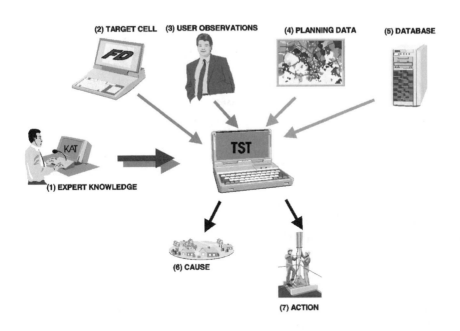

Figure 12.25 Troubleshooting process.

- Cause identification: automatic reasoning mechanisms to identify the cause of the problems and the best sequence of actions to solve them.
- Problem solving: execution of the actions to solve the problems.

Hereafter when speaking about TS it will be understood that we refer to cause identification, keeping in mind that TS is a wider process. Figure 12.25 explains the cause identification procedure. First, the knowledge of the troubleshooters has to be transferred to the system using a knowledge acquisition tool (1). The fault detection subsystem indicates which are the cells with problems and the type of problem (2). The TST chooses the proper model and makes the reasoning based on user observations (e.g. antenna down tilt) (3), planning data (e.g. frequency reuse) (4) and indicators from databases (e.g. measured interference level, hardware (HW) alarms, etc.) (5). The output of the TST is a list of possible causes with the probability associated with each cause (6). The TST also recommends the most cost-efficient action to solve the problem (7).

Decision support systems

When there is a problem in the network, e.g. high amount of dropped calls, TS experts analyse some key performance indicators that show the status of the network. Based on their previous experience, they can find the cause of the problem and decide how to proceed. One significant issue is that the number of variables to be taken into account can be very high and so the process can take a long time and a lot of effort. Besides, there are only very few specialists in the company who actually have the knowledge to determine the right cause. Therefore, the TS performance can be very much dependent on these experts, who may not always be available for TS or might also change jobs.

Traditional expert systems for fault diagnosis try to extract the knowledge of one or several experts to build a tool that can be used by non-specialists. This allows newcomers to effectively analyse and solve the majority of faults. Furthermore, the knowledge from different experts is stored in the tool in an organised manner, not in a diffuse set of rules that only a few people know and understand.

An expert system consists of two parts: a knowledge base and an inference engine. The knowledge base contains the domain-specific knowledge, which should be transferred from the experts to the system (e.g. knowledge about mobile networks TS, medical knowledge of diseases and their symptoms, etc.). The inference engine performs reasoning to obtain conclusions based on the knowledge base and the domain observations. It is important to point out that these expert systems are actually decision support systems, which do not try to completely replace experts, but to support less experienced people.

Rule based systems

The first expert systems were based on rules. Rules have the form: 'IF A THEN B', where A is an assertion and B can be an assertion or an action. Some rules for TS could be: 'IF there are HW alarms THEN there is a HW fault' or 'IF number of HOs failures is too high THEN check whether calls are being handled by expected adjacencies'. The set of rules forms the knowledge base and the inference engine can select the rules relevant to the particular case under consideration and suggest further assertions or deductions.

In several cases operators have flowcharts describing their TS procedure, which can be directly converted into rules. The main advantage of rule based systems is that they provide a logical way of reasoning for human beings, and therefore it is easy to build models based on the flowcharts. But rule based systems also have several drawbacks: they are not suitable for large and complex domains, they are only valid for deterministic cases, they model the TS procedure rather than the domain, etc.

Furthermore, normally experts have some uncertainty in their reasoning, that is, conclusions are not usually completely certain, e.g. 'IF there are HW alarms THEN *very probably* there is a HW fault'. There are many schemes for treating uncertainty in rule based systems (certainty factors, fuzzy logic, etc.), but normally combining uncertainties from different rules is very complex, so that the resulting systems may lead to wrong conclusions.

Bayesian networks

One alternative to rule based expert systems are Bayesian networks (BNs) [17,18], which use classical probability calculus and decision theory to cope with uncertainty. BNs got their name from the eighteenth-century Reverend Thomas Bayes, a mathematician who first used probability inductively and established a mathematical basis for probability inference. He formulated the Bayes' theorem, which in the BN is used to infer in the opposite direction of the causal one. For example, if there is an interference problem, very probably there will be a high number of interference HOs. Thus, there is a causal relation from interference problem to number of interference HOs. But it is also possible to do reasoning in the reverse direction. For instance, if there is a high number of interference HOs, then the probable cause is interference in the network. Using Bayes' theorem it is possible to calculate the second probability from the first one.

A BN consists of a set of variables or nodes (which have different exclusive states) and a set of arrows between these variables. Arrows normally reflect cause–effect relations within the domain. The strength of an effect is modelled as a probability. For example, Figure 12.26 shows a very simple Bayes model. If there is an interference problem in the cellular network, this will cause HO failures and the interference level will probably be high. In this case, there are three variables with the causal relations shown in the figure. The states of the variables are interference problem: yes/no; interference level: high/medium/low; and HO failure: yes/no. The meaning of these states should be defined, for example, high interference level can be specified so that for 95% of the calls the interference level is higher than − 85 dBm.

The conditional probabilities associated with the links have to be specified too, that is, the probability of the interference level being high, medium or low given that there is an interference problem and likewise the probability of the

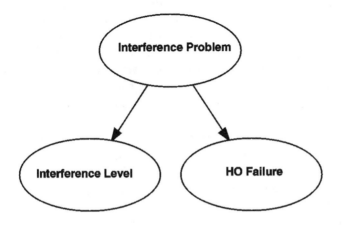

Figure 12.26   Example of Bayesian network.

HO failure being in state 'yes' or 'no' given that there is an interference problem. Furthermore, it is necessary to define the prior probability of having an interference problem.

A BN has several advantages compared with rule based expert systems:

- reasoning can be made even if not all the required information is available, which is the most common situation in mobile networks;
- it can cope not only with deterministic cases, as it was the case with rule based systems, but also with uncertainty using probability theory;
- it models the domain, whereas rule based systems modelled the expert's way of reasoning;
- it can learn with each new case;
- it can learn from data available from the domain.

A clear drawback of a BN is that it is relatively difficult to build models, especially if several probabilities have to be specified.

Bayesian network models

When creating a BN there should be a compromise between accuracy and simplicity. On one hand, the Bayesian model should reflect the domain as accurately as possible in order to assure that the right cause of the problem is found in each case. On the other hand, if the model is too complex the knowledge acquisition will be extremely difficult and the algorithms to find the causes will be far more complicated. Indeed one of the major problems when trying to build models for mobile networks has been obtaining knowledge from the experts. Therefore, the aim is to define a model as simple as possible, but sufficiently accurate.

In order to create a model, the important variables have to be identified and their exclusive states have to be defined. Then, it is necessary to specify the dependency between the variables. Finally, probabilities have to be assigned.

For simplicity, variables will be divided in three types: causes, evidences and actions. *Causes* are the possible faults that are causing problems in the network. *Evidences*, which are variables related to the causes, can be divided in two classes: symptoms and conditions. *Symptoms* are indications of the causes. *Conditions* are variables that could have caused the problem or could have an impact on some cause. *Actions* are the activities that may solve the problem once the cause is known.

Figure 12.27 shows an example to illustrate these definitions. For example, frequency reuse may have an influence on having interference, i.e. if the frequency reuse is very tight it is more probable to have an interference problem than if co-channel cells are further apart. On the other hand, if there is an interference problem, very probably the interference level will be high and there will be HO failures.

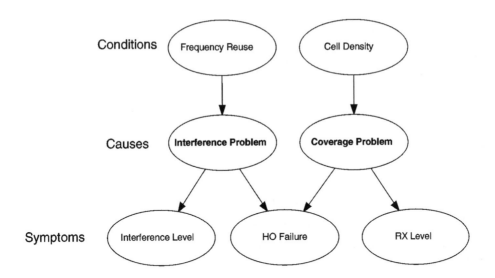

Figure 12.27   Example of Bayesian model for troubleshooting.

Once the variables, their states and the complete structure are defined, then the probabilities have to be specified. When a node has several parents this becomes a cumbersome task because probabilities for each combination of the parent nodes have to be set. For example, even for an expert it is very difficult to define which is the probability of having a HO failure given that there is an interference problem and, at the same time, a coverage problem but not a HW problem.

In order to simplify, two plain models are proposed: naïve model and noisy-OR model.

The *naïve model* shown in Figure 12.28 has been used in many diagnostic systems [17]. When this model has been used in medicine, C represented a set of alternative diseases and A_i were potential symptoms of the diseases. This model can also be used for TS, C would be the possible causes of the problem and A_i the possible symptoms, conditions and actions.

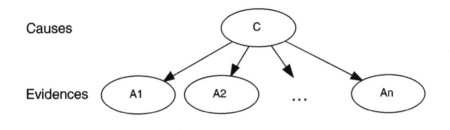

Figure 12.28   Naïve Bayes model.

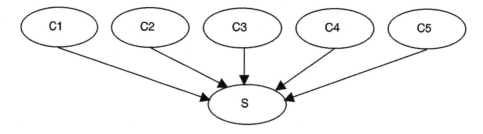

Figure 12.29   Scenario for Noisy-OR.

The following assumptions are implicitly inherent in this model:

- Single fault assumption: the states of C are mutually exclusive, that is, only one cause can be present at a time.
- The evidences A_i are independent given the cause C. It implies that once the cause C is known, information about some evidence A_i is of no relevance to predict the value of the other A_i.

To overcome the single fault restriction, which is the main limitation of the naïve model, the *Noisy-OR model* or other *causal-independence models* can be used instead [19]. As was explained before, one of the major problems in building a Bayesian model with multiple faults is the difficulty of providing conditional probabilities for a symptom given each possible combination of the causes. For example, in Figure 12.29 the probability of node S should be specified for each possible value of the causes C_i. That means specifying 32 probabilities, whereas only 6 probabilities have to be defined if Noisy-OR is assumed.

Noisy-OR supposes that C_i causes S unless an inhibitor prevents it. For example, if there is either a coverage problem or an interference problem, then there will be a high number of quality HOs unless something prevents it. That means, that almost certainly there will be a high number of quality HOs, although there is a slight possibility that the symptom is not present even if any of the causes is occurring. Considering that the inhibitors are independent, the probabilities are very easy to calculate.

Knowledge acquisition

Converting the expert knowledge into a BN can be a complex process. If the domain expert not only knows about TS mobile networks, but also understands BNs, he would be able to build accurate models. But this is not usually the case, normally a knowledge engineer should work with the expert and try to reflect his knowledge in the form of BNs. Frequently this is a troublesome task because of the different languages used by the BN and the TS expert, and it can take a long time to build even a simple model.

The TST will never be able to work better than the expert who transferred his knowledge to the system. This means that the knowledge acquisition is a crucial task because the TS will only give accurate results if the experts have a deep understanding of the domain and if this is correctly converted into BN models.

Eliciting knowledge from experts should include:

- Variables: what are the important variables to consider? What are the states of these variables?
- Structure: what are the causal relations between the variables? Are there any dependencies or independencies between variables?
- Parameters: what are the probabilities needed in the defined structure?
- Utilities: what is the usefulness of the different possible actions?

These stages can take place iteratively, being a difficult and time-consuming process. In this context it would be very useful to have a tool acting as the knowledge engineer, i.e. it should guide the experts through a sequence of questions and it should automatically translate the answers into Bayesian models. This tool will be called the knowledge acquisition tool (KAT).

Nevertheless, it can be extremely complex for KAT to build any BN model. In order to make it simpler, some assumptions can be made, for example regarding the complexity of the BN structure. In that case, only questions concerning some basic relations, variables, parameters and utilities are required. Besides, if it is assumed that the Bayesian model is any of the previously explained, i.e. naïve or noisy-OR model, only a limited subset of probabilities is needed.

For the simple case shown in Figure 12.27, KAT would ask about the causes (interference, coverage problem), symptoms related to each cause (interference level, HO failure, RXLevel) and conditions related to each cause (frequency reuse, cell density). It would also inquire about the different states of each variable and some probabilities, e.g. probability of each symptom given the related causes. Based on the answers to the questions it would automatically build a model like the one shown in Figure 12.27, a naïve model or a noisy-OR model.

When dealing with TS experts it has been realised that even specifying probabilities for a given structure is quite difficult and that different experts can provide completely different values. This can lead to inaccurate results. Normally, operators have system databases containing the history of most variables in the network, which can be used to train the BN (*learning from data*). This means that it is possible to train the BN based on previous problem cases and the probabilities can then be obtained from those previous experiences instead of being elicited from experts. Furthermore, probabilities can be updated every time there is a new case. This presents the additional value that the system can adapt to different environments, e.g. probabilities can be different for a rural and urban area.

In conclusion, learning from data is an important feature when using BNs. But not only learning parameters, learning the structure of the BN is also possible (although more complex). Therefore, it is not necessary to make simplifications such as done in the naïve model, but the structure is the one that best matches the existing data, and also it adapts to changes in the TS domain.

Troubleshooting sequence

It is important to point out that the target of TS is not only to determine what is wrong, i.e. what the causes are, but also to solve the problems. At any stage of the process there are several possible evidences that can be analysed and actions that can be performed. Because these evidences and actions take time and have a cost, it is required to find a sequence of steps (evidences or actions) that minimise cost (time/money) of solving the problems.

A very simple TS procedure collects all evidences and then makes reasoning based on them, ordering the possible causes according to their probabilities and carrying out the action associated with the most probable cause. This is not an optimum procedure because all evidences are collected, when probably not all of them are required to find out the cause. Furthermore, sometimes it could be more efficient to perform an action different to the one associated with the most probable cause. For example, suppose that the most probable cause was a HW problem, and the next probable one was a configuration problem. In that case, it is perhaps more cost efficient to make a simple parameter change first to see if the problem was in configuration. Only if it fails to solve the problem, someone has to be sent to the site to change the problematic HW unit, which is always a very expensive and time-consuming action. Without an optimised TS sequence, the HW would have been changed before trying a simple change of a parameter.

An optimum procedure would order the steps (evidences and actions) in a sequence minimising the cost (money/time). In general, some actions will be carried out before collecting all evidences and very probably the problem will be solved by one of these actions, saving time and money. For example, if after collecting some evidence the probabilities of faulty HW and bad configuration are very high, the parameter change will be proposed and if the problem is solved, it is not required to collect the other evidences or carry out the other actions anymore.

The procedure consisting of identifying causes of problems and solving them through a sequence of TS steps is called decision-theoretic troubleshooting. Some algorithms to choose the best sequence of steps are proposed in [17,20,21].

Advanced features

Some issues to be taken into account when modelling with a BN are briefly reviewed in this section.

First, *sensitivity analysis* examines how sensitive are the conclusions to minor changes in probabilities and evidences. As was previously explained, specifying accurate probabilities is a difficult task for the domain experts. Thus, it is important to assure that minor deviations of these probabilities will not lead to big errors in the conclusions (probabilities of the causes). On the other hand, it is important to notice which are the more important evidences, e.g. knowing that the interference level is high may lead to the conclusion that there is an interference problem, independently of the value of the other evidences.

In some cases, it can be required to change some probabilities in a model that is already built. *Tuning* explains how the parameters in the model can be changed to meet the new requirements. For example, let's suppose that there is a high number of quality HOs and this knowledge is incorporated into the model, obtaining the posterior probability of having a coverage problem. But, the expert may think that the obtained probability is not right and prefer to manually set a different value. In that case, all the conditional probabilities previously defined in the model should be tuned in order to obtain the probability required by the expert. Tuning may be especially useful when there is a previously existing model, and then one expert, who may be different to the one that initially built the model, wants to modify certain probability.

Another important topic is verifying the chosen model, i.e. once the model is selected and the probabilities are specified, it is important to check the validity of the model. This can be done by processing several cases and examining that the conclusions obtained are the right ones.

Interaction with the network management system

The network management system (NMS) is an essential part of cellular networks providing operational tools for the operator to run the network efficiently. The TS expert system can work independently from the NMS, but most of the benefits are achieved when it is an integrated part of it. This integrated solution will provide direct access to information required in fault analysis as well as access to the operators' fault management system. Also, some operators can have special automatic fault-detection tools in their NMS that can highlight cells that are performing badly. With an integrated solution, it is possible to start up TS analysis automatically based on the output of the fault-finding tool.

An integrated solution is also beneficial in case of TS of multi-vendor networks as well as with networks where new network system layer is added, e.g. extension from 2G network to 3G. The TS expert system based on BN methods is system independent itself, so it is flexible to support multi-vendor networks and 3G networks with minor modifications

The access to the fault management system in the NMS guarantees that work done with the TST is very well synchronised with the whole fault management process of the operator. Thus, all relevant TS cases can be automatically directed to the TS expert system, and if it finds the solution, the case is cleared, reported and filed. If the problem is not found by the expert system it can be redirected to the specialists for further analysis and the final conclusions can be incorporated into the knowledge of the expert system.

References

[1] Okumura T., Ohmori E., Fukuda K., 'Field Strength and its Variability in VHF and UHF Land Mobile Services', Rev. Elect. Commun. Lab., Vol. 16, No. 9–10, September-October 1968, pp. 825–873.

[2] Walfish J., Bertoni H. L., 'A Theoretical Model of UHF Propagation in Urban Environments', IEEE Trans. Antenna Propagation, Vol. AP-36, October 1988, pp. 1788–1796.

[3] Athanasiadou G. E., Nix A. R., McGeehan J. P., 'A Ray Tracing Algorithm for Microcellular and Indoor Propagation Modelling', ICAP'95, IEE Conference Publication No. 407, pp. 231–235.

[4] Brown P. G., Constantinou C. C., 'Prediction of Radiowave Propagation in Urban Microcell Environments using Ray Tracing Methods', ICAP'95, IEE Conference Publication No. 407, pp. 373–376.

[5] Timmer Y., Bergenlid M., 'Estimating the Inter Cell Dependency Matrix in a GSM Network', VTC'99, pp. 3024–3028.

[6] GSM Specification 05.08 v8.4.0, Section 8.4.1.

[7] Mouly M., Pautet M. B., 'The GSM System for Mobile Communications', Cell & Sys., 1992.

[8] Barco R., Cañete F. J., Díez L., Ferrer R., Wille V., 'Analysis of Mobile Measurement-based Interference Matrices in GSM Networks', IEEE Vehicular Technology Society Fall Conference, VTC 2001.

[9] Kuurne A., 'Mobile Measurement Based Frequency Planning in GSM Networks', Master's thesis, Helsinki University of Technology, Laboratory of Systems Analysis, 2001.

[10] Nielsen T. T., Wigard J., 'Performance Enhancements in a Frequency Hoping GSM Network', Kluwers Academic Publishers, Dordrecht, 2000.

[11] Okumura–Hata propagation model in 'Prediction Methods for the Terrestrial Land Mobile Service in the VHF and UHF Bands', ITU-R Recommendation P.529-2, ITU, Geneva, 1995, pp. 5–7.

[12] Control Group, http://www-control.eng.cam.ac.uk/.

[13] Ogata Katsuhiko, Modern Control Engineering.

[14] Astrom Karl, Albertos Pedro, Blanke Mogens, Isidori Alberto, Schaufelberger Walter, Sanz Ricardo, Control of Complex Systems.

[15] Kourtis S., Tafazzolli R., 'Adaptive Handover Boundaries: A Proposed Scheme for Enhanced System Performance', IEEE Proc. Vehicular Technology Conference, 51st, Tokyo, Vol. 3, 2000, pp. 2344–2349.

[16] Kim J., et al., 'Determining Velocity Threshold for Handoff Control in Hierarchically Structured Networks', IEEE Personal, Indoor and Mobile Radio Commun., Vol. 1, 1998,

pp. 350–354.

[17] Jensen F. V., 'Bayesian Networks and Decision Graphs', UCL Press, 2001.

[18] Cowell R., Dawid A. P., Lauritzen S. L., Spiegelhalter D., 'Probabilistic Networks and Expert Systems', Springer-Verlag, Berlin, 1999.

[19] Kim, Pearl, 'A Computational Model for Causal and Diagnostic Reasoning in Inference Engine', Proc. 8th Int. Joint Conf. on AI, 1983, pp. 190–193.

[20] Heckerman D., Breese J. S., Rommelse K., 'Decision-theoretic Troubleshooting', Commun. ACM, Vol. 38, No. 3, pp. 49–56.

[21] Skaanning C., Jensen F., Kjaerulff U., 'Printer Troubleshooting using Bayesian Networks', Proc. 13th Int. Conf. on Ind. and Eng. of AI and Expert Systems, AIE, 2000.

13

IMT-2000 3G Radio Access Technologies

Juan Melero, Antti Toskala, Petteri Hakalin and Antti Tolli

This chapter provides an introduction to the different third-generation (3G) radio technologies, providing a technical benchmark of their voice and data capabilities and presenting the concept of UMTS multi-radio networks.

13.1 IMT-2000 3G technologies and evolution paths

Third-generation radio access technologies aim to provide the mass-market with high quality, efficient and easy-to-use wireless mobile multimedia services. IMT-2000 defines a set of technical requirements for the realisation of such target, which can be summarised as follows:

- high data rates: 144 kbps in all environments and 2 Mbps in low-mobility and indoor environments
- symmetrical and asymmetrical data transmission
- circuit switched and packet switched based services
- speech quality comparable to wire-line quality
- improved spectral efficiency
- several simultaneous services to end users, for multimedia services
- seamless incorporation of second-generation cellular systems
- global roaming
- open architecture for the rapid introduction of new services and technology

Several technologies have been approved by the International Telecommunication Union (ITU) and included in the IMT-2000 3G radio access technologies family. Out of these technologies, wideband code division multiple access (WCDMA), enhanced data rates for global evolution (EDGE) and cdma2000 have gained substantial backing from vendors and operators. Therefore, this chapter will focus on these radio technologies:

- *EDGE.* Enhanced data rates for GSM evolution. This technology provides an enhanced air interface which includes the adoption of octagonal phase shift keying (8-PSK) modulation, the use of adaptive modulation and coding schemes and incremental redundancy. This technology, fully integrated within the existing GSM radio access network, is an integral part of GSM/EDGE radio access network (GERAN), which, together with UMTS terrestrial radio access network (UTRAN), constitutes the UMTS 3G multi-radio network. The first two parts of this book describe in depth the characteristics of this radio technology, its evolution and its associated performance.
- *WCDMA (FDD/TDD).* Wideband code division multiple access. It has two modes of operation, frequency division duplex (FDD) and time division duplex (TDD). TDD will not be part of this chapter's analysis due to its expected late introduction and the lack of current market interest. WCDMA FDD is a code division multiple access (CDMA) based technology, specifically designed for the efficient support of wideband multimedia services and implementing an advanced set of CDMA related functionality. WCDMA is the radio access technology of UTRAN. The next section will introduce the main technical principles of this technology.
- *cdma2000.* Code division multiple access 2000. This radio technology is the natural evolution of IS-95 (cdmaOne). It includes additional functionality that increases its spectral efficiency and data rate capability, as described in Section 13.3.

These radio access technologies are grouped under two main categories or evolution paths, which are standardised by different bodies:

- *UMTS multi-radio evolution path.* 3GPP (3rd Generation Partnership Project) is a global project aiming to develop open standards for the UMTS 3rd Generation Mobile System based on evolved GSM core networks. This multi-radio mobile system comprises two different 3G radio access networks, GERAN and UTRAN, which are based on different radio access technologies, GSM/EDGE and WCDMA, respectively. Initially, UMTS defined, within the scope of UTRAN standardisation, a new radio access network architecture, with protocols, interfaces and quality of service architectures specifically designed for the efficient provision of 3G multimedia services. As presented in Chapters 1 and 2, GERAN has adopted all these,

Figure 13.1   UMTS multi-radio network.

hence becoming an integral part of the UMTS 3G frame. Furthermore, 3GPP standards ensure that an efficient integration between UTRAN and GERAN can be accomplished so they can be merged under a single UMTS multi-radio network. This concept is illustrated in the Figure 13.1.

*   *cdma2000 evolution path.* 3GPP2 (3rd Generation Partnership Project 2) forms a collaborative effort from the organisational partners to put together the standardisation of the cdma2000 based 3G radio access technology, which is the natural evolution of the IS-95 (cdmaOne) 2G radio access technology.

13.2 3G technology support of licensed frequency bands

An important factor in the 3G technology evolution path strategy is the support, in the spectrum licensed to operators, of the different radio technologies. Depending on the spectrum licensed for cellular use in a certain region, local operators may have a limited technology choice. Unfortunately, frequency licensing has not been historically uniform across different regions worldwide. Figure 13.2 shows the spectrum allocation in different regions of the ITU IMT-2000 recommended new bands. The recommended ITU spectrum for new bands for 3G deployment has been allocated in some regions for other purposes.

Therefore, operators in different regions will have to try to accommodate the preferred radio access technology in the specific licensed band. Table 13.1 shows the supported technologies in the main bands allocated worldwide for cellular networks use. Both the UMTS 3G multi-radio (GSM/EDGE and WCDMA) and the cdma2000 technologies are supported in most of the

Table 13.1   3G RAT support in different bands[a]

Bands (MHz)	MHz	Main deployment regions	GSM/EDGE	WCDMA	cdma2000
450	450–467	Europe	x		x
480	478–496	Europe	x		
800	824–894	America	x		x
900	880–960	Europe/APAC	x		x
1700	1750–1870	Korea			x
1800	1710–1880	Europe/APAC	x	x	x
1900	1850–1990	America	x	x	x
2100	1885–2025 and 2110–2200	Europe/Japan/Korea/ APAC		x	x

[a] Some of the bands supported in the standards do not have commercial products available.

bands. The standardisation of WCDMA support in the 1800 and 1900 bands is being finalised. With this standardisation wireless operators in America will be able to deploy WCDMA in existing or about to be licensed 1800/1900 cellular bands since new 3G bands have not been licensed yet in this region.

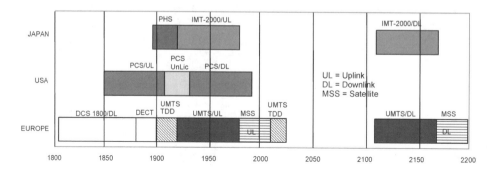

Figure 13.2   Spectrum allocation in different regions of the IMT-2000 recommended new bands for 3G.

13.3 3G radio access technologies introduction

As mentioned before there are several different radio access technologies defined within ITU, based on either CDMA or TDMA technology. This section will give a technical overview about the CDMA based technologies, multi-carrier CDMA (cdma2000) and WCDMA. Out of these two radio technologies, WCDMA is the one used in UTRAN, which together with GERAN make up the UMTS 3G multi-radio network. A more detailed description of these CDMA based technologies, with specific focus on WCDMA FDD, can be found in [1].

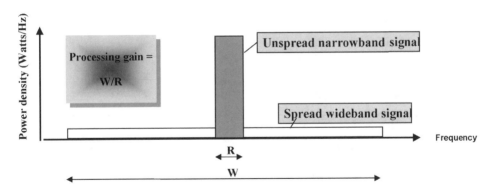

Figure 13.3   Concept of processing gain in CDMA.

WCDMA basics

Introduction

WCDMA is based on DS-CDMA (direct sequence – code division multiple access) technology where user information bits are spread over a wide bandwidth (much larger than the information signal bandwidth) by multiplying the user data with spreading code. The chip (symbol rate) rate of the spreading sequence is 3.84 Mcps, which in the WCDMA system deployment is used together with the 5 MHz carrier spacing. The processing gain term refers to the relationship between the signal bandwidth and the information bandwidth as illustrated in Figure 13.3. Thus the name wideband is derived to differentiate it from the 2G CDMA (IS-95) with has a chip rate of 1.2288 Mcps. In a CDMA system all users are active at the same time on the same frequency and are separated from each other with the use of user specific spreading codes.

The wide carrier bandwidth of WCDMA allows supporting high user data rates and also has certain performance benefits, such as increased multipath diversity. The actual carrier spacing to be used by the operator may vary on a 200 kHz grid between approximately 4.4 and 5 MHz, depending on spectrum arrangement and interference situation.

WCDMA supports highly variable user data rates, with the concept of obtaining bandwidth on demand (BoD) very well supported. Each user is allocated frames of 10 ms duration, during which the user data rate is kept constant. However, the data rate among the users can change from frame to frame. This fast radio capacity allocation (or the limits for variation in the uplink) is controlled and coordinated by the radio resource management functions in the network to achieve optimum throughput for packet data services and to ensure sufficient quality of service (QoS) for circuit switched users.

WCDMA supports two basic modes of operation: frequency division duplex (FDD) and time division duplex (TDD), as illustrated in Figure 13.4. In the FDD

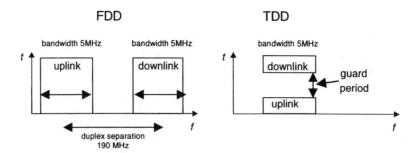

Figure 13.4   FDD and TDD operation principles.

mode, separate 5 MHz carrier frequencies with duplex spacing are used for the uplink and downlink, respectively, whereas in TDD only one 5 MHz carrier is time-shared between uplink and downlink. The TDD mode is based heavily on FDD mode concepts and was added in order to leverage the basic WCDMA system also for the unpaired spectrum allocations of the ITU for the IMT-2000 systems. The TDD mode is not expected to be used in the first phase of WCDMA network deployment. The TDD mode is described in more detail in [1] and is not considered further in this book.

WCDMA operates with asynchronous base stations, without the need for chip level timing source. From the signal detection point of view WCDMA uses coherent detection based on the pilot symbols and/or common pilot. WCDMA allows many performance enhancement methods to be used, such as transmit diversity or advanced CDMA receiver concepts. Adaptive antennas can also be efficiently deployed even though its full support with the performance requirements for the terminals and measurements to support radio resource management (RRM) is not included in Rel'99 and will be part of the later WCDMA 3GPP standard releases.

As WCDMA is going to be mostly deployed in the current GSM footprint, the support for handovers between GSM and WCDMA is part of the first standard version. This means that all multi-mode WCDMA/GSM terminals will support measurements from the one system while camped on the other one. This allows networks using both WCDMA and GSM to balance the load between the networks and base the handovers on actual measurements from the terminals for different radio conditions in addition to other criteria available. All WCDMA terminals, with the exception of those targeting the Japanese market, are expected to be multi-mode, supporting both WCDMA and GSM.

Table 13.2 summarises the main WCDMA parameters.

Table 13.2 Main WCDMA parameters

Multiple access method	DS-CDMA
Duplexing method	Frequency division duplex/time division duplex
Base station synchronisation	Asynchronous operation
Chip rate	3.84 Mcps
Frame length	10 ms
Service multiplexing	Multiple services with different quality of service requirements multiplexed on one connection
Multirate concept	Variable spreading factor and multicode
Detection	Coherent using pilot symbols or common pilot
Multiuser detection, smart antennas	Supported by the standard, optional in the implementation

WCDMA *radio access network architecture*

The WCDMA technology has shaped the WCDMA radio access network architecture due to soft handover requirements in any cellular CDMA system with reuse 1. Without support for soft handover, one or more cells would be suffering from the near–far problem when terminals operate near the cell border. In Figure 13.5 the basic WCDMA/UTRAN architecture is shown indicating the connections to the core network for both circuit switched traffic (Iu-cs) and packet switched traffic (Iu-ps). The existence of the open, standardised Iur interface between radio network controllers (RNCs) is essential for proper network operation including soft handover support in a multi-vendor environment.

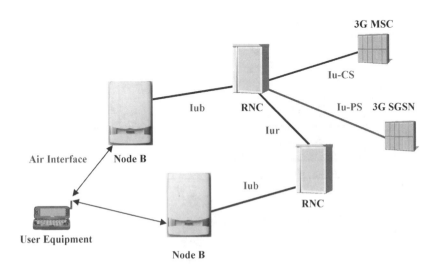

Figure 13.5 The WCDMA radio access network architecture.

One focus area in the architecture standardisation work in the coming releases is the multi-radio operation. From a service point of view there is going to be a single entity from a core network to provide the service to the GSM/WCDMA dual mode terminal either via GSM or WCDMA radio access. Therefore, it is obvious that optimising the GSM and WCDMA radio access networks (GERAN and UTRAN) jointly is the key for cost efficient multi-radio network operation. The multi-radio architecture concept will be further explored in the last section of this chapter.

WCDMA receiver structure
In WCDMA standardisation the receiver type to be used is not specified. As in the GSM standards, only minimum performance requirements in particular test cases are specified. In the case of CDMA, the basic receiver solution, known for over 50 years, is the RAKE receiver, which is illustrated in Figure 13.6. The receiver consists of multiple fingers, each one of them being able to receive a multipath component or a different downlink transmission in case soft handover is used. Typically, a RAKE receiver has a searcher, implemented with matched filter, which tracks the delay components from one or more base stations and then allocates the RAKE fingers to different propagation delays. The signal is then combined coherently together for maximum performance. In WCDM, the common pilot and pilot symbols allow coherent combining. RAKE receivers have been used in 2G already in IS-95 terminals. The first version of IS-95 uplink was non-coherent and thus the base station receiver principle was slightly different. The later versions of IS-95 standard also used coherent uplink.

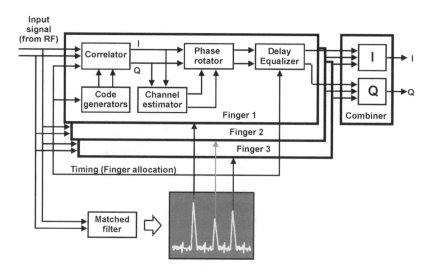

Figure 13.6   Example WCDMA receiver structure.

Soft handover + fast TPC

In WCDMA, since all users share the same time and frequency space, the separation is done in the code domain. The processing gain indicates how much the despreading process can suppress interference present in the whole of the carrier. As the processing gain is limited and reduces further when the data rate increases, the power control is the essential element for a CDMA system operation, especially in the uplink direction. As the path loss between a terminal and base station varies a lot, it was necessary to use a large dynamic range for the uplink power control operation. In WCDMA the typical power class is 21 dBm with minimum requirement for lowest power level of -50 dBm. This results in a 71 dB power control range in the uplink.

The practical power control operation consists of a simple up/down command sent by the base station and terminal at the rate of 1500 Hz. This causes the transmitter in the normal case to increase or decrease the TX power by 1 dB. This principle is illustrated in Figure 13.7, which shows the uplink power control operation. Fast power control is used in WCDMA downlink as well as performance improvements, though in the downlink the near–far problem is not that critical like in the uplink.

The operation in soft handover is another important feature for all CDMA systems, including WCDMA (Figure 13.8). All base stations have their own power control procedure, which ensures that none of them suffers from the near–far problem. In the uplink direction all base stations will decode the same data and the RNC will then select (based on the error indication from the base station) which frames to forward to the core network. Soft handover support is the key function of the Iur interface, as otherwise the near–far problem would emerge in the border of RNC areas. As WCDMA is designed to operate in asynchronous fashion, the timing information from the measurements done by the terminal is provided to the target cells. This allows the target cell to adjust the timing of a new radio link and enables the terminal to combine the links in the RAKE receiver without excessive time difference between the links. In addition to solving the near–far problem, the use of soft handover

Figure 13.7   Fast closed loop power control principle.

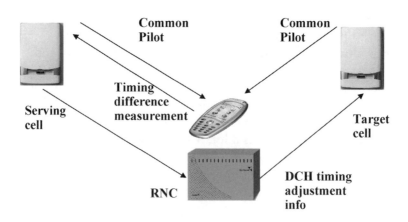

Figure 13.8   Asynchronous WCDMA soft handover operation.

provides also diversity benefits, especially in the uplink where there is no additional transmission but the same signal is decoded by multiple base stations. In the downlink direction each new transmission from additional cells adds also interference and does not always have a positive impact on the capacity.

WCDMA packet data operation

For packet data transmission in WCDMA downlink, several alternatives exist, especially in the downlink. There are basically three different channels in Rel'99/Rel'4 WCDMA specifications, completed December 1999 and March 2001, respectively, that can be used for downlink packet data:

- The dedicated channel (DCH) can be used basically for any type of service, and has a fixed spreading factor (SF) in the downlink. Thus, it reserves the capacity according to the peak rate for the connection. For example, with adaptive multi-rate codec (AMR) speech service and packet data, the DCH capacity reserved is equal to the sum of the highest rate used for the AMR speech and the highest rate allowed to be sent simultaneously with full-rate AMR. This can be even used up to 2 Mbps but obviously reserving the code tree for a very high peak rate with low actual duty cycle is not very efficient use of code resources. The DCH is power controlled and may be operated in macro diversity as well. In the uplink the DCH SF is variable for the data part while the control channel part has fixed SF.
- The downlink shared channel (DSCH) has been developed to operate always together with a DCH. This allows defining the channel properties to be best suited for packet data while leaving the data with tight delay budget to be carried by DCH. The DSCH in contrast with DCH (or forward access channel (FACH)) has dynamically varying SF informed to the terminal on a frame-by-frame basis with physical layer signalling carried on the

DCH. This allows dynamic multiplexing of several users to share the DSCH code resource and thus optimising the orthogonal code resource usage in the downlink. The DSCH is power controlled relative to the DCH power level, and the DSCH is not operated in soft handover. However, the associated DCH can be in soft handover.

- The forward access channel (FACH) can be used for downlink packet data as well. The FACH is operated on its own, and it is sent with a fixed SF and typically with rather high power level to reach all users in the cell. There is no fast power control or soft handover for FACH. The uplink counterpart for FACH is random access channel (RACH), which is intended for short duration 10/20 ms packet data transmission in the uplink. There exists also another uplink option, named common packet channel (CPCH), but that is not foreseen to be part of the first phase WCDMA network deployment.

From the terminal capability point of view, all terminals obviously support DCH and FACH/RACH, but the maximum data rates supported vary. The DSCH support is a terminal optional capability, and the reference classes have optional DSCH up to 384 kbps class in Rel'99. In Rel'4 DSCH is part of 384 kbps reference class. CPCH support is not part of any reference classes.

Transmit diversity
Rel'99 of WCDMA contains a powerful performance enhancement feature, transmit diversity, which is applicable for all downlink transmission. There are two methods included, the open loop transmit diversity coding, as shown in Figure 13.9, where the information is space time coded to be sent from two antennas, and the feedback mode transmit diversity, where the signal is sent from two antennas based on the feedback information from the terminal. The feedback mode uses phase, and in some cases also amplitude, offsets between the antennas. The use of these two antennas is based on the feedback from the terminal, transmitted in the feedback (FB) bits in the uplink DCH.

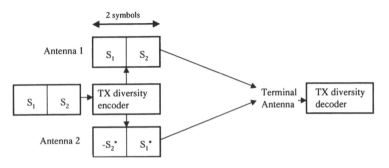

Figure 13.9   Open loop transmit diversity encoding.

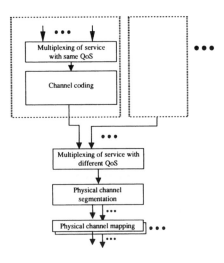

Figure 13.10   WCDMA service multiplexing principle.

Service multiplexing
The simplified WCDMA service multiplexing principle is illustrated in Figure 13.10. In the actual multiplexing chain there are additional steps such as rate matching or interleaving functions that are necessary as well for proper system operation. A single physical connection (channel) can support more than one service, even if the services have different quality requirements. This avoids for example multicode transmission in case there are two services running simultaneously as they can be dynamically multiplexed on a single physical resource.

WCDMA downlink packet data evolution with HSDPA
In WCDMA Rel'5 specifications, scheduled March 2002, the biggest improvement is the high speed downlink packet access (HSDPA) concept. In brief the feature introduces base station (or Node B in 3GPP terminology) based scheduling for the downlink packet data operation and uses similar methods like GSM evolution such as higher order modulations (e.g. 16QAM in WCDMA) and hybrid automatic repeat request (ARQ) (incremental redundancy and chase combining) to improve packet data performance. In 3GPP, performance improvements up to 100% have been reported compared with Rel'99 architecture packet data performance. The HSDPA principle with Node B based scheduling is illustrated in Figure 13.11. The HSDPA capable Node B will take care of the scheduling functionality and retransmission control which in a Rel'99/Rel'4 based network is handled solely by the RNC. Terminal or user equipment (UE) in 3GPP terminology, will send positive or negative acknowledgement to Node B which initiates retransmission at a suitable moment taking into account the channel condition and data in the buffer for different terminals.

Figure 13.11   WCDMA HSDPA operation.

Multi-carrier CDMA (cdma2000) fundamentals

Introduction

Multi-carrier CDMA (ITU-R term) or cdma2000 (3GPP2 standards term) is designed to be deployed together with the existing IS-95 (cdmaOne) systems in the same frequency band. With multi-carrier mode of cdma2000, the downlink transmission consists of multiple (up to three in the IMT-2000 version) parallel narrowband CDMA carriers. The bandwidth in each narrowband carrier is equal to the bandwidth of a single IS-95 carrier.

The multi-carrier mode of cdma2000 is often denoted as 3X to differentiate it from its other mode, the evolved narrowband version of IS-95, 1X. The 3X uplink chip rate is three times the 1X chip rate while in the downlink three parallel 1.2288 Mcps carriers are used. This is illustrated in Figure 13.12.

When comparing with WCDMA, the uplink principles are quite similar while the downlink is totally different due to its multi-carrier nature. It seems however that 1X is the focus for practical deployments and thus work on cdma2000 standardisation has been also lately focusing on 1X.

Figure 13.12   Multi-carrier CDMA (3X) and 1X cdma2000 modes.

Table 13.3   Key differences between WCDMA and cdma2000 1X

	WCDMA	1X
Carrier spacing	5 MHz	1.25 MHz
Chip rate	3.84 Mcps	1.2288 Mcps
Power control frequency	1500 Hz, both uplink and downlink	Uplink: 800 Hz, downlink: 800 Hz
Base station synchronisation	Not needed	Yes, typically obtained via GPS
Inter-frequency handovers	Yes, measurements with slotted mode	Possible, but measurement method not specified
Efficient radio resource management algorithms	Yes, provides required quality of service	Not needed for speech only networks
Packet data	Load-based packet scheduling	Packet data transmitted as short circuit switched calls
Downlink transmit diversity	Mandatory support of closed and open loop methods	Optional support of open loop method

When comparing WCDMA with 1X there is a number of key differences starting from the synchronous network operation of IS-95 based networks. 1X introduces certain improvements over IS-95 such as downlink fast power control and optional terminal support for downlink transmit diversity. Some of these key differences are summarised in Table 13.3. From the service provisioning point of view, the WCDMA has a more flexible service multiplexing capability although cdma2000 should be able to support an equivalent set of services.

Packet data operation with cdma2000
The basic concept of cdma2000 packet data operation with data rates above 14.4 kbps relies on the concept of using a fundamental and supplemental channel. The counter parts in WCDMA are DCH and DSCH, respectively. In cdma2000 the fundamental channel has a fixed SF (as in WCDMA downlink) but it does not contain physical layer information on a frame-by-frame basis as in WCDMA. Instead the existence of the supplemental channel carrying the packet data is informed by higher layer signalling on the supplemental channel. Thus the operation is similar from the conceptual level but the WCDMA resource sharing process is more dynamic on a 10 ms frame basis.

cdma2000 network architecture and operation
From the radio access network architecture point of view, there is no specification adopted for the radio access network architecture. Depending on the vendor the level of functionality in the base station will vary and in some implementations the base station can be connected even directly to the mobile switching centre (MSC) while in some others there exists a RNC type of entity. There are no open interfaces between different network elements as in WCDMA.

The network design is also based on the use of synchronised base stations. Terminals have been designed accordingly to search for the different base stations with identical timing but different code phase. The timing element needs to be accounted for in the network planning process.

cdma2000 downlink packet data evolution
The evolution of cdma2000 packet data operation is using similar methods as those used in EDGE or WCDMA evolution, such as higher order (adaptive) modulation methods or the combining of retransmissions with hybrid ARQ (HARQ) and faster scheduling based on the feedback from the terminal. In the first phase, 1X-EV-DO (DO = data only), expected to be first deployed during 2002, the technology requires packet data to be operated on a separate carrier but the next version, 1X-EV-DV (DV = data and voice), is expected to allow packet data to be combined with speech on the same carrier like in EDGE or WCDMA HSDPA. The target peak data rate of 1X-EV-DV is around 5 Mbps, whereas in WCDMA HSDPA work the peak rate targeted is 10 Mbps. The 1X-EV-DV standard is scheduled for publication 1Q/02 from 3GPP2.

13.4 3G radio access technology (RAT) performance benchmark

This section will benchmark the performance of the relevant 3G radio access technologies, GSM/EDGE and WCDMA as part of the UMTS multi-radio and cdma2000. Different areas will be analysed, such as link budget performance, technology data capabilities and spectral efficiency both of voice and data services. The scenario selected for this analysis represents a 10 MHz (10 MHz in downlink + 10 MHz in uplink) deployment in the 1900 band. For GSM/EDGE technology all the results presented are those resulting from the thorough performance analysis presented in this book. For cdma2000, available information from different sources is used. Finally, for WCDMA, the same methodology and results presented in [1] have been used, adapting the relevant parameters to match the same conditions used in the cdma2000 studies. Future technology evolution (beyond 2002) has been previously mentioned, although in order to carry out a fair analysis, in this section each technology will be analysed according to the functionality that can be realistically deployed during 2002:

- for GSM/EDGE, AMR and dynamic frequency and channel allocation (DFCA) are considered for speech and EGPRS Rel'99 for data
- WCDMA uses Rel'99 functionality
- cdma2000 uses 1X release 0 and 1X-EV-DO

Voice performance

Coverage. Link budget analysis

The coverage analysis is performed using the link budget calculation. Link budget is defined as the maximum path loss (dB) between the base transceiver station (BTS) and mobile station (MS). Since the CDMA based technologies are coverage limited in uplink, this is the link analysed. Table 13.4 summarises the uplink link budget for speech services for the three radio technologies deployed using the 1900 band.

This analysis shows there are no significant differences between the different technologies, with all of them having an approximate maximum path loss of 155 dB for speech services.[1]

Spectral efficiency analysis

This section will analyse the spectral efficiency of the different technologies. As mentioned before, only functionality realistically available in 2002 has been considered in this analysis. As mentioned before a 10 MHz band (10 MHz DL + 10 MHz UL) in the 1900 band has been selected as a deployment scenario. The results are presented in terms of users (voice paths) per sector since this is the common methodology used for the CDMA based technologies. In order to transform these numbers to Erlangs of carried traffic, Erlang B tables should be used.

In Chapters 5, 6 and 8 the baseline performance of GSM and the gains associated with the deployment of AMR and DFCA were presented. In Chapter 5, the achievable GSM baseline performance of the hopping layer was defined to be

Table 13.4   1900 band speech uplink link budget comparison

	cdma2000	GSM/EDGE	WCDMA
MS TX power [dBm]	23	30	21
BS sensitivity [dBm]	− 124	− 110	− 124
BS ant diversity gain [dB]	3	4	3
Interference margin [dB]	3	0	3
Fast fading margin (incl. SHO gain) [dB]	4	0	2
BS antenna gain [dBi]	15	15	15
Cable loss incl. MHA gain	0	0	0
Body loss for speech terminal [dB]	3	3	3
MS antenna gain [dB]	0	0	0
Maximum path loss [dB]	155	156	155

---

[1] GSM/EDGE pathloss calculation has not taken into account the benefits associated from AMR (Chapter 6) or Transmit Diversity techniques (Chapter 10)

12% effective frequency load (EFL). In Chapter 6, the hopping layer gain with the introduction of AMR was quantified to be 150%. Finally, an additional 40% hopping layer gain from DFCA was described in Chapter 8. With this set of functionality the total GSM capacity reaches 16.6 users/MHz/sector.

The WCDMA capacity has been calculated using the methodology described in [1]. The following equation is utilised to compute the capacity:

$$\eta_{DL} = \sum_{k=1}^{K} \frac{(E_b/N_0)_k}{W/R_k} \times [(1 - \alpha_k) + i_k] \times v_k$$

where

AMR speech codec	R_k = 7.95 and 12.2 kbps	4.75–12.2 kbps supported in AMR
Effective voice activity	v_k = 58% (12.2 kbps)	50% activity + DPCCH overhead
		SF = 128 and slot format 8
E_b/N_0 12.2 kbps	E_b/N_0 = 7.0 dB (12.2 kbps)	Typical macro cell based on 3GPP
		Performance requirements [2]
Orthogonality	α = 0.5	Typical macro cell
Other cell interference	i = 0.65	Typical three-sector macro cell
Load factor	η_{DL} = 80%	80% of pole capacity
Common channels	15%	15% of maximum BTS power
Number of users	k	

In the case of 7.95 kbps speech codec and assuming an Eb/No requirement of 7.5 dB, the capacity reaches 88 users per WCDMA carrier. Finally transmit diversity, which is a standard functionality supported by all terminals, provides an additional 0.5 dB gain, which brings the total capacity to 19.75 users/MHz/ sector.

The cdma2000 1X capacity has been calculated from the available market sources. According to the CDMA Development Group (CDG) [3] the maximum capacity per cdma2000 1X carrier is 33 users. This figure is calculated over the one-path radio channel propagation environment. The equivalent number for the Vehicular B channel profile, which has been selected for all the presented studies since it reflects much better a realistic macrocellular propagation environment, is 24 users per carrier. Since seven carriers can be deployed in 10 MHz so the capacity reaches 16.8 users/MHz/sector.

WCDMA, even though based on the same CDMA principles, outperforms cdma2000 1X due its wideband nature. WCDMA has a higher multipath diversity gain thus reducing the required fast fading margin, a higher effective power due to the reduced BTS power variance associated with the higher number of users per carrier and an important trunking gain in the common channels since there are less carriers for the same bandwidth, which leads to less power required to broadcast the common control channels.

Figure 13.13 displays the speech spectral efficiency results in terms of total

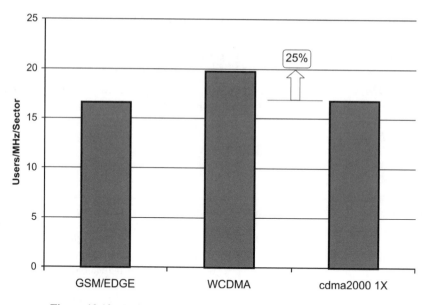

Figure 13.13   Voice spectral efficiency. 10 MHz case in 1900 band.

number of users per sector. GSM/EDGE and cdma2000 1X have an equivalent performance while WCDMA displays the best performance with 25% higher spectral efficiency.

Another aspect to consider is the hardware configuration required to support the maximum spectral efficiency associated to each technology. This can impact the cost of the Capital Expenditure (CAPEX) required to deploy different capacity configurations. Figure 13.14 displays the TRX per sector and cabinet per site configurations needed to support the maximum spectral efficiency capabil-

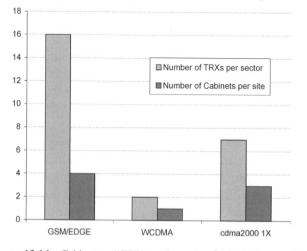

Figure 13.14   Cabinet and TRX configurations[a].   10 MHz case in 1900 band.

[a]It assumes 12 TRXs GSM/EDGE cabinets, 9 TRXs cdma2000 cabinets and 6 TRXs WCDMA cabinets

ities of each technology. The most important element is the cabinet configuration per site. WCDMA is clearly the most efficient technology for very high capacity configurations requiring a single cabinet per site to support its maximum traffic capacity. On the other hand, and due to the lower cost of GSM/EDGE cabinet and TRX units and its granular scalability, this technology will be the most cost efficient for relatively low capacity configurations.

Data performance

Data capabilities

This section gives an overview of the different areas regarding data capabilities, such as theoretical realistic and cell edge throughputs, spectral efficiency and QoS capabilities. All the data presented will refer to downlink capabilities since this is the limiting link in the case of data services.

Each technology has a theoretical peak throughput that can be achieved under certain conditions. However, these peak throughputs will be unlikely experienced in the field due to the demanding conditions required, the maximum level of interference that can be generated in the systems, or simply due to terminal limitations. In CDMA based technologies, i.e. WCDMA and cdma2000, the maximum throughput is going to be limited by the maximum interference that can be maintained across the network, which can be potentially exceeded by a single user. In EDGE, the peak throughputs are going to be limited by the terminal timeslot configurations supported, which, during 2002, are likely going to be limited to Type 1 configurations. This is due to the cost associated with higher configurations, which would require duplication of the radio architecture of the terminals since they would need to transmit and receive simultaneously. The supported multi-slot configurations are presented in Chapter 1. The maximum multi-slot configurations (DL + UL) associated in Rel'99 to Type 1 terminals are 3 + 2, 4 + 1 and 6 + 0. This configurations can be dynamically supported according to the connection requirements. For interactive services such as web browsing, 4 + 1 is the most adequate configuration, providing in downlink a maximum of 236.8 kbps (4 TSL*59.2 kbps/TSL) throughput.

In terms of average throughputs expected in the cell, these are going to be fully dependant on the network load. Table 13.5 displays typical average throughputs for medium loaded networks. The difference between WCDMA and cdma2000 1X throughput values is mostly due to the higher chip rate of WCDMA. These results are in line with the simulation results presented in Figure 13.17 and several cellular operators publicly stated expectations and measured performance.[2]

---

[2]cdma2000 1X measured/expected average throughput: SKT (Korea) and Bell Mobility (Canada) 60-80 kbps, Sprint (US) 60-70 kbps, Verizon (US) 40-60 kbps. WDMA measured average throughput: NTT DoCoMo (Japan) 270 kbps.

Table 13.5 Maximum, average and cell edge throughputs

	EGPRS	WCDMA	cdma2000 1X
Absolute maximum (standard)	473.6 kbps	2 Mbps	307 kbps
Average throughput	128 kbps	200 kbps	80 kbps
Cell edge throughput	128 kbps	384 kbps	128 kbps

The data services link budget is conditioned by the selected data rates and, in the case of GSM/EDGE, of the MS timeslot configuration capability. A simple approach to this complex problem is to determine what is the available through-put in the cell edge, defining cell edge as the maximum path loss from the speech link budget calculation. Using the Table 13.4 link budget results, and considering the lack of body loss in the data link budget calculation, the selected pathloss has been 152 dB. According to the results presented in Chapter 7, EGPRS 4+1 term-inals supports 128 kbps at the cell edge. WCDMA supports 384 kbps assuming loaded surrounding cells, while cdma2000 1X would support 128 kbps under the same conditions. The CDMA technologies, WCDMA and cdma2000 1X, are capacity limited, i.e. in order to support these throughputs at the cell edge a single connection would use all the sector capacity.

Table 13.5 illustrates the maximum throughputs defined in the standards, the typpical average throughputs and the achievable throughputs at cell edge.

Best-effort data spectral efficiency
This section presents the capabilities of different technologies in terms of spec-tral efficiency for data only deployments. The analysis is carried out for packet switched best-effort data with a simple round robin scheduling algorithm, which will deliver different data rates to different connections according to their signal to interference ration. The analysis is based on existing simulation results and does not consider application layer throughput impacts.

As presented in Chapter 7, in a loaded network, the TSL capacity of 1/3 reuse converge to 26.25 kbps, which is equivalent to a spectral efficiency of 350 kbps/MHz/sector. The evolution of spectral efficiency with load is illustrated in Figure 13.17. For the BCCH TRX, in the case of reuse 12, the TSL capacity reaches 47 kbps, which is equivalent to 137 kbps/MHz/sector. In a 10 MHz deployment the total throughput per sector would be 2989 kbps.

According to the methodolgy presented [1], and using the latest 3GPP perfor-mance requirements, the spectral efficiency of WCDMA reaches 161 kbps/MHz for guaranteed throughput power scheduling. The following conditions have been considered for evaluation of the simulations:

E_b/N_0	$E_b/N_0 = 4.5$ dB	Typical macro cell
Orthogonality	$\alpha = 0.5$	Typical macro cell
Other cell interference	$i = 0.65$	Typical three-sector macro cell
Load factor	$\eta_{DL} = 80\%$	80% of the pole capacity
Common channels	15%	20% of max. BTS power

The use of round robin best-effort data rates scheduling[3] introduces a gain of 40% in the spectral efficiency, increasing its value to 226 kbps per MHz. In a 10 MHz deployment, the total throughput per sector would reach 2260 kbps.

According to [4], for equivalent conditions, cdma2000 1X reaches a capacity of 284.75 kbps per carrier for omni-directional antenna sites. In sectorised antenna site scenarios equivalent simulations provide a higher capacity that reaches 230.4 kbps per carrier. As with WCDMA, an equivalent 40% increase is applied for best-effort data scheduling which brings the capacity up to 322.5 kbps per carrier. In the selected 10 MHz scenario seven carriers can be deployed with a maximum throughput of 2258 kbps per sector. The spectral efficiency for 1X-EV-DO is presented in [5] under ideal conditions. According to [6], the introduction of realistic interference conditions, the spectral efficiency of 1X-EV-DO reaches 337 kbps per carrier. In the selected scenario this yields a total throughput per sector of 2359 kbps.

Figure 13.15 summarises the spectral efficiency results for all technologies. It assumes data only configurations. WCDMA and cdma2000 1X have an equivalent performance. EGPRS is clearly the best performing technology for best-effort data, with a 35% higher capacity than WCDMA or cdma2000 1X. The main reason for this is that EGPRS includes adaptive modulation and coding schemes (AMCS) as well as type II hybrid ARQ (HARQ) retransmissions combining. These advanced data functionalities will be only deployed for WCDMA and cdma2000 1X when their data technology evolution (HSDPA and 1X-EV-DV, respectively), which are being currently standardised, are available in commercial products (realistically not earlier than 2004), providing an estimated capacity increase of 75% [7].

1X-EV-DO is expected to be available during 2002 and most likely will be first deployed in Korea. Even though it provides high peak data rates and competitive spectral efficiency, its main disadvantage is that it can't be mixed with voice in the same carrier so the system trunking efficiency will be significantly affected, as displayed in Figure 13.16. Additionally, its introduction requires one full 1.25 MHz carrier to be deployed to support 1X-EV-DO even though initially there will likely be a marginal number of 1X-EV-DO capable terminals.

---

[3] This can be accomplished either assigning same power to all data users or by using the Dynamic Shared Channel (DSCH)

Figure 13.15   Best-effort data spectral efficiency. 10 MHz case (10 MHz DL + 10 MHz UL).

QoS capabilities and impact in performance

As presented in Chapter 3, Rel'99 defines the UMTS traffic classes and basic QoS architecture, which allows for differentiated QoS requirements for various service and user profiles. The GSM/EDGE and WCDMA radio access networks, GERAN and UTRAN, may utilise the available RRM mechanisms (link adaptation, power control, channel allocation and scheduling) in order to ensure that such differentiated requirements are fulfilled. On the other hand, cdma2000 1X release 0 and 1X-EV-DO do not support QoS differentiation for packet switched data and are effectively best-effort only data technologies. The

Figure 13.16   1X-EV-DO trunking efficiency limitation.

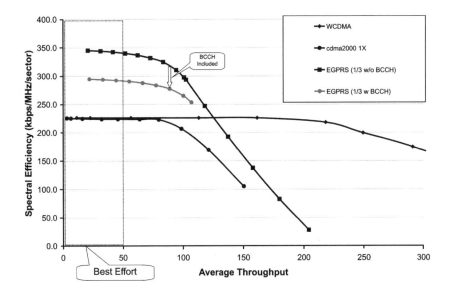

Figure 13.17   Spectral efficiency (load) vs. average throughput. 10 + 10 MHz data only deployment.

cdma2000 1X support for an equivalent QoS architecture will come with release A, unlikely to be deployed before 2002.

The spectral efficiency of the radio network is going to be conditioned by the required QoS to be provided to the end users in terms of user throughputs. Figure 13.15 displayed the best effort Spectral Efficiency for the different technologies under analysis. However, the best effort figures will decrease as the user throughput requirements get higher, as displayed in Figure 13.17. This figure display the EGPRS simulation results presented in Chapter 7 and Matlab based static simulations for the CDMA based technologies. The spectral efficiency degradation with high throughputs is however significantly different for each technology. Such degradation starts at a certain throughput, which is determined by terminal capabilities. GSM/EDGE is the most efficient technology for average data rates below 128 kbps, while WCDMA becomes more efficient for higher data rates. The combination of GSM/EDGE for relatively low data rates and WCDMA for high data rates provides the most efficient data networks performance.

Conclusions

The speech link budget analysis, which generally characterises the dimensioned network coverage, showed an equivalent performance for the 3G technologies under analysis. According to the presented results, in terms of voice spectral efficiency, WCDMA has a 25% higher performance than GSM/EDGE and cdma2000 1X. Similarly, in terms of best-effort data, GSM/EDGE has a 35% higher performance than WCDMA and cdma2000 1X. The spectral efficiency characterisation for different average throughputs show that WCDMA is the most robust technology for throughputs over 128 kbps, due to its wideband nature. Finally, both GSM/EDGE and WCDMA Rel'99 offer full support of QoS differentiation, whereas cdma2000 1X release 0 and 1X-EV-DO are best effort only packet switched data capable technologies.

The combination of GSM/EDGE and WCDMA under a seamlessly integrated UMTS multi-radio network should yield the best possible radio performance both for speech and data services. Provided multiple bands availability, GSM/EDGE is the most efficient technology in the already deployed cellular network bands (800, 900, 1800 and 1900) and WCDMA in the new 2100 band (IMT-2000) or other coming bands over 2 GHz (as it is likely the case in the US). GSM/EDGE can support efficiently speech and narrow band services and WCDMA can offer further capabilities, supporting wide band data services (high quality music or video and high speed internet access) and conversational services (video calls). For operators licensed with multiple bands a joint operation of GSM/EDGE and WCDMA could be the optimum scenario with a UMTS multi-radio deployment. The next section analyses how these technologies can be efficiently integrated.

13.5 UMTS multi-radio integration

Introduction

With the GSM evolution presented in this book, GSM/EDGE effectively becomes a new UMTS radio access technology since it can get directly connected to the UMTS core network (CN) supporting the same QoS architecture. The UMTS multi-radio networks will make use of two different radio technologies within a single 3G network, GSM/EDGE in existing bands and WCDMA mostly in the IMT-2000 recommended new bands. The advantage of this solution is that these technologies can be seamlessly integrated in order to deliver services through the most efficient radio access network based on QoS management. The service request and QoS negotiation take place between the terminal and the 3G CN, so the radio access network becomes just the means to deliver the requested services to the end user.

In Rel'5, a full integration is achieved with the support of the Iu interfaces which allow GERAN to be connected to the UMTS CN. Following this principle, other radio technologies could be 'plugged' into the UMTS multi-radio network (UTRAN TDD, WLAN ...) as long as they can be connected as well to the UMTS CN.

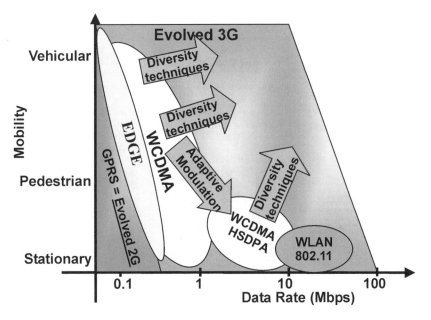

Figure 13.18 Evolution towards a seamless broadband network.

Figure 13.18 illustrates the current technology evolution towards a seamless broadband network where multiple radio technologies coexist together as part of the same network and are used according to the requirements of the service to be delivered.

UMTS multi-radio evolution

In a typical scenario, the integration of GSM/EDGE and WCDMA will take place smoothly. Figure 13.19 shows a typical multi-radio network evolution scenario. The first stages of the multi-system network are characterised by relatively small multi-mode mobile penetration and low initial coverage of the WCDMA network. The level of integration required in this initial scenario is limited to the availability of coverage based emergency handovers.

As the WCDMA coverage grows and the dual-mode mobile penetration increases, the integration of the networks should increase as well providing load and interference balancing between both networks. Load based handovers with traffic differentiation based on service profile, are required at this phase. Mature multi-radio networks with high multi-mode mobile penetration should provide seamless service and RRM integration with the technology selection based on service QoS requirements, radio technology capabilities and resource availability.

Figure 13.19   Multi-radio evolution.

Mechanisms for UMTS multi-radio integration

This section provides an overview of the most important mechanisms for traffic management between GSM/EDGE and WCDMA.

Initial cell selection

When the multi-mode terminal is switched on, or following recovery from lack of coverage, the terminal scans the bands for supported radio access technologies (GERAN and UTRAN) and finds the available PLMNs. The terminal reads the system information from the strongest cell, in order to find out which PLMN the cell belongs to. It selects the PLMN based on the priority ordered list of PLMN/access technology combinations (GERAN and UTRAN) provided by the subscriber identity module (SIM)/universal subscriber identity module (USIM). In manual PLMN selection mode, the user may select the desired PLMN. Finally, the terminal initiates registration on this PLMN using the access technology chosen or using the highest priority available access technology for that PLMN.

Idle mode

In general, the traffic management can be divided into idle and active mode traffic management. In idle mode the terminal is performing the cell selections autonomously based on system information parameters broadcasted on the

Table 13.6 Supported integration mechanisms between A/Gb and Iu GERAN modes and UTRAN

Source/target	GERAN A/Iu-cs modes	GERAN Gb/Iu-ps modes	UTRAN CS	UTRAN PS
GERAN A/Iu-cs modes	HO	–	Inter-RAT HO	–
GERAN Gb/Iu-ps modes	–	CRS	–	Inter-RAT CRS
UTRAN CS	Inter-RAT HO	–	HO	–
UTRAN PS	–	Inter-RAT CRS	–	CRS

control channels. Thus, the traffic management flexibility in idle mode is limited as compared with that in active mode where more information is available. Idle mode parameters can, however, be used to keep idle terminals camped on the preferred technology.

Active mode
Both basic inter-system handover (for circuit switched services) and network controlled re-selection (for packet switched services) between GSM/EDGE and WCDMA systems have been standardised in order to provide the tools for an efficient active mode interworking. Table 13.6 summarises the different possible handovers (HO) and cell reselections (CRS) between different radio access technologies.

Trunking efficiency benefits from multi-radio integration

One immediate effect of multi-radio integration is the trunking gain resulting from all radio resources being managed as a single pool of resources. Table 13.7 presents the load sharing gains that could be achieved by ideal multi-radio resource management of two separate systems, WCDMA and GSM EDGE. In practice, networks will carry a mixture of high and low data rate services, so the overall gain would be a weighted mixture of these gains. Common multi-radio resource management provides more gain at higher bit rates.

Table 13.7 Trunking gains with 2% blocking for 5 MHz WCDMA bandwidth and equivalent GSM/EDGE capacity

	WCDMA/EDGE capacity	WCDMA and EDGE separately	Common channel pool	Trunking gain
Speech	49.7 Erl	99.4 Erl	107.4 Erl	8%
64 kbps	5.1 Erl	10.2 Erl	13.1 Erl	28%
144 kbps	1.6 Erl	3.2 Erl	5.1 Erl	60%

The efficient load balance between the available radio technologies will provide as well an optimum interference distribution, which enhances the overall network performance.

QoS based multi-radio integration

The final level of integration comes when the UMTS multi-radio network is able to direct the incoming call or handover to the most appropriate cell and radio access technology according to its QoS requirements. At the same time, the system should try to maintain the uniform distribution of traffic and interference

Table 13.8 Possible inputs for candidate cell prioritisation algorithm

Terminal inputs	QoS requirements (RAB parameters), terminal capabilities, measurement report, terminal location
Cell inputs	Cell capability, available maximum capacity, throughput, current UL/DL load for RT, delay for NRT services, interference levels, statistics collected from the idle/active terminals
O&M inputs	Operator preferences, cell capabilities, parameters and settings for algorithms, limits/thresholds, statistics

between different cells, layers and radio systems. The system should evaluate the required QoS and both mobile station and base station capabilities and then select the most adequate cell/sector and technology.

There can be many information inputs and measurements available from the network to be utilised in the cell/technology selection algorithms. In Table 13.8 a few possible inputs from network elements for the cell selection algorithms are presented.

For real-time services (conversational and streaming) the most important indicator, apart from the received signal strength, is the amount of load currently occupied in the cell of interest. The amount of resources dedicated to non-real-time services without any guaranteed bandwidth does not have much effect on new requests of real-time services. NRT traffic is usually scheduled via time-shared channels, and because of lower priority the amount of resources dedicated to interactive and background traffic can be flexibly scaled down if requested by real-time services. Therefore, the occupation of physical resources by NRT traffic is not an optimum criterion for candidate cell prioritisation as new NRT service request is considered. Instead, a measure of achieved throughput or average delay per NRT connection could be used.

	GSM/EDGE	WCDMA	TDD	WLAN	....
Network operations	OSS	OSS	OSS	OSS	
Radio Resource Management	BSC	RNC	RRM	RRM	
Base Stations	BS	BS	BS	BS	

Figure 13.20 Functional layer view of network architecture.

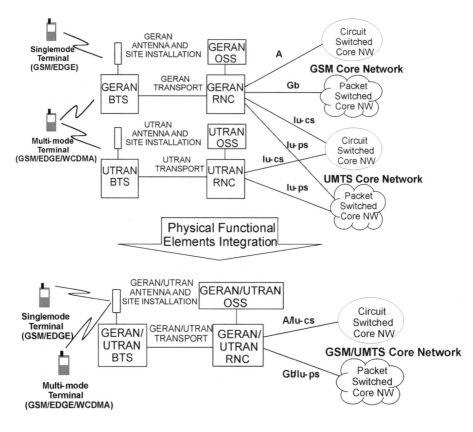

Figure 13.21   UMTS multi-radio network architecture integration.

Architecture integration

Each radio access network has its own functional and physical elements. As the number of radio technologies connected to the UMTS CN increases, the associated complexity and deployment cost of the UMTS multi-radio network increases as well. Figure 13.20 illustrates the main functional entities of each technology.

In order to design an efficient multi-radio solution, these functional and physical entities must be efficiently integrated. As previously mentioned, both radio access technologies, GSM/EDGE and WCDMA, constitute two physical and logically independent radio access networks connected to the same UMTS CN. The required level of architectural integration is achieved as both logical networks merge together into a single physical UMTS radio access network, i.e. all the physical elements of the network support both technologies providing the cellular operator a seamless operation of multiple technologies. This concept is illustrated in Figure 13.21.

References

[1] Holma Toskala, 'WCDMA for UMTS', Wiley, Chichester, 2000.

[2] 3GPP Technical Specification 25.101. UE Radio Transmission and Reception (FDD).

[3] CDG, www.cdg.org. '3rd Generation Mobile Wireless', December 2001.

[4] Lin Ma, Zhigang Rong, 'Capacity Simulations and Analysis for cdma2000 Packet Data Services', IEEE VTC 2000.

[5] Viterbi A. J., et al., 'CDMA/HDR: A Bandwidth-efficient High-speed Wireless Data Service for Nomadic Users', IEEE Communications Magazine, July 2000.

[6] SKT, www.sktelecom.com. R&D 1X-EV-DO Technology Features.

[7] 3GPP TR 25.848 V4.0.0. Technical Report 3rd Generation Partnership Project; Technical Specification Group Radio Access Network; Physical Layer Aspects of UTRA High Speed Downlink Packet Access.

14

3G Technology Strategy and Evolution Paths

Juan Melero

14.1 3G multimedia services

Operators business impact

The wireless communications business has experienced an exceptional growth during the last 10 years. This growth has been particularly remarkable in those markets where GSM was the dominant standard, Europe and Asia Pacific. GSM economies of scale and its fundamental characteristics such as global roaming and competitive short message services (SMS) helped this boost. Current global volumes were not expected in any existing forecast. Figure 14.1 displays the evolution of the cellular subscribers worldwide from 1991 and shows several forecasts for the coming years, which indicate that the cellular subscriber growth rate is expected to continue growing during the next 10 years.

During the 1990s, with the introduction of second-generation (digital) standards, voice telephony went wireless. The 1990s was as well the decade of the Internet take off, with Internet based multimedia services becoming increasingly popular and the web becoming the 'de facto' world information database. Second-generation (2G) cellular standards were however not designed to support these new Internet based services and applications. Third-generation

(3G) standards are meant to become the vehicle for Internet based multimedia and other data services to go wireless as well. Figure 14.2 shows the projected evolution of the number of devices accessing the Internet. It is expected that the number of wireless devices with access to the Internet will exceed that of fixed devices during 2002.

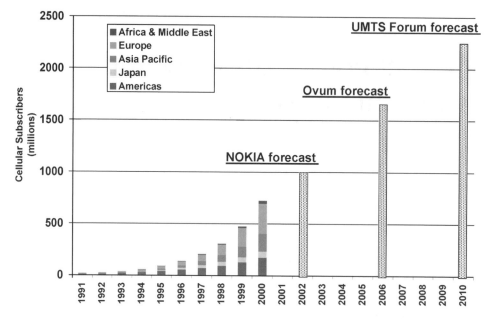

Figure 14.1   Global cellular subscriber evolution.

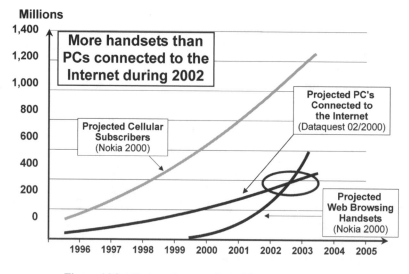

Figure 14.2   Projected penetration of Internet connections.

Figure 14.3 Evolution of the global mobile service revenues [1].

The take off of these new 3G multimedia services will drive a major transformation in the cellular operators' business case. With these services boosting the compound annual growth rate (CAGR) of the global mobile service revenues, the business opportunities for the 3G operators are enormous. Figure 14.3 shows the estimated evolution from 2000 to 2006 of the total global mobile service revenues and the revenue split in terms of different services. Today, voice services usually account for well over 90% of the total cellular operator revenue, but in the next few years data services are expected to account for 40% of the revenues, boosting the total global cellular business revenues to a value three times its current volume.

During the last years the growth experienced, both by the number of subscribers and the voice service usage, was enough for cellular operators to attain a solid revenue growth. However, as these factors tend to stabilise, and the price erosion impacts the margins, the monthly average revenue per user (ARPU) will tend to decrease and the operators' total revenue growth will start to slow down. Figure 14.4 displays some examples of the impact the introduction of value added data services had on the operators' monthly ARPU. The introduction of the coming 3G multimedia services is expected to have a major impact on the revenue per user, thus allowing cellular operators to continue with a sustained revenue growth.

3G technologies requirements

3G radio access networks (RAN) should efficiently deliver the coming services providing the quality of service (QoS) expected by the end user. The main characteristic to consider when describing data services is the required data rate. Figure 14.5 shows the data rates required by different 3G services. It is important to highlight that most of the foreseen services are well supported with data rates below 128 kbps. As presented in Chapter 7, enhanced general packet

Source HSBC, Nov 2000

Figure 14.4 Examples of monthly ARPU with the introduction of value added data services.

radio system (EGPRS) (Rel'99) can already efficiently deliver such data rates throughout the whole cell print by effectively controlling the QoS of the ongoing connections.

QoS management is a key element for the radio access technology to efficiently deliver 3G multimedia services, since such services will be fundamentally packet switched based. Distinct services and user profiles will demand an important differentiation in terms of the QoS to be provided. Therefore, the RAN should be able to control the QoS provision and ensure certain guaranteed data rates can be provided throughout the system by dynamically assigning the radio access resources. As presented in Chapter 3, UMTS terrestrial radio access network (UTRAN) and GSM EDGE radio access network (GERAN) utilises

Figure 14.5 3G service throughput and QoS requirements.

the same UMTS QoS architecture, which was specifically designed for the efficient support of the expected 3G service mix. Even though cdma2000 first release only supports best-effort data, ongoing standardisation will provide this technology of equivalent QoS architecture.

14.2 Globalisation

3G is about service delivery, and the competitive differentiation operators will seek in order to increase their revenues and market share will be based on when, how and what is offered in terms of services to the targeted end user segment. Provided different technologies are able to deliver a similar service set, the technology utilised in the RAN will be, to a certain extent, transparent to the end user perception. There are however a number of aspects of vital strategic importance related to the selection of the right technology, which are linked with the globalisation process of the cellular business.

Technology globalisation

Today's telecommunications market segment is becoming increasingly global and major players are emerging in this market struggling to take a dominant position. The cellular business is not an exception and through acquisitions, mergers and strategic alliances most operators are linked to one of these global players. This increasingly global business environment has been driving, and will continue doing so, the globalisation of technologies, and in particular, radio access technologies. Therefore, the selection of technology is likely to have an impact on the global strategy of each particular cellular operator. Figure 14.6 displays the technology globalisation evolution and how the number of existing technologies is continuously decreasing with a clear trend to reach a unique global radio technology in the near future.

Figure 14.6 Technology globalisation.

Economies of scale

In this global scenario in which operators, technologies, services and marketing strategies are becoming increasingly global, the selection of the right 3G technology is critical to remain competitive. Chapter 13 introduced a technical comparison between the different 3G radio access technologies. Such analysis showed certain advantages of the UMTS 3G multi-radio access technologies (GSM/EDGE/WCDMA) over cdma2000. But there are other aspects associated with the globalisation trend that will likely be determining factors for the selection of technology. These aspects are related to the benefits associated with 'economies of scale'. The basic principle is related to global business volumes and market shares. The higher the volumes the lower the technology cost, the faster the technology and its supported services evolves and the easier to deploy true global market strategies.

- *Network infrastructure and mobile handset costs.* The cost of the network infrastructure and mobile terminals is heavily dependent on economies of scale. Higher competitiveness and larger volumes will naturally bring costs down. Hence, the cost impact is likely to be significant if there is a significant difference in the global market share of the technologies. Additionally, the fact that GSM has enjoyed open interfaces and shared intellectual proprietary rights (IPR) has had a significant influence on the infrastructure and terminals costs. This cost impact may become particularly important in the terminal side. Figure 14.7 shows the wholesale prices of GSM and cdmaOne (IS-95) 2G terminals sorted in ascending order, each point representing a terminal price.

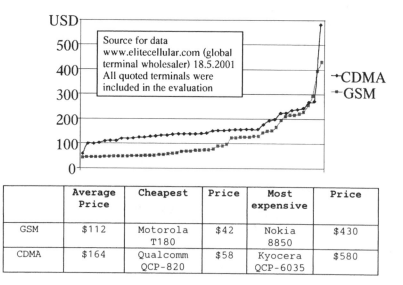

	Average Price	Cheapest	Price	Most expensive	Price
GSM	$112	Motorola T180	$42	Nokia 8850	$430
CDMA	$164	Qualcomm QCP-820	$58	Kyocera QCP-6035	$580

Figure 14.7 Comparison of unsubsidised wholesale terminal prices for 43 cdmaOne and 54 GSM terminal models [2].

The impact is of higher relative significance in the low-end section, where unsubsidised prices are roughly double for cdmaOne terminals.

- *Speed of technology development.* Industry analysts estimate vendors are allocating 200 billion US dollars in R&D in order to carry out the specification, design and manufacturing of 3G infrastructure only. The R&D investment in the two differentiated available 3G paths (UMTS 3G multi-radio and cdma2000) will likely be proportional to their respective global market share. Therefore, the choice of the right technology will ensure a faster, more advanced and efficient technology development. Today's products portfolio is a good example of this factor. For instance, in the case of terminals, the number of available GSM models on the market (312) is roughly double that of cdmaOne (159) [3].

- *Roaming and service continuity.* Traditionally roaming has been one of the competitive advantages that has made GSM technology highly attractive, and recently is becoming increasingly important in parallel with the globalisation process. In the past, GSM subscribers have been able to roam into other GSM networks worldwide. Today, global players are starting to deploy worldwide service continuity strategies, in order to ensure their customers always stay in their global networks when roaming outside the particular customers' 'home network'. A single global technology will be required in order to realise these targets.

- *Available service portfolio.* Finally, since 3G is, as described before, about delivery of 3G multimedia services, the technology with the broader global footprint and pioneering the introduction of true 3G services will have an advantage in terms of service portfolio available. In September 2000, SK Telecom launched its first commercial cdma2000 network, which positioned cdma2000 as the first International Telecommunication Union (ITU) approved 3G technology to be commercially launched worldwide. However, true 3G multimedia services, such as video telephony or simultaneous speech and Internet access, with efficient QoS management and data rates up to 384 kbps, were first successfully launched 1 October 2001 in Tokyo by NTT DoCoMo using UMTS WCDMA technology.

All these items define the competitive benefits coming from economies of scale, and are likely to be determining factors of the selection of 3G technologies. Economies of scale is synonymous of volumes. The current 3G technology paths are based on the evolution of 2G standards and hence the future 3G technologies volumes will be highly influenced by the current 2G technologies volumes and their natural evolution to 3G. The following section will analyse the global trends of the technology evolution.

14.3 3G technology evolution paths. UMTS multi-radio and cdma2000

Analogue technologies were dominant in the cellular market up to 1997, when their global market share was exceeded by that of 2G digital technologies. As presented before, the last 10 years have seen a phenomenal growth of the cellular penetration worldwide and has defined the shape of the 2G technologies global market share as it is displayed in Figure 14.8. GSM technology market share shows a sustained growth and today it has become the global 2G standard, deployed by more than 400 operators around the world and accounting for more than two-thirds of the total number of cellular subscribers. On the other hand, despite an important growth during its first years of deployment, cdmaOne market share has stabilised. During 2001, 81% of all the new subscriptions globally were GSM whereas only 13% where cdmaOne [4]. The current subscription trend will broaden the already significant relative market share gap between GSM and CDMA.

In terms of revenue generation, GSM SMS has been the most successful 2G data service and has defined a clear frame for the expected highest revenue generator 3G service, the enhanced 3G multimedia service (MMS). According to the GSM Association, 15 billion SMS text messages were sent worldwide during December 2000, which indicates a five-fold increase in the SMS volume in the past year. During October 2001 the GSM Association announced 1 billion SMS were sent daily worldwide. In the service revenue forecast presented

Figure 14.8   Technology global market share evolution [4].

in Figure 14.3, the different messaging services, grouped under the communication category, account in 2006 for €164 billion a year, which represents a market size equivalent to almost half of the total global market revenues of 2000.

As GSM technology has become by far the most successful 2G standard, UMTS, its 3G multi-radio evolution, made up of the seamless integration of GERAN and UTRAN, is likely not only to match its predecessor but naturally increase its dominance as UMTS multi-radio becomes the evolution path of other 2G technologies.

From 2G to 3G

Today, the convergence of 2G technologies towards the UMTS multi-radio 3G evolution path is clear. The entities, such as operators, global associations and standardisation bodies, which are representing and driving the evolution of three out of the four current most representative 2G technologies, have endorsed the UMTS multi-radio evolution path:

- The GSM Association, representative of the GSM operators worldwide, is a market representative of the 3rd Generation Partnership Project (3GPP) and fully supports the UMTS multi-radio 3G evolution path.
- The Universal Wireless Communications Consortium (UWCC), which is the international consortium driving the US-TDMA (IS-136) operators evolution, endorses the UMTS multi-radio evolution in which US-TDMA technology can inter-operate with GSM, GPRS, EDGE and WCDMA thanks to the standardised GAIT functionality (GSM/ANSI-136 Interoperability Team). GAIT terminals are dual mode, effectively supporting US-TDMA and a combination of GSM/GPRS/EDGE/WCDMA radio technologies. With the introduction of such terminals the US-TDMA operators have a smooth path available to converge into the UMTS multi-radio evolution. Figure 14.9 shows the combined TDMA/GSM/EDGE architecture based on GAIT interworking.
- In Japan, the two most important PDC operators (NTT DoCoMo and J-Phone), which on October 1st, 2001 covered 75.7% of the Japanese subscriber base, have selected UMTS WCDMA technology for their 3G evolution [4].

Table 14.1 summarises the evolution paths associated with the existing 2G technologies.

The alternative 3G evolution path to UMTS is the one endorsed by the CDMA Development Group (CDG), which supports an evolution to 3G based on cdmaOne technology. cdmaOne was initially developed in the US and soon adopted in other regions. Figure 14.10 shows the market share cdmaOne technology holds in different global regions [4].

Figure 14.9   TDMA/GSM/EDGE GAIT architecture.

Although cdmaOne has been introduced in most of the regions worldwide, only in America, especially in the US, and to some extent in Asia Pacific, fundamentally in Korea, has it developed an important momentum.

South Korea represents 70% of the cdmaOne Asia Pacific customer base. This country accounted for 28.1 million cdmaOne subscribers in June 2001, which was roughly one-third of the cdmaOne subscriber base worldwide. After several acquisitions and mergers during 2001, the three remaining operators defined their 3G evolution path. Two of these operators (SKT and KT) opted for WCDMA hence converging into the UMTS evolution path. These operators hold over 85% of the Korean market share [4]. Therefore Korea, the main cdmaOne stronghold in Asia, is effectively converging as well towards the UMTS evolution path.

In the US, Verizon Wireless, with 19 million cdmaOne subscribers, accounts for over 40% of the total US cdmaOne subscribers and well over 50% of the

Table 14.1   Evolution paths associated with the existing 2G technologies

2G standard	3G standard	Evolution path
GSM	GSM/EDGE/WCDMA	UMTS
PDC (Japan)	WCDMA[a]	UMTS
US-TDMA (IS-136)	TDMA/GSM/EDGE/WCDMA	UMTS
cdmaOne (IS-95)	cdma2000 or GSM/EDGE/WCDMA	cdma2000 or UMTS

[a] The two Japanese operators (NTT DoCoMo and J-Phone) that have selected UMTS WCDMA technology for their 3G evolution represent 92.4% of the existing PDC market [4].

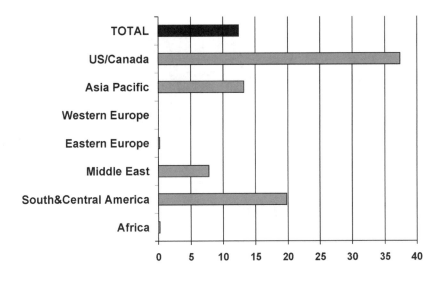

Figure 14.10 cdmaOne worldwide regional market share (cdma2000 terminals included).

potential subscribers if its analogue customer base is counted [4]. Verizon Wireless, which is 44% owned by Vodafone, the largest global wireless operator with 100 million subscribers worldwide [5], has publicly announced the likelihood of its UMTS 3G choice by the adoption of WCDMA radio technology. If Verizon Wireless UMTS selection finally takes place, over 40% of the current global cdmaOne customer base (counting SKT, KT and Verizon Wireless cdmaOne subscribers) would evolve towards the UMTS 3G evolution path.

Additionally, most current cdmaOne operators are analysing the possible migration paths to UMTS. Such migration can be done through the migration to GERAN (GSM/EDGE) with the later integration of UTRAN (WCDMA) or it can be done through the direct introduction of UTRAN. This last option is however dependent on the cellular operator licensed bandwidth, since WCDMA technology is currently not supported in the 800 MHz band, the future availability of dual mode cdmaOne/WCDMA terminals and the integration effectiveness of the technologies. Figure 14.11 summarises the global evolution paths to 3G.

At the end of 2001, the year that UMTS WCDMA was first launched in Tokyo, cdmaOne based technology covered 11.8% of the global market share whereas the technologies to converge to UMTS[1] covered 83.5% of the global market [4]. This constitutes an initial relative advantage of 7 to 1 for the UMTS 3G multi-radio evolution path. With the globalisation dynamics, this

---

[1] Including GSM, PDC and US-TDMA.

Figure 14.11   3G technology evolution paths.

gap will surely broaden, possibly ending up in an effective single universal 3G wireless access technology, as it was devised when UMTS (universal mobile telecommunications system) was first proposed.

References

[1] Nokia, Nokia Mobile Internet Conference, November 2001.
[2] Elite Cellular (global terminal wholesaler). www.elitecellular.com
[3] Wireless Tomorrow. www.wirelesstomorrow.com
[4] EMC Database, January 1st, 2002.
[5] Vodafone. www.vodafone.com.

Appendix A

MAIO Management Limitations

Angel Rivada

When random radio frequency (RF) hopping is used and fractional load is introduced, i.e. there are more hopping frequencies than transceivers (TRXs) per sector, there is a certain degree of interference control available in order to avoid undesired co-channels and adjacent channel interference between the sectors of each site. This is achieved by the mobile allocation offset index (MAIO) management functionality, which basically controls the offsets in the Mobile Allocation (MA) list of each TRX within a synchronised site. However, the gains of this functionality get limited as the effective reuse decreases. When the effective reuse is below 6, adjacencies in the hopping band within sites start to appear. Subsequently, for effective reuse lower than 3, adjacencies within the sector itself and co-channels between the site sectors are unavoidable. Although this is a general problem, narrowband scenarios suffer this limitation for relatively low TRX configurations. This appendix studies the performance degradation as the effective reuse decreases.

A.1 *MAIO management limitations and planning*

RF hopping networks usually use lists of frequencies (MA lists) that are shared by different sectors to hop over. In order to reduce the impact of interference from sectors using the same MA lists, the usage of the frequencies can be randomised (random hopping). The hopping sequence number (HSN) is a parameter that defines the pseudo-random sequence used. Within one site using a single MA list and HSN, both co-channel and adjacent channel interference can be avoided by controlling the frequencies used in each TRX within the site. To achieve this, the MAIO associated with each TRX must be controlled.

However, the degree of control provided by the MAIO management functionality is conditioned by the site effective reuse (number of frequencies in the MA list of the site/average number of TRXs per sector of the site). If the MA list is conformed by consecutive frequencies (1/1 hopping reuse) the degrees of control would be:

• Effective reuse ≥ 6	No co-channel or adjacent channel interference within the site
• $6 >$ Effective reuse ≥ 3	Adjacent channel interference between site sectors
• $3 >$ Effective reuse ≥ 1	Co-channel and adjacent channel interference between site sectors and adjacent interference within the sectors
• $1 >$ Effective reuse	Co-channel and adjacent channel interference both between site sector and within the sectors

Therefore, the MAIO management limitations are not load dependent but only configuration dependent (effective reuse dependent). The different available deployment strategies when the effective reuse goes under the mentioned thresholds are either to maintain the MAIO management in all the sectors within the site, which causes permanent interference between certain TRXs, or start using different HSNs per sector, which increases the overall interference but randomises its occurrence. Figure A.1 shows an example of the performance of different strategies for the case of a single MA list of 12 frequencies using 1/1 hopping reuse with an average of four hopping TRXs per sector, i.e. effective reuse of 3. In this example adjacencies within the sites are unavoidable. Additionally, in this figure the ideal case with 12 separate frequencies without adjacent interference (reference best performing case) has been added. The 1/1 with MAIO management in all sectors outperforms the other configurations even though existing adjacent channel interference will degrade the performance compared with the interference-free (within the site) case. The same exercise can be applied to higher number of TRXs, where co-channel interference within the site starts to be a problem. In those cases, the co-channel interference should be handled assigning different HSNs between co-channel interfering sectors and applying MAIO management when possible (for example with effective reuse 2, MAIO management could be applied in two sectors to avoid co-channels and a different HSN in the third one).

Figure A.1   Performance for four hopping TRXs and different configurations (typical urban 3 km/h (TU3), 3000 m inter-site distance; 65-deg antennae beamwidth; DL displayed).

A.2 MAIO *management limitations for different effective reuses and antenna beamwidth*

Figure A.2 displays the performance of the network when using different effective reuses. The best configuration cases are displayed for each effective reuse. Only hopping layer and downlink (DL) results from simulations are displayed. The degradation when going from effective reuse 6 to 4 and 2 can be observed. This is due to the impact of MAIO management degradation. With reuse 4, adjacent channels within sectors of the same site appear; and with reuse 2 co-channels within sites are unavoidable. For a 65-deg antennae a degradation of

Figure A.2   MAIO management limitations for different antenna beamwidth (DL performance, TU3).

around 5% can be detected when going form reuse 6 to reuse 4, and 10% when going to reuse 2.

Additionally, the effect of using narrower antenna beamwidth is also presented in the same figure. Basically, according to the simulations, in a regular-grid scenario the use of narrower beamwidth decreases the interference level produced towards sectors of the same site, improving the performance of the interference limited scenarios. Finally, the MAIO management degradation presented in Figure A.1 will be higher in the case of 90-deg antennas, as presented in Figure A.2.

Appendix B

Hardware Dimensioning Studies

Salvador Pedraza

This appendix includes different studies on hardware dimensioning in cases where the Erlang B formula cannot be used. The first section covers mixed full and half-rate voice scenarios while the second deals with packet data-traffic connection sharing the system resources.

B.1 Blocking probability for half- and full-rate speech modes

This section is a study about blocking probability (BP) in the GSM system in which two different types of connections can be established: full-rate (FR) connections using one timeslot (TSL) and half-rate (HR) connections using half a TSL. This study is very useful for hardware dimensioning when a percentage of the calls can be converted to HR in order to save resources. Note that this study is purely based on hardware (HW) considerations and does not consider interference issues.

The Erlang B formula

The Erlang B formula has been used for a very long time to dimension the number of transceivers (TRXs) with FR calls. This problem was first studied by Erlang in 1899 [1], who derived the well-known and much used Erlang B formula for blocking probability (BP):

$$BP_{\text{Erlang B}}(T, N) = \frac{\dfrac{T^N}{N!}}{\displaystyle\sum_{i=0}^{N} \frac{T^i}{i!}}$$

where

T = traffic generated in the cell
N = number of available channels

Call arrival is assumed to be a Poisson distributed stochastic process with exponentially distributed call duration. This formula provides the BP in a cell with N channels for a given traffic (T). It is important to note that T is *potential* traffic and not real carried traffic in the cell, which can be calculated as follows:

$$CT = T(1 - BP_{\text{Erlang B}}(T, N)).$$

Although if BP is low it can be assumed that carried traffic is approximately equal to potential traffic

$$CT \approx T \qquad \text{if BP is low.}$$

The Erlang B formula is the steady response of the stochastic process of arriving and serving calls assuming the following hypothesis:

- Traffic is originated from an infinite number of sources.
- The number of channels in the system is limited (N).
- Call requests arrive independently of each other.
- The number of call requests in a time unit is Poisson distributed with parameter λ.
- The call duration is exponentially distributed with parameter τ.

Where

λ = call request arriving rate (e.g. $\lambda = 0.1$ calls/s)
τ = average call duration (e.g. $\tau = 120$ s)

The call termination rate is $\mu = 1/\tau$. Little's formula connects μ and λ:

$$T = \frac{\lambda}{\mu}.$$

Blocking probability for HR/FR scenario

Let's assume a combined scenario with a ratio of γ HR connections. There are two factors that may contribute to γ: HR mobile penetration and radio conditions (connections with bad radio conditions can't be converted to HR). N is the number of available channels (e.g. $N = 15$ for a two TRX cell). The call arrival rate is λ (e.g. $\lambda = 0.1$ calls/s = 1 call/10 s) and the call termination rate is μ (e.g. $\mu = 0.05$ calls/s = 1 call/20 s) or equivalently $\tau = 1/\mu$ is the average call duration.

Simple case: two timeslots

For the two timeslot (TSL) case the problem can be modelled with the state diagram shown in Figure B.1. Here (P*f*F*h*H) represent the state in which there are *f* FR connections and *h* HR connections. The arrows describe the possible state transitions and the label gives the relative probability for that transition.

It is possible to derive the steady state equations for Figure B.1, in terms of offered traffic (T):

$$0 = \text{P0F1H} + \text{P1F0H} - T\,\text{P0F0H}$$

$$0 = T\gamma\text{P0F0H} + 2\text{P0F2H} + \text{P1F1H} - (T + 1)\,\text{P0F1H}$$

$$0 = T\gamma\text{P0F1H} + 3\text{P0F3H} + \text{P1F2H} - (T + 2)\,\text{P0F2H}$$

$$0 = T\gamma\text{P0F3H} - 4\text{P0F4H}$$

$$0 = T(1 - \gamma)\text{P0F0H} + \text{P1F1H} + 2\text{P2F0H} - (T + 1)\,\text{P1F0H}$$

$$0 = T(1 - \gamma)\text{P0F0H} + \text{P1F1H} + 2\text{P2F0H} - (T + 1)\,\text{P1F0H}$$

$$0 = T\gamma\text{P1F0H} + T(1 - \gamma)\text{P0F1H} + 2\text{P1F2H} - (T\gamma + 2)\,\text{P1F1H}$$

$$0 = T\gamma\text{P1F1H} + T(1 - \gamma)\text{P0F2H} - 3\text{P1F2H}$$

$$0 = T(1 - \gamma)\text{P1F0H} - 2\text{P2F0H}.$$

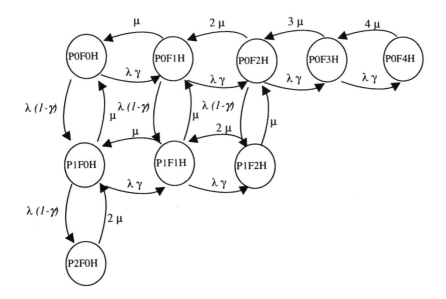

Figure B.1 State diagram for two TSL case.

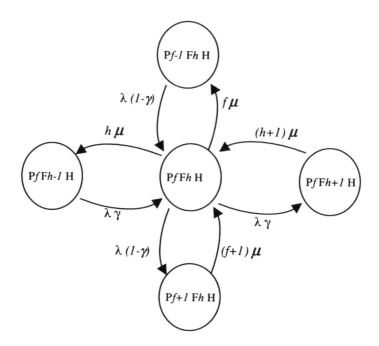

Figure B.2   General node state diagram for N TSL case.

The solution for BP for two TSLs is

$$\mathrm{BP}(T, \gamma, 2) =$$

$$
\left(T^2 \left(\begin{array}{l} T^4\gamma^4 + 72(\gamma - 1)^2(2\gamma + 1) + T^3\gamma(6 + 6\gamma - 8\gamma^2 + \gamma^3) \\ -\, 2T^2(-6 - 12\gamma + 14\gamma^3 + \gamma^4) - 12T(-5 + 9\gamma^2 - 9\gamma^3 + 5\gamma^4) \end{array} \right) \right)
$$

$$
\Big/ \left(\begin{array}{l} 144 + 264T + T^6\gamma^4 + T^2(216 + 48\gamma - 48\gamma^2) - 12T^3(-7 - 5\gamma - \gamma^2 + 4\gamma^3) \\ +\, T^5\gamma(6 + 6\gamma - 8\gamma^2 + 5\gamma^3) + 2T^4(6 + 18\gamma + 12\gamma^2 - 26\gamma^3 + 9\gamma^4) \end{array} \right).
$$

General case: N TSLs

In a general case there is a similar state diagram with $N^2 + 2N + 1$ states, where a general node has the state transition diagram shown in Figure B.2.

Similar to the two TSL case the following linear equation can be written describing the steady state behaviour of Figure B.2:

$$0 = T\gamma PfF(h - 1)H + T(1 - \gamma)P(f - 1)FhH + (h + 1)PfF(h + 1)H$$

$$+ (f + 1)P(f + 1)FhH - fPfFhH \; \mathrm{Exist}(P(f - 1)FhH)$$

$$-T\gamma PfFhH \text{ Exist}(PfF(h+1)H)$$

$$-T(1-\gamma)PfFhH \text{ Exist}(P(f+1)FhH)$$

$$-hPfFhH \text{ Exist}(PfF(h-1)H)$$

where Exist(PfFhH) is a function equal to 1 if the state exists in the state diagram, and is equal to 0 otherwise. Two terms contribute to the BP equation:

The sum of those states in which the system is totally full:

$$\sum_{f=0}^{N} PfF(2N-f)H.$$

and when the system has 1/2 TSL free, so that the system is full for FR mobiles which are $1-\gamma$ of total. So for this case:

$$(1-\gamma)\sum_{f=0}^{N-1} PfF(2(N-f)-1)H.$$

In conclusion BP can be written as

$$BP = \sum_{f=0}^{N} PfF(2N-f)H + (1-\gamma)\sum_{f=0}^{N-1} PfF(2(N-f)-1)H.$$

Solving the $N^2 + 2N + 1$ dimensional system of equations, BP can be obtained as a function of total traffic generated in the cell (T), capability of HR mobiles (γ) and number of available channels (N):

$$BP = BP(T, \gamma, N).$$

A general solution for this equation is not easy to find as in the two TSL case. The next section presents numerical solutions in the form of tables.

Numerical solutions
Similar to the Erlang B formula, it is more useful for dimensioning purposes to have a table with traffic for a certain grade of service (GoS = blocking probability). Tables B.1 and B.2 have been generated for 1% of GoS and 2% of GoS as a function of the HR capability and number of channels (= TSL).

Notice that the first column ($\gamma = 0\%$) corresponds to the value obtained with the Erlang B formula for n TSLs and the last column ($\gamma = 100\%$) corresponds to values obtained with the Erlang B formula for $2n$ TSLs. For example, with eight TSLs the offered traffic at 2% GoS is 3.62 Erlang if no HR calls can be done ($\gamma = 0\%$), while with 100% HR capability 9.82 Erlang can be offered for the same GoS, the same as with 16 TSLs.

Table B.1 Traffic (Erlangs) for 1% GoS with different ratio of HR connections

	Percentage of HR connections										
TSL	0%	10%	20%	30%	40%	50%	60%	70%	80%	90%	100%
1	0.0101	0.0102	0.0105	0.0111	0.012	0.0133	0.0158	0.0143	0.0279	0.0696	0.1526
2	0.1526	0.1554	0.1613	0.1722	0.19	0.2144	0.255	0.3097	0.4132	0.5913	0.8694
3	0.4555	0.4652	0.4875	0.5258	0.5838	0.6615	0.7724	0.9163	1.1335	1.4544	1.909
4	0.8694	0.8908	0.9398	1.019	1.1312	1.2746	1.4651	1.7013	2.0326	2.4958	3.1277
5	1.3608	1.3993	1.4826	1.6109	1.7843	1.9964	2.2678	2.5982	3.0452	3.6518	4.4614
6	1.909	1.9705	2.0931	2.2744	2.5121	2.7939	3.1472	3.5737	4.1381	4.8893	5.8761
7	2.5009	2.5928	2.7556	2.9877	3.2875	3.6435	4.0839	4.6082	5.2906	6.1866	7.3517
8	3.1276	3.2512	3.4595	3.7488	4.1152	4.5419	5.0646	5.6873	6.4895	7.5319	8.8753
9	3.7825	3.9414	4.1969	4.5439	4.977	5.4747	6.0807	6.8034	7.7269	8.9165	10.437
10	4.4612	4.6604	4.962	5.3627	5.8593	6.4327	7.1275	7.9511	8.9957	10.332	12.031
11	5.1599	5.4014	5.7503	6.2056	6.7663	7.4159	8.1999	9.1247	10.291	11.774	13.652
12	5.876	6.1616	6.5584	7.0681	7.6928	8.4196	9.2938	10.32	11.608	13.239	15.296
13	6.6072	6.9376	7.3835	7.9485	8.6362	9.434	10.394	11.528	12.945	14.728	16.96
14	7.3517	7.7287	8.2237	8.8434	9.5947	10.469	11.519	12.757	14.299	16.231	18.642
15	8.108	8.5325	9.077	9.7517	10.567	11.518	12.658	13.998	15.662	17.744	20.339
16	8.875	9.3477	9.9421	10.672	11.551	12.579	13.81	15.259	17.05	19.281	22.051
24	15.295	16.16	17.174	18.371	19.787	21.453	23.422	25.687	28.456	31.879	36.109
32	22.048	23.307	24.767	26.466	28.443	30.714	33.386	36.49	40.267	44.904	50.589
40	29.007	30.689	32.58	34.747	37.256	40.155	43.567	47.538	52.343	58.198	65.322
48	36.109	38.227	40.546	43.168	46.196	49.719	53.88	58.77	64.646	71.724	80.221
56	43.315	45.882	48.621	51.682	55.213	59.355	64.281	70.151	77.148	85.451	95.237

Table B.2 Traffic (Erlangs) for 2% GoS with different ratio of HR connections

TSL	Percentage of HR connections										
	0%	10%	20%	30%	40%	50%	60%	70%	80%	90%	100%
1	0.0204	0.0207	0.0213	0.0224	0.0243	0.0268	0.032	0.034	0.0559	0.1138	0.2235
2	0.2236	0.2278	0.2366	0.2527	0.2789	0.315	0.3722	0.4515	0.5828	0.7889	1.0923
3	0.6024	0.6155	0.6456	0.6965	0.772	0.8729	1.0118	1.1897	1.4419	1.795	2.2759
4	1.0927	1.1204	1.1828	1.2819	1.4197	1.5938	1.8195	2.0963	2.468	2.9674	3.6271
5	1.6578	1.7065	1.8084	1.9628	2.1683	2.4169	2.7292	3.1061	3.5987	4.2451	5.084
6	2.2769	2.3526	2.4985	2.7109	2.986	3.3096	3.7089	4.1874	4.8022	5.5967	6.6147
7	2.9367	3.0448	3.2375	3.509	3.8538	4.2532	4.7404	5.3216	6.0596	7.003	8.2003
8	3.6287	3.7733	4.0146	4.3457	4.7601	5.2359	5.8119	6.4968	7.359	8.4517	9.8284
9	4.3468	4.5311	4.8222	5.2131	5.6969	6.2502	6.9161	7.7053	8.692	9.9344	11.491
10	5.0864	5.3128	5.6548	6.1058	6.6593	7.2905	8.0469	8.9409	10.053	11.445	13.182
11	5.8443	6.1143	6.5082	7.0199	7.6432	8.3528	9.2001	10.199	11.436	12.979	14.896
12	6.6178	6.9206	7.3662	7.9445	8.6452	9.4345	10.372	11.477	12.84	14.534	16.631
13	7.405	7.7521	8.2508	8.8911	9.6629	10.532	11.561	12.772	14.261	16.105	18.383
14	8.204	8.5963	9.1486	9.8511	10.694	11.643	12.764	14.082	15.697	17.693	20.15
15	9.0137	9.4517	10.058	10.823	11.738	12.766	13.981	15.405	17.147	19.294	21.932
16	9.8328	10.317	10.978	11.806	12.792	13.901	15.208	16.739	18.609	20.907	23.725
24	16.636	17.497	18.601	19.943	21.515	23.274	25.329	27.724	30.615	34.129	38.392
32	23.729	24.978	26.522	28.367	30.516	32.932	35.753	39.044	42.985	47.729	53.428
40	30.998	32.618	34.611	36.982	39.735	42.819	46.408	50.589	55.575	61.548	68.688
48	38.387	40.358	42.8	45.709	49.082	52.839	57.201	62.278	68.315	75.519	84.1
56	45.863	48.161	51.051	54.511	58.519	62.955	68.092	74.073	81.164	89.602	99.624

Effective factor

The effective factor (EF) is defined as the ratio between the effective number of channels (the equivalent number of channels required to offer the same traffic at the same GoS with FR mobiles only) and the actual number of channels

$$EF = \frac{\text{Effective number of channels}}{\text{Actual number of channels}} = \frac{\text{Number of channels with } \gamma = 0}{\text{Number of channels for a certain } \gamma}.$$

This factor shows the resources that can be saved due to HR capability. Logically, the larger the capability (γ), the larger the EF. Tables B.3 and B.4 show the EF for 1 and 2% GoS for different number of TSLs.

From the results with different number of channels it can be noticed that the EF is not dependent on the number of channels (N) when N is greater than 16. Additionally, the difference between 1 and 2% GoS is relatively small (less than 2%).

Figure B.3 shows the EF for 4, 8, 16 and 56 channels with 2% GoS. It can be noticed that for $N > 16$ the curves are very similar, so that the curve for $N = 56$ can be considered as describing the behaviour for a cell with more than two TRXs ($N > 16$).

Also the respective curves for 1% (not shown here) and 2% GoS are very similar.

From Figure B.3 it can be seen that trunking gain increases with the number of channels. Although this gain practically disappears for $N > 16$. Additionally it can be concluded that EF is not a linear function of γ and is always below the diagonal.

Figure B.3   Effective factor for $N = 4, 8, 16, 56, 2\%$ GoS.

Table B.3 Effective number of channels and efficiency factor for 1% GoS

TSL	Percentage of HR connections										
	0%	10%	20%	30%	40%	50%	60%	70%	80%	90%	100%
1	1.0	1.0013	1.0037	1.0084	1.0168	1.0274	1.0494	1.0362	1.151	1.4747	2.0
2	1.0	1.0078	1.0234	1.0517	1.0959	1.1516	1.2338	1.3264	1.4578	1.6189	2.0
3	1.0	1.006	1.0195	1.0417	1.0748	1.1234	1.2153	1.3676	1.5189	1.726	2.0
4	1.0	1.0118	1.0384	1.0804	1.138	1.2088	1.2995	1.4081	1.5537	1.7479	2.0
5	1.0	1.0148	1.0462	1.0937	1.1562	1.2304	1.3229	1.4318	1.5743	1.7607	2.0
6	1.0	1.0179	1.0531	1.1042	1.1697	1.2457	1.3384	1.4476	1.588	1.7693	2.0
7	1.0	1.0214	1.059	1.1116	1.1783	1.2559	1.3497	1.459	1.5977	1.7753	2.0
8	1.0	1.024	1.064	1.1187	1.1868	1.2646	1.3581	1.4673	1.605	1.780	2.0
9	1.0	1.0263	1.0683	1.1244	1.1934	1.2714	1.3647	1.4739	1.6109	1.7837	2.0
10	1.0	1.0288	1.0719	1.1286	1.1977	1.2763	1.3701	1.4794	1.6156	1.7867	2.0
11	1.0	1.0309	1.0751	1.1321	1.2014	1.2805	1.3746	1.4839	1.6195	1.7892	2.0
12	1.0	1.0327	1.0778	1.1351	1.2044	1.284	1.3784	1.4877	1.6228	1.7912	2.0
13	1.0	1.0343	1.0802	1.1377	1.2069	1.2862	1.3804	1.4903	1.6255	1.7933	2.0
14	1.0	1.0357	1.0823	1.1399	1.2091	1.2886	1.383	1.4929	1.6279	1.7948	2.0
15	1.0	1.037	1.0841	1.1419	1.2109	1.2907	1.3852	1.495	1.6295	1.7958	2.0
16	1.0	1.0381	1.0857	1.1436	1.2126	1.2925	1.3872	1.4973	1.6318	1.7973	2.0
24	1.0	1.0435	1.094	1.1534	1.2231	1.3044	1.3998	1.5086	1.6405	1.8021	2.0
32	1.0	1.0457	1.0984	1.1594	1.2299	1.3105	1.4047	1.5133	1.6446	1.8047	2.0
40	1.0	1.0477	1.1011	1.1619	1.232	1.3126	1.4069	1.5161	1.6475	1.8066	2.0
48	1.0	1.0492	1.1029	1.1633	1.2328	1.3132	1.4078	1.5184	1.6506	1.809	2.0
56	1.0	1.0505	1.1042	1.1639	1.2326	1.3128	1.4078	1.5204	1.6541	1.812	2.0

Table B.4 Effective number of channels and efficiency factor for 2% GoS

TSL	0%	10%	20%	30%	40%	50%	60%	70%	80%	90%	100%
						Percentage of HR connections					
1	1.0	1.0018	1.0052	1.0119	1.0238	1.0389	1.0694	1.0812	1.2082	1.5153	2.0
2	1.0	1.0084	1.0252	1.0551	1.1014	1.1608	1.2454	1.3462	1.4822	1.7051	2.0
3	1.0	1.0103	1.0332	1.071	1.125	1.1939	1.2836	1.3941	1.5445	1.7434	2.0
4	1.0	1.0133	1.0423	1.0876	1.1487	1.2234	1.3176	1.4294	1.5744	1.7618	2.0
5	1.0	1.0166	1.0505	1.1008	1.1663	1.2438	1.3387	1.4502	1.5919	1.7726	2.0
6	1.0	1.0199	1.0574	1.111	1.1791	1.2577	1.3526	1.4639	1.6037	1.7799	2.0
7	1.0	1.023	1.0632	1.1189	1.1885	1.2677	1.3626	1.4737	1.612	1.7851	2.0
8	1.0	1.0258	1.068	1.1252	1.1956	1.2753	1.3702	1.4811	1.6183	1.789	2.0
9	1.0	1.0283	1.0721	1.1302	1.2012	1.2812	1.3761	1.4869	1.6232	1.7921	2.0
10	1.0	1.0304	1.0756	1.1344	1.2057	1.286	1.3809	1.4916	1.6272	1.7945	2.0
11	1.0	1.0323	1.0785	1.1379	1.2095	1.2899	1.3849	1.4954	1.6305	1.7966	2.0
12	1.0	1.0326	1.0796	1.1401	1.2126	1.2934	1.3882	1.4987	1.6332	1.7983	2.0
13	1.0	1.0339	1.0818	1.1426	1.2153	1.2962	1.3911	1.5014	1.6356	1.7998	2.0
14	1.0	1.0351	1.0836	1.1448	1.2176	1.2986	1.3935	1.5038	1.6376	1.801	2.0
15	1.0	1.0361	1.0852	1.1467	1.2196	1.3008	1.3957	1.5059	1.6394	1.8022	2.0
16	1.0	1.037	1.0866	1.1484	1.2214	1.3027	1.3976	1.5078	1.641	1.8032	2.0
24	1.0	1.0413	1.0937	1.1569	1.2306	1.3124	1.4074	1.5175	1.6493	1.8084	2.0
32	1.0	1.0435	1.0968	1.1601	1.2336	1.3158	1.4112	1.5219	1.6538	1.8116	2.0
40	1.0	1.0441	1.0981	1.1621	1.2361	1.3186	1.4142	1.525	1.6566	1.8134	2.0
48	1.0	1.044	1.0984	1.163	1.2376	1.3204	1.4161	1.5271	1.6586	1.8147	2.0
56	1.0	1.0434	1.098	1.1633	1.2386	1.3216	1.4175	1.5287	1.660	1.8158	2.0

Since the EF does not depend on the number of channels when N is large ($\text{EF}(N, \gamma) \approx \text{EF}(\gamma)$), it is possible to use the Erlang B formula to calculate BP or traffic replacing N with a function of capability (γ) as

$$\text{BP}(T, \gamma, N) = \text{BP}_{\text{Erlang B}}(T, N(\gamma)).$$

And

$$N(\gamma) \approx N^* \text{EF}(\gamma).$$

Dimensioning factor

A dimensioning factor (DF) can be defined as the rate between the traffic offered for a certain GoS with a capability (γ) and the traffic offered with only FR connections ($\gamma = 0\%$)

$$\text{DF} = \frac{\text{Traffic offered for a certain } \gamma}{\text{Traffic offered for } \gamma = 0}.$$

Tables B.5 and B.6 show the DF for 1 and 2% GoS for different number of TSLs.

DF shows the increase of the offered traffic for a certain number of channels as a function of HR capability. For example, for 16 TSLs and 2% GoS the offered traffic is increased by a factor of 1.41 for 50% of HR capability and by 2.41 if all calls are done as HR ($\gamma = 100\%$).

Figure B.4 shows the DF for different number of channels and 2% GoS. It can be seen that offered traffic increases non-linearly and surprisingly, the lines for

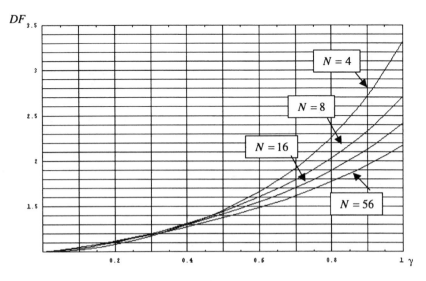

Figure B.4 Dimensioning factor for $N = 4, 8, 16, 56$, 2% GoS.

Table B.5 Dimensioning factor for 1% GoS

TSL	0%	10%	20%	30%	40%	50%	60%	70%	80%	90%	100%
						Percentage of HR connections					
1	1.0	1.0147	1.0419	1.0959	1.1914	1.3126	1.5644	1.4133	2.759	6.893	15.107
2	1.0	1.0187	1.057	1.1282	1.2454	1.4051	1.6709	2.0293	2.7081	3.8751	5.6977
3	1.0	1.0213	1.0702	1.1544	1.2817	1.4524	1.6957	2.0116	2.4886	3.193	4.1912
4	1.0	1.0246	1.0809	1.1721	1.3011	1.4661	1.6851	1.9568	2.3379	2.8707	3.5974
5	1.0	1.0283	1.0895	1.1838	1.3113	1.4671	1.6666	1.9093	2.2378	2.6836	3.2785
6	1.0	1.0322	1.0964	1.1914	1.3159	1.4635	1.6486	1.872	2.1677	2.5611	3.0781
7	1.0	1.0367	1.1018	1.1946	1.3145	1.4569	1.6329	1.8426	2.1154	2.4737	2.9396
8	1.0	1.0395	1.1061	1.1986	1.3158	1.4522	1.6194	1.8184	2.0749	2.4082	2.8378
9	1.0	1.042	1.1096	1.2013	1.3158	1.4474	1.6076	1.7986	2.0428	2.3573	2.7594
10	1.0	1.0447	1.1123	1.2021	1.3134	1.4419	1.5977	1.7823	2.0164	2.316	2.6969
11	1.0	1.0468	1.1144	1.2026	1.3113	1.4372	1.5891	1.7684	1.9943	2.2819	2.6459
12	1.0	1.0486	1.1161	1.2029	1.3092	1.4329	1.5817	1.7564	1.9755	2.2531	2.6032
13	1.0	1.05	1.1175	1.203	1.3071	1.4278	1.5732	1.7448	1.9592	2.229	2.567
14	1.0	1.0513	1.1186	1.2029	1.3051	1.424	1.5668	1.7353	1.9449	2.2077	2.5358
15	1.0	1.0524	1.1195	1.2027	1.3032	1.4205	1.5612	1.7265	1.9317	2.1885	2.5085
16	1.0	1.0533	1.1202	1.2025	1.3015	1.4173	1.5561	1.7193	1.9211	2.1725	2.4846
24	1.0	1.0566	1.1228	1.2011	1.2937	1.4026	1.5313	1.6795	1.8605	2.0843	2.3608
32	1.0	1.0571	1.1233	1.2004	1.29	1.3931	1.5142	1.655	1.8263	2.0366	2.2944
40	1.0	1.058	1.1232	1.1979	1.2844	1.3843	1.5019	1.6388	1.8045	2.0063	2.2519
48	1.0	1.0587	1.1229	1.1955	1.2794	1.3769	1.4922	1.6276	1.7903	1.9863	2.2217
56	1.0	1.0593	1.1225	1.1932	1.2747	1.3703	1.484	1.6196	1.7811	1.9728	2.1987

Table B.6 Dimensioning factor for 2% GoS

2% GoS	Percentage of HR connections										
	0%	10%	20%	30%	40%	50%	60%	70%	80%	90%	100%
1	1.0	1.0148	1.0421	1.0965	1.1923	1.3154	1.5664	1.664	2.7409	5.5765	10.95
2	1.0	1.0194	1.0588	1.131	1.248	1.4098	1.6657	2.0206	2.6082	3.5301	4.8878
3	1.0	1.0221	1.0721	1.1566	1.2819	1.4496	1.6802	1.9756	2.3944	2.9808	3.7792
4	1.0	1.0258	1.0829	1.1736	1.2998	1.4592	1.6658	1.9192	2.2596	2.7167	3.3207
5	1.0	1.0298	1.0913	1.1844	1.3085	1.4585	1.6469	1.8744	2.1716	2.5617	3.0679
6	1.0	1.0337	1.0978	1.1912	1.312	1.4542	1.6297	1.8399	2.11	2.4591	2.9064
7	1.0	1.0373	1.1029	1.1954	1.3129	1.4489	1.6149	1.8129	2.0643	2.3857	2.7936
8	1.0	1.0403	1.1068	1.1981	1.3124	1.4436	1.6024	1.7912	2.0289	2.3302	2.7098
9	1.0	1.0429	1.1099	1.1999	1.3112	1.4386	1.5918	1.7735	2.0006	2.2865	2.6448
10	1.0	1.045	1.1123	1.201	1.3099	1.434	1.5828	1.7586	1.9773	2.2512	2.5927
11	1.0	1.0467	1.1141	1.2017	1.3084	1.4299	1.5749	1.746	1.9578	2.2219	2.55
12	1.0	1.0462	1.1136	1.201	1.307	1.4263	1.5681	1.7351	1.9412	2.1972	2.5142
13	1.0	1.0474	1.1147	1.2012	1.3055	1.4229	1.562	1.7256	1.9268	2.176	2.4837
14	1.0	1.0483	1.1156	1.2013	1.3041	1.4198	1.5566	1.7173	1.9143	2.1576	2.4573
15	1.0	1.0491	1.1164	1.2013	1.3028	1.417	1.5517	1.7098	1.9032	2.1415	2.4342
16	1.0	1.0497	1.1169	1.2012	1.3016	1.4144	1.5473	1.7032	1.8933	2.1271	2.4139
24	1.0	1.0521	1.1185	1.1992	1.2937	1.3994	1.523	1.6671	1.8409	2.0522	2.3085
32	1.0	1.0528	1.1179	1.1957	1.2863	1.3881	1.507	1.6457	1.8118	2.0118	2.252
40	1.0	1.0523	1.1166	1.1931	1.2819	1.3814	1.4972	1.632	1.7929	1.9856	2.2159
48	1.0	1.0512	1.1148	1.1906	1.2784	1.3763	1.4899	1.6222	1.7794	1.9671	2.1906
56	1.0	1.0498	1.1128	1.1882	1.2756	1.3723	1.4843	1.6146	1.7692	1.9532	2.1716

different number of channels cross for low γ. Additionally, it can be observed how for $\gamma < 50\%$ there is not much difference in DF for different number of channels.

B.2 (E)GPRS HW dimensioning analysis

In (E)GPRS, the resources (i.e. TSLs) are shared between the users being served, and consequently, the throughput is reduced when the number of active users increases. The throughput experienced by a user can be calculated as (see throughput reduction factor in Section 5.2).

User throughput = Number of allocated TSLs × TSL capacity × Reduction factor

    or

User throughput per allocated TSL = TSL capacity × Reduction factor

where TSL capacity is the average throughput per fully utilised TSL. It represents the average throughput from the network point of view. It basically depends on the network carrier/interference (C/I) distribution. Reduction factor (RF) takes into account the user throughput reduction due to TSL sharing with other users. Figure B.5 represents how the EGPRS peak throughput is reduced by interference and TSL sharing. In some networks TSL capacity can be the most limiting factor while in others TSL sharing can be the dominant effect.

2.1. Reduction factor calculation

Reduction factor is a function of number of TSLs assigned in the connection (Nu), number of TSLs in the system (Ns) and average utilisation of resources in the system (U). Note that traffic (Data Erlangs) is U^*Ns. More precisely, the

Figure B.5   Effect of interference and HW dimensioning on average user throughput.

reduction factor is a function of Ns/Nu (N), since, for instance, a process in which users take 3 TSLs in each connection in a 24 TSL system can be modelled with a system with 8 TSLs and users taking 1 TSL per connection. Therefore

$RF = f(Ns/Nu, U) = f(N, U)$.

Following the hypothesis used by Erlang in his well-known blocking probability formula:

- Traffic is originated from an infinite number of sources.
- The number of channels in the system is limited (Ns).
- Call requests arrive independently of each other.
- The number of call requests in a time unit is Poisson distributed with parameter λ.
- The call duration is exponentially distributed with parameter τ.
- The system resources (Ns) are shared evenly for all connections.

Some of these assumptions are difficult to hold for certain kind of packet switched (PS) traffic. Therefore, different results might be expected with different kinds of traffic mix. In any case, the results of this simple modelling are still useful for coarse (E)GPRS HW dimensioning. The assumption that system resources are shared in the best possible way is, as well, optimistic considering realistic implementation of channel allocation algorithm and TSL allocation limitations, therefore, the analysis represents upper performance limits. The effect of channel allocation limitations is more noticeable with low network load.

Model formulation
Let X be a random variable which measures the reduction factor in a certain system state:

$$\begin{cases} X = 0, & \text{if } n = 0 \\ X = 1, & \text{if } 0 < n \le N \\ X = N/n, & \text{if } N < n \end{cases}$$

where n is the instantaneous number of connections in the system.

Then, the reduction factor can be calculated as the average of X over the states with at least one connection, as

$$\text{Reduction factor} = \sum_{n=0}^{\infty} X \cdot \frac{P(X = n)}{P(X > 0)} = \sum_{n=0}^{\infty} X \cdot \frac{P(X = n)}{\displaystyle\sum_{i=1}^{\infty} P(X = i)}$$

where $P(X = n)$ is the probability function (i.e. the probability of having n connections in the system).

Under the assumptions considered above, the probability function can be expressed as

$$
\begin{cases}
P(X = n) = \dfrac{\dfrac{(U \cdot N)^n}{n!}}{\displaystyle\sum_{i=0}^{N} \frac{(U \cdot N)^i}{i!} + \sum_{i=N+1}^{\infty} \frac{(U \cdot N)^i}{N! \cdot N^{(i-N)}}}, & \text{if } 0 \le n \le N \\[3em]
P(X = n) = \dfrac{\dfrac{(U \cdot N)^n}{N! \cdot N^{(n-N)}}}{\displaystyle\sum_{i=0}^{N} \frac{(U \cdot N)^i}{i!} + \sum_{i=N+1}^{\infty} \frac{(U \cdot N)^i}{N! \cdot N^{(i-N)}}}, & \text{if } N < n
\end{cases}
$$

And finally

$$
\text{Reduction factor} = \frac{\displaystyle\sum_{i=1}^{N} \frac{(U \cdot N)^i}{i!} + \sum_{i=N+1}^{\infty} \frac{(U \cdot N)^i}{N! \cdot N^{(i-N)}} \left(\frac{N}{i} \right)}{\displaystyle\sum_{i=1}^{N} \frac{(U \cdot N)^i}{i!} + \sum_{i=N+1}^{\infty} \frac{(U \cdot N)^i}{N! \cdot N^{(i-N)}}}.
$$

Figure B.6 represents graphically the above equation. Numerical tables can be found in Table B.7. In tables N (Ns/Nu) values have been calculated with combinations using $Nu = \{1, 2, 3, 4\}$ and $Ns = \{1, 2, 3, 4, 5, 6, 7, 8, 9, 10, 11, 12, 13, 14, 15, 16, 20, 24, 28, 32, 36, 40, 48, 56, 64, 72, 80, 88, 96\}$.

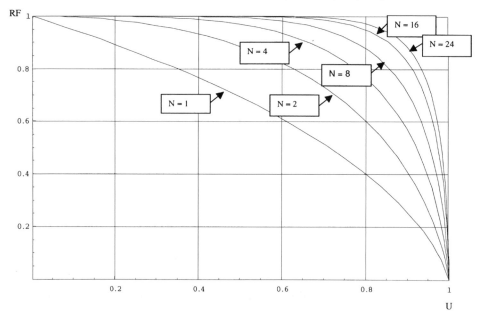

Figure B.6 Reduction factor curves.

Table B.7 Utilisation (u) for different reduction factor and N (Ns/Nu)

Ns/Nu	Reduction factor								
	0.95	0.9	0.8	0.7	0.6	0.5	0.4	0.3	0.2
1	0.0967	0.1871	0.35	0.4911	0.6123	0.7153	0.8019	0.8731	0.9301
5/4	0.1707	0.2831	0.4548	0.5861	0.6911	0.7763	0.8457	0.9016	0.9458
4/3	0.1947	0.3112	0.4826	0.6101	0.7104	0.791	0.8561	0.9084	0.9496
3/2	0.2398	0.3615	0.53	0.65	0.7421	0.8148	0.8729	0.9193	0.9556
5/3	0.2807	0.4046	0.5686	0.6818	0.7669	0.8333	0.8859	0.9277	0.9602
7/4	0.2996	0.4239	0.5854	0.6953	0.7773	0.8411	0.8914	0.9312	0.9622
2	0.3502	0.4737	0.6274	0.7288	0.8031	0.8601	0.9047	0.9397	0.967
9/4	0.3928	0.5141	0.6603	0.7545	0.8226	0.8744	0.9147	0.9462	0.9706
7/3	0.4056	0.5259	0.6697	0.7618	0.8281	0.8785	0.9175	0.948	0.9716
5/2	0.429	0.5473	0.6866	0.7748	0.838	0.8857	0.9225	0.9513	0.9734
8/3	0.4501	0.5663	0.7014	0.7862	0.8465	0.8919	0.9269	0.9541	0.9749
11/4	0.4599	0.575	0.7081	0.7913	0.8504	0.8947	0.9289	0.9553	0.9756
3	0.4866	0.5985	0.7261	0.805	0.8607	0.9022	0.9341	0.9587	0.9775
13/4	0.5097	0.6186	0.7413	0.8166	0.8693	0.9085	0.9384	0.9615	0.9791
10/3	0.5168	0.6247	0.7459	0.82	0.8719	0.9104	0.9397	0.9623	0.9795
7/2	0.5301	0.6361	0.7545	0.8265	0.8767	0.9139	0.9422	0.9639	0.9804
11/3	0.5423	0.6466	0.7623	0.8323	0.881	0.917	0.9444	0.9653	0.9812
15/4	0.5481	0.6515	0.7659	0.8351	0.8831	0.9185	0.9454	0.966	0.9816
4	0.5642	0.6651	0.776	0.8426	0.8887	0.9226	0.9482	0.9678	0.9826
13/3	0.5831	0.681	0.7877	0.8514	0.8952	0.9273	0.9515	0.9699	0.9838
9/2	0.5917	0.6882	0.793	0.8554	0.8982	0.9294	0.953	0.9708	0.9843
14/3	0.5998	0.6949	0.798	0.8591	0.9009	0.9314	0.9543	0.9717	0.9848
5	0.6146	0.7072	0.807	0.8658	0.9059	0.935	0.9568	0.9733	0.9857
16/3	0.6278	0.7182	0.815	0.8717	0.9103	0.9382	0.959	0.9747	0.9865
11/2	0.6339	0.7233	0.8187	0.8745	0.9123	0.9396	0.96	0.9754	0.9868
6	0.6507	0.7371	0.8287	0.8819	0.9178	0.9436	0.9627	0.9771	0.9878
13/2	0.6654	0.7492	0.8374	0.8883	0.9225	0.947	0.9651	0.9786	0.9886
20/3	0.6699	0.7529	0.8401	0.8903	0.924	0.948	0.9658	0.979	0.9888

Table B.7 (continued)

Ns/Nu	Reduction factor								
	0.95	0.9	0.8	0.7	0.6	0.5	0.4	0.3	0.2
7	0.6784	0.7599	0.8451	0.894	0.9267	0.95	0.9671	0.9799	0.9893
15/2	0.6901	0.7694	0.852	0.8991	0.9304	0.9526	0.9689	0.981	0.9899
8	0.7007	0.7781	0.8582	0.9036	0.9337	0.955	0.9705	0.982	0.9905
9	0.7193	0.7932	0.8689	0.9114	0.9393	0.959	0.9732	0.9838	0.9914
28/3	0.7249	0.7977	0.8721	0.9137	0.941	0.9601	0.974	0.9843	0.9917
10	0.7352	0.806	0.8779	0.9179	0.944	0.9623	0.9755	0.9852	0.9922
32/3	0.7445	0.8135	0.8831	0.9217	0.9468	0.9642	0.9768	0.986	0.9926
11	0.7489	0.817	0.8856	0.9235	0.948	0.9651	0.9774	0.9863	0.9928
12	0.761	0.8266	0.8922	0.9282	0.9514	0.9675	0.979	0.9873	0.9934
13	0.7716	0.835	0.8981	0.9324	0.9544	0.9695	0.9804	0.9882	0.9938
40/3	0.7749	0.8377	0.8999	0.9337	0.9553	0.9702	0.9808	0.9885	0.994
14	0.7812	0.8426	0.9032	0.936	0.957	0.9713	0.9816	0.9889	0.9942
15	0.7898	0.8494	0.9079	0.9393	0.9593	0.9729	0.9826	0.9896	0.9946
16	0.7976	0.8555	0.912	0.9422	0.9613	0.9743	0.9836	0.9902	0.9949
18	0.8113	0.8661	0.9191	0.9472	0.9648	0.9768	0.9851	0.9911	0.9954
56/3	0.8154	0.8693	0.9212	0.9487	0.9658	0.9775	0.9856	0.9914	0.9956
20	0.8229	0.8751	0.9251	0.9513	0.9677	0.9787	0.9864	0.9919	0.9958
64/3	0.8297	0.8803	0.9286	0.9537	0.9694	0.9799	0.9872	0.9924	0.9961
22	0.8329	0.8827	0.9302	0.9548	0.9701	0.9804	0.9875	0.9926	0.9962
24	0.8416	0.8894	0.9345	0.9578	0.9722	0.9818	0.9885	0.9932	0.9965
80/3	0.8516	0.897	0.9395	0.9612	0.9745	0.9834	0.9895	0.9938	0.9968
28	0.8561	0.9003	0.9416	0.9627	0.9755	0.9841	0.9899	0.9941	0.9969
88/3	0.8602	0.9035	0.9437	0.9641	0.9765	0.9847	0.9903	0.9943	0.9971
32	0.8677	0.9091	0.9473	0.9665	0.9781	0.9858	0.9911	0.9947	0.9973
36	0.8773	0.9162	0.9518	0.9696	0.9802	0.9872	0.992	0.9953	0.9976
40	0.8854	0.9222	0.9556	0.9721	0.9819	0.9883	0.9927	0.9957	0.9978
44	0.8923	0.9273	0.9588	0.9742	0.9833	0.9893	0.9933	0.9961	0.998
48	0.8983	0.9317	0.9615	0.976	0.9845	0.9901	0.9938	0.9964	0.9982
56	0.9082	0.9389	0.9659	0.9789	0.9865	0.9913	0.9946	0.9969	0.9984

Table B.7 (*continued*)

		Reduction factor							
	0.95	0.9	0.8	0.7	0.6	0.5	0.4	0.3	0.2
Ns/Nu									
64	0.916	0.9446	0.9694	0.9811	0.988	0.9923	0.9952	0.9972	0.9986
72	0.9225	0.9492	0.9721	0.9829	0.9892	0.9931	0.9957	0.9975	0.9987
80	0.9279	0.953	0.9744	0.9844	0.9901	0.9937	0.9961	0.9978	0.9989
88	0.9325	0.9563	0.9764	0.9856	0.9909	0.9943	0.9965	0.998	0.999
96	0.9365	0.9591	0.978	0.9867	0.9916	0.9947	0.9967	0.9981	0.9991

Example: consider one broadcast control channel (BCCH) TRX fully dedicated for (E)GPRS connections ($Ns = 7$) and a target reduction factor of 0.9. Assuming three TSL average mobile station (MS) capability ($Nu = 3$), $N = 7/3$. Looking at Table B.7 the utilisation with 0.9 reduction factor corresponds to 0.5259.

Blocking probability

In GPRS, there is no blocking as in circuit switched connections, if a new temporary block flow (TBF) stablisment is requested and there are already M users per TSL, the communication is queued in the system to be established later. M is the maximum number of simultaneous TBFs per TSL that (E)GPRS can handle. In this sense, blocking probability means the probability of TBF stablisment queuing and this happens when there are more than $M*N$ active TBFs in the system.

Blocking probability can be calculated as

$$\text{Blocking probability} = \frac{\displaystyle\sum_{i=M\cdot N+1}^{\infty} \frac{(U\cdot N)^i}{N!\cdot N^{(i-N)}}}{\displaystyle\sum_{i=0}^{N} \frac{(U\cdot N)^i}{i!} + \sum_{i=N+1}^{\infty} \frac{(U\cdot N)^i}{N!\cdot N^{(i-N)}}}.$$

Figure B.7 represents graphically the above equation using the typical maximum number of users sharing a TSL, $M = 9$. Numerical tables can be found in Table B.8. This table can be used in a similar way as described in the previous section.

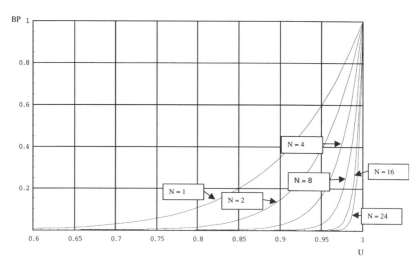

Figure B.7 Blocking probability curves.

Table B.8 Utilisation (u) for different blocking probability and N (Ns/Nu)

	Blocking probability							
	1%	2%	3%	5%	10%	20%	30%	40%
1	0.631	0.6762	0.7042	0.7411	0.7943	0.8513	0.8866	0.9124
5/4	0.6846	0.7247	0.7493	0.7815	0.8273	0.8759	0.9056	0.9273
4/3	0.6991	0.7378	0.7614	0.7922	0.8361	0.8823	0.9106	0.9312
3/2	0.7245	0.7605	0.7823	0.8108	0.8511	0.8934	0.9191	0.9378
5/3	0.7459	0.7795	0.7999	0.8263	0.8635	0.9025	0.9261	0.9433
7/4	0.7554	0.7879	0.8076	0.8331	0.869	0.9065	0.9292	0.9456
2	0.78	0.8096	0.8275	0.8506	0.8831	0.9167	0.937	0.9517
9/4	0.8	0.8273	0.8437	0.8648	0.8944	0.9249	0.9433	0.9565
7/3	0.8059	0.8325	0.8484	0.869	0.8976	0.9273	0.9451	0.9579
5/2	0.8167	0.8419	0.8571	0.8765	0.9036	0.9316	0.9484	0.9605
8/3	0.8263	0.8504	0.8648	0.8832	0.909	0.9354	0.9513	0.9627
11/4	0.8308	0.8543	0.8683	0.8863	0.9114	0.9372	0.9526	0.9637
3	0.8428	0.8648	0.8779	0.8947	0.918	0.9419	0.9562	0.9665
13/4	0.8533	0.8739	0.8862	0.9019	0.9237	0.946	0.9593	0.9689
10/3	0.8565	0.8766	0.8887	0.9041	0.9254	0.9472	0.9603	0.9696
7/2	0.8624	0.8818	0.8934	0.9081	0.9286	0.9495	0.962	0.9709
11/3	0.8679	0.8866	0.8977	0.9119	0.9315	0.9516	0.9636	0.9722
15/4	0.8705	0.8888	0.8997	0.9137	0.9329	0.9526	0.9643	0.9727
4	0.8776	0.895	0.9053	0.9185	0.9368	0.9554	0.9664	0.9743
13/3	0.886	0.9023	0.9119	0.9242	0.9412	0.9585	0.9688	0.9762
9/2	0.8898	0.9055	0.9149	0.9268	0.9432	0.96	0.9699	0.977
14/3	0.8933	0.9086	0.9176	0.9292	0.9451	0.9613	0.9709	0.9778
5	0.8997	0.9141	0.9227	0.9335	0.9485	0.9637	0.9727	0.9792
16/3	0.9054	0.919	0.9271	0.9374	0.9515	0.9658	0.9743	0.9804
11/2	0.908	0.9213	0.9291	0.9391	0.9529	0.9668	0.9751	0.981
6	0.915	0.9273	0.9346	0.9438	0.9565	0.9694	0.977	0.9825
13/2	0.921	0.9325	0.9393	0.9479	0.9597	0.9716	0.9787	0.9837
20/3	0.9228	0.934	0.9407	0.9491	0.9606	0.9723	0.9792	0.9841
7	0.9262	0.937	0.9433	0.9513	0.9624	0.9736	0.9801	0.9849
15/2	0.9308	0.9409	0.9468	0.9544	0.9648	0.9752	0.9814	0.9858
8	0.9348	0.9443	0.95	0.9571	0.9668	0.9767	0.9825	0.9867
9	0.9416	0.9502	0.9552	0.9616	0.9703	0.9792	0.9844	0.9881
28/3	0.9436	0.9518	0.9567	0.9629	0.9714	0.9799	0.9849	0.9885
10	0.9471	0.9549	0.9595	0.9653	0.9732	0.9812	0.9859	0.9892
32/3	0.9502	0.9576	0.9619	0.9673	0.9748	0.9823	0.9867	0.9899
11	0.9517	0.9588	0.963	0.9683	0.9755	0.9828	0.9871	0.9902
12	0.9555	0.962	0.9659	0.9708	0.9775	0.9842	0.9882	0.991
13	0.9587	0.9648	0.9684	0.973	0.9791	0.9854	0.989	0.9916
40/3	0.9597	0.9657	0.9692	0.9736	0.9796	0.9857	0.9893	0.9918
14	0.9616	0.9672	0.9706	0.9748	0.9806	0.9864	0.9898	0.9922
15	0.964	0.9693	0.9725	0.9764	0.9818	0.9873	0.9905	0.9927
16	0.9662	0.9712	0.9741	0.9779	0.9829	0.988	0.991	0.9932
18	0.9698	0.9743	0.9769	0.9802	0.9848	0.9893	0.992	0.9939
56/3	0.9708	0.9752	0.9777	0.9809	0.9853	0.9897	0.9923	0.9941
20	0.9727	0.9768	0.9791	0.9822	0.9863	0.9904	0.9928	0.9945

Ns/Nu (row label at left, beside 9/2 row)

Table B.8 (*continued*)

		Blocking probability							
		1%	2%	3%	5%	10%	20%	30%	40%
	64/3	0.9744	0.9782	0.9804	0.9832	0.9871	0.991	0.9932	0.9948
	22	0.9751	0.9788	0.981	0.9837	0.9875	0.9912	0.9934	0.995
	24	0.9771	0.9805	0.9825	0.9851	0.9885	0.9919	0.994	0.9954
	80/3	0.9793	0.9824	0.9842	0.9865	0.9896	0.9927	0.9946	0.9959
	28	0.9803	0.9832	0.985	0.9871	0.9901	0.9931	0.9948	0.996
	88/3	0.9812	0.984	0.9856	0.9877	0.9905	0.9934	0.995	0.9962
	32	0.9827	0.9853	0.9868	0.9887	0.9913	0.9939	0.9954	0.9965
	36	0.9846	0.9869	0.9882	0.9899	0.9923	0.9946	0.9959	0.9969
Ns/Nu	40	0.9861	0.9882	0.9894	0.9909	0.993	0.9951	0.9963	0.9972
	44	0.9873	0.9892	0.9903	0.9917	0.9936	0.9956	0.9967	0.9975
	48	0.9884	0.9901	0.9911	0.9924	0.9942	0.9959	0.9969	0.9977
	56	0.99	0.9915	0.9924	0.9935	0.995	0.9965	0.9974	0.998
	64	0.9912	0.9925	0.9933	0.9943	0.9956	0.9969	0.9977	0.9982
	72	0.9922	0.9934	0.994	0.9949	0.9961	0.9973	0.998	0.9984
	80	0.993	0.994	0.9946	0.9954	0.9965	0.9975	0.9982	0.9986
	88	0.9936	0.9946	0.9951	0.9958	0.9968	0.9978	0.9983	0.9987
	96	0.9941	0.995	0.9955	0.9962	0.9971	0.9979	0.9985	0.9988

References

[1] Allen A. O., 'Probability, Statistics, and Queuing Theory. With Computer Science Application'.

Appendix C

Mapping Link Gain to Network Capacity Gain

Pablo Tapia

C.1 Introduction

This appendix presents a simple method to estimate the network capacity associated with link level enhancing functionality. The method converts a link level gain (in dB) to an estimated network capacity increase. This is done by means of a function that has been constructed from extensive network level simulation analysis.

This gain estimation method can be very useful when there is no time or possibilities to model the specific feature in a network simulator, which is usually a highly time-consuming task. However, if resources are available, the best way to achieve reliable capacity numbers is always the feature modelling.

The results presented in this appendix have been obtained from network level simulations. The modelling of the link gain is very simple. It consists of a fixed dB offset that is added systematically after every carrier/interference (C/I) calculation:

New C/I (burst n) = Real C/I (burst n) + Link offset (dB).

This offset corresponds to the studied link level gain and generates the same result in the resulting frame error rate (FER) distribution as a link level enhancing functionality with a link offset (dB) gain.

C.2 Theoretical analysis

In principle, it could be expected that a 3 dB gain (3 dB higher interference tolerated for the same FER) would lead to a capacity increase of up to 100% (since the network can tolerate double the interference). With the same reasoning, a 6 dB gain would translate to a maximum of 300% increase. However, mobile networks are complex systems and not easy to predict, and these numbers can be considered as *ideal* ones. As it will be shown later, there are several factors that have to be taken into account that will limit the capacity increase associated with certain link level gains. Theoretically, a method could be developed assuming an ideal mobile network to calculate the capacity increase. Considering a generic cellular system with an ideal hexagonal layout, the mean CIR in the network can be easily estimated based on geometrical considerations (assuming that interference is coming from the first ring of co-channel cells) [1].

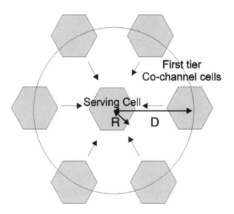

Assuming an exponential propagation law, the pathloss factor is:

$$Lb = k \cdot r^\gamma$$

(k is a constant, r is the distance, and γ the propagation factor)
    Considering CIR calculated at the edge of the cell:

$$CIR = \frac{k_a \cdot R^{-\gamma}}{\sum\limits_{i\ \text{Neighbours}} k_i \cdot r_i^{-\gamma}}$$

Taking into account the relationship between the cell radius (R) and reuse distance (D) in a hexagonal grid:

Reuse distance $(D) = \sqrt{3K} \cdot R$

where K is a rhombic number $(K = 1, 3, 4, 7, 9, ...)$.

A relationship can be established between the frequency reuse factor (K) and the mean CIR in the network. According to [1] the final expression can be simplified to the following formula, for an omni-directional layout:

$$CIR = \frac{1}{2} \cdot \frac{1}{\left(\sqrt{3K}-1\right)^{-\gamma}+\left(\sqrt{3K}\right)^{-\gamma}+\left(\sqrt{3K}+1\right)^{-\gamma}}.$$

In the case of sectorised antennas, there is no good simplification for the CIR, and it should be calculated for every specific reuse pattern by geometrical calculations (taking into account also the beamwidth of the antennas: 120°, 65°, etc.). In this theoretical approach we will just consider omni-directional antennas.

Comparing two different networks, A and B, with the same traffic conditions and different reuse schemes, the capacity difference between them is obtained from the following expression:

$$\text{Capacity increase } (\%) = 100 \times \left(\frac{\text{Reuse A (loose)}}{\text{Reuse B (tight)}} - 1 \right).$$

This capacity increase means that network B is using fewer of the resources than network A, while keeping the same amount of traffic in both systems. This will cause the CIR distribution in network B to be degraded, which will be perceived as a degradation in the quality of the system.

In order to achieve a certain capacity increase by tightening the reuse scheme, it will be necessary to improve the link level performance of the network by a certain factor. This factor will be called *link level gain*, and has to be chosen so that it compensates the CIR degradation caused by the reduction of the reuse:

$$\text{Link level gain} = \frac{CIR_A}{CIR_B} = \left(\frac{\left(\sqrt{3K_B}-1\right)^{-\gamma}+\left(\sqrt{3K_B}\right)^{-\gamma}+\left(\sqrt{3K_B}+1\right)^{-\gamma}}{\left(\sqrt{3K_A}-1\right)^{-\gamma}+\left(\sqrt{3K_A}\right)^{-\gamma}+\left(\sqrt{3K_A}+1\right)^{-\gamma}} \right).$$

Table C.1 shows some values calculated as an example.

These results show that introducing some physical limitation (in this case, the propagation factor) reduces the ideal capacity figures. In order to achieve these ideal figures (100% for 3 dB and 300% for 6 dB), a propagation factor of $\gamma = 2$ should be considered, that is, assuming space-free propagation conditions.

Table C.1

Link level gain (dB)	Capacity increase (theoretical, $\gamma = 3.5$)
3	43%
6	110%

It should be noted that these values have been obtained by simplified theoretical calculations. Also, user mobility hasn't been considered, as well as many other features existing in real networks: frequency hopping, power control, discontinuous transmission (DTX), etc. As a result, this method could only be suitable to calculate the capacity increase on a broadcast control channel (BCCH) transceiver (TRX), under certain circumstances: good network layout, optimal frequency plan, high reuse factors, etc.

C.3 Simulations

In the previous section a theoretical study presented how the link level gain could be translated into capacity increase of cellular networks. However, real cellular systems are not easy to analyse because of their dynamic behaviour. There are many factors to consider, and the behaviour of each of them separately is not necessary the same as when they are combined with other effects. With this complexity the resolution of this problem requires the use of a dynamic network simulator.

The results of the network simulations will be presented here. The simulations have been run over an ideal hexagonal layout, assuming typical urban 3 km/h (TU3) propagation model. The curves present the evolution of FER when effective frequency load (EFL) is increased, calculating the capacity difference between the EFL points with the same quality.

BCCH layer performance

Next simulations were configured to emulate BCCH TRX performance. Since BCCH is studied, only one TRX per cell is modelled, assuming full power transmission in all slots, no power control and no frequency hopping. Sectorised antennae with 65° beamwidth were used.

In Figure C.1, three curves are presented: the first one ('no offset') corresponds to a reference network; the second one ('3 dB offset') emulates the performance of a network with 3 dB better link level performance than the reference; and the third one ('6 dB offset') shows the behaviour when the link level gain is 6 dB. The Y-axis shows the percentage of bad quality FER samples, i.e. those whose FER is above 4%.

The different points in the curve have been obtained by changing the reuse pattern (reducing the number of frequencies). It is the only way to observe differences in the performance on this layer, due to the full time transmission of the broadcast channels.

When the trend of the curves becomes stable, roughly at 8% outage, it can be seen that the capacity increases for 3 and 6 dB link gain are 40 and 90%,

Figure C.1   Capacity increase in the BCCH layer.

respectively. These values are in line with the predicted capacity increase of the theoretical method, which indicated 43 and 110% gain, respectively.

Hopping layer

In this book we have considered the baseline speech global system for mobile communications (GSM) network configuration to include frequency hopping, DTX and power control in the non-BCCH TRXs. The combination of these factors causes the interference to be more equally distributed in the network. This section will include the capacity gain analysis for both hopping and non-hopping configurations in the non-BCCH TRXs.

Non-hopping network
The simulation results in Figure C.2 apply for a non-hopping network, with power control and DTX enabled.

Looking at the results at 5% outage it can be seen that the capacity increases are 90 and 190% for 3 and 6 dB link level gain, respectively. It can be concluded that power control effect is taking better advantage of the potential gains introduced by link level enhancements. This effect will be further analysed in Section entitled *Effect of power control* in C.3.

Figure C.2   Capacity increase in the TCH layer (non-hopping, power control, DTX).

Hopping network
The simulations in Figure C.3 provide the values for a typical GSM network configuration on the traffic channel (TCH) layer. In this case, frequency hopping as well as power control and DTX are switched on. As in the previous section, power control parameters have been tuned to achieve the maximum capacity in every case.

Also, the capacity increase values are now slightly better than in the non-hopping network case: 100% for 3 dB and 215% for 6 dB (at 2% outage).

Figure C.3   Capacity increase in the TCH layer (frequency hopping, power control, DTX).

Figure C.4 CIR distribution of BCCH and hopping layer.

Effect of power control

It has been shown in previous results how the use of power control could be applied to improve the capacity gain due to a link level improvement. While in a network with no power control the capacity increase with 3 dB was 50%, when power control is activated, this capacity increase rises to 100%.

Power control algorithms are an effective method to limit the waste of spare resources in the network, thus leaving space for new incoming users. As power control reduces the power of some users, there will effectively be more resources that can be utilised to improve the FER of mobiles with poorer quality. This effect can be observed by examining Figure C.4.

Figure C.5 represents the *pseudo CIR* that the user with enhanced link level would experience, if the link level gain was translated into an effective reduction of the interference. In reality, gains at link level are provided at the receiver chain, however, these representations can help to understand the effect of power control.

Figure C.4 shows the CIR distribution for the BCCH and hopping layer, both for the reference case and for a 3 dB link level gain. The figure on the left can be taken as an example of a non-power control network, while the one on the right is using power control.

It can be observed that, when no power control is applied, the shape of the distribution remains almost unchanged (the effect is a right shift of the whole curve). When power control is used, the shape of the distribution changes, and the potential gain from a better link level is used to effectively share the quality in the network; this can be observed by a narrower shape in the 3 dB gain case, which is placed almost completely in the *good CIR zone* of the X-axis (from 15 dB onwards). Thus, both distributions have different impact on the capacity since the difference of the low CIR values is broader when power control is used (see Figure C.4).

Figure C.5   Link gain vs. capacity gain figures (BCCH, TCH layer).

C.4 Final results and conclusions

The conclusions of this study can be summarised in Figure C.5, which presents values of link gain vs. capacity gain for both typical BCCH and hopping configurations.

Figure C.5 can be used to calculate the approximate capacity increase of a link level enhancement. Usage examples may be adaptive multi-rate (AMR) codecs, antenna hopping, delay diversity, etc.

For instance, keeping in mind that the link level gain of the AMR 5.90 codec with frequency hopping is 4.5 dB, the capacity increase for this codec would be 150% according to Figure C.5. Figure C.6 shows the results when the codec is modelled in the simulator, where the gain achieved is around 140%.

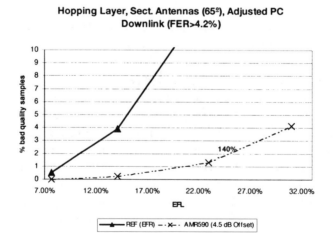

Figure C.6   Simulated performance of AMR590 codec.

Finally, note that the values from these figures should only be taken as an *estimation* of the real capacity increase associated with the feature under study since the link level gain is usually not constant but dependent on the propagation environment and the instantaneous C/I. If accurate capacity studies are needed, the feature should be modelled using a dynamic network simulator.

References

[1] "Wireless Communications, Principles & Practice", Rappaport, T.S. Prentice Hall Communications Engineering and Emerging Technologies series copyright 1996, Edition 1999.

Appendix D

Interference between GSM/EDGE and other cellular radio technologies

Kari Heiska

D.1 Introduction

This appendix describes the main interference sources and the network planning principles associated with the adjacent channel operation of different technologies. The interference sources and basic mechanisms are typically technology independent but the relative importance of various interference mechanisms is dependent on the implementation of different network elements, locations of the interfered and interfering sites with respect to each other as well as type and size of the cells/sectors. Typically, in order to control the effect of the adjacent system interference, guardbands between the adjacent systems are deployed. The size of these guardbands has to be carefully selected in order to maximise the overall performance.

The interference between global system for mobile communications (GSM)/ enhanced data rates for global evolution (EDGE) and other cellular technologies is an important issue to be carefully considered in scenarios when these technologies are deployed in adjacent frequency bands. Typical reasons for these multi-technology deployments would be:

- Technology migration. If cellular operators using IS-136 or IS-95 decide to migrate technology to GSM/EDGE, during the migration period, both technologies are likely to be deployed within the same band and inter-system interference has to be considered for the deployment.
- GSM/EDGE and wideband code division multiple access (WCDMA) multiradio operation within the same band. GSM/EDGE and WCDMA can be seamlessly integrated and deployed together using the same 1800 and/or 1900 band. In this deployment scenario, the inter-system interference has to be taken into account.
- Licensed bands using different technologies. In the case different technologies are deployed in adjacent bands by different operators, there will be certain inter-system interference to consider. National regulations, such as Federal Communications Commission (FCC) requirements in the US [1,2], will ensure this interference is controlled.

In the first two scenarios, the technologies will typically be co-sited since the same operator deploys them. These cases are under the control of operators' design criteria and will be the focus of this appendix.

The interference sources in the mobile telecommunication systems can be divided into controllable and non-controllable interference. The controllable interference refers to the interference from the sources that can be controlled by the operator itself, i.e. the operator is able to move the interfering sites, change the equipment of the site and to reconfigure the network parameters. Non-controllable interference from other operator devices or from background noise is thus more difficult to eliminate.

The easiest way to decrease the inter-system interference is to co-site the interfering systems. By co-siting it is possible to avoid the near–far effect which means, for example, that when the mobile of operator 1 is close to operator 2's site and far away from its own site there could be significant uplink interference from operator 1's mobile to operator 2's base station receiver. Also, when the operator 2 mobile terminal is close to the operator 1 base station there can be a large downlink interference component from operator 1 to operator 2 in downlink. Similar interference components occur from operator 2 to operator 1, respectively.

Figure D.1 shows the example frequency allocation case where different systems and different operators have been allocated into the same frequency band. In the general situation operator 1 and operator 2 are not co-sited and their network structures are not congruent. This means, for example, that when operator 1 utilises micro cells to cover a certain part of the urban area, operator 2 utilises macro cells. The average cell size for these two operators can be thus rather dissimilar, which increases the inter-system interference.

Figure D.1 Frequency allocation example. Operator 1 utilises two different technologies in its frequency band and the adjacent operator 2 uses its own technology. Δf_{gn} refers to the guard between operators (non-coordinated case) and Δf_{gc} refers to the guardband between technologies (coordinated case).

D.2 Interference mechanisms

This section describes the most dominant interference mechanisms, which take place in the adjacent channel operation.

Adjacent channel power

The adjacent channel interference (ACI) refers to the power coming to the receiver due to finite receive filter attenuation outside the band of interest. Even with the ideal transmitter emission mask there is interference coming from adjacent channels, which depends on the implementation of analogue and digital filtering. Additionally, the ACI is dependent on the power of the interfering system as well as the frequency offset between the interferer and the interfered systems. The effect of the adjacent channel interference decreases rapidly outside the receive band so the ACI can be eliminated with an adequate guardband aside the band of interest.

Out-of-band emission refers to all the interference power outside the wanted channel of the interfering system and can be measured with the spectrum analyser from the output port of the transmitter. When the power density of the transmitter is $p_e(f)$ [W/Hz] its total emitted power, P_e, is defined as

$$P_e = \int_{-\infty}^{\infty} p_e(f)\, \mathrm{d}f. \tag{1}$$

The adjacent channel interference ratio (ACIR) is the key parameter describing both the effect of non-ideal receive filtering and the out-of-band emission of the transmitter. L_{ACIR} is then given by

$$L_{\text{ACIR}}(\Delta f) = \frac{\displaystyle\int_{-\infty}^{\infty} p_e(f) \cdot L_{\text{filter}}(f - \Delta f)\, df}{\displaystyle\int_{-\infty}^{\infty} p_e(f)\, df} \tag{2}$$

$$= \frac{1}{P_e} \int_{-\infty}^{\infty} p_e(f) \cdot L_{\text{filter}}(f - \Delta f)\, df \tag{3}$$

where $L_{\text{filter}}(f)$ is the attenuation of the receive filtering at frequency f. Typically, the emission spectrum values have been measured over some finite bandwidth and in some predefined frequency grid. When the carrier separation between two systems is Δf_k the L_{ACIR} is then given by

$$L_{\text{ACIR}}(\Delta f_k) = \frac{1}{\displaystyle\sum_{k=-M}^{M} p_e(f_k)} \sum_{k=-M}^{M} (p_e(f_k) \cdot L_{\text{filter}}(f_k - \Delta f_k)) \tag{4}$$

where the emission spectrum, $p_e(f_k)$, is negligible if $|f_k| > f_M$. The effect of adjacent channel interference on the system performance has been studied in [3,4,5].

Intermodulation distortion (IMD)

Because of the non-linearity of the receiver, the receiver's output spectrum may contain frequency components that are not present in the input signal. These new signal components may fall into the band of interest causing intermodulation distortion. The non-linearity takes place in every component of the receiver if the signal level is high enough. The high power, unwanted signal may drive the receiver of the mobile to its non-linear part. When two or more signals are added together in the non-linear element the resulting outcome from the element includes in addition to the wanted signal frequency also other frequency components. In the radio frequency (RF) design the third-order intermodulation is particularly problematic, because of its strength and because it falls close to the band of interest. In the case of two interfering signals on frequencies f_1 and f_2, in the proximity of the desired signal, the third-order IMD products are those falling on frequencies $2f_1 - f_2$ and $2f_2 - f_1$. Also higher order IMD products exist but they are usually not that strong. The strength of the third-order intermodulation power is given by

$$P_{\text{IMD}}^{\text{in}} = 3 \cdot P_i - 2 \cdot \text{IIP}_3 \tag{5}$$

assuming that the incoming signal powers at frequencies f_1 and f_2 are equal. P_i (dBm) is the power at the input of the non-linear component and IIP_3 (dBm) is the third-order input intercept point of the same. The strength of the IMD depends on the input power of the interferer as well as the linearity of the

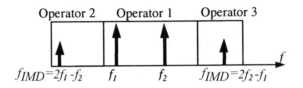

Figure D.2 Third-order intermodulation components falling from operator 1 into the band of operators 2 and 3.

receiver. P_{IMD} is proportional to the third power of P_i so that it is large when the receiver is close to the interfering source but decreases rapidly as the path loss increases.

Figure D.2 shows the example of the intermodulation effects between two operators. Operator 1 has two strong signal powers at frequencies f_1 and f_2. These can be originating from one site. If this jamming signal from the adjacent band is high enough, causing saturation at the receiver, the mobiles of operators 2 and 3 might observe intermodulation distortion power at frequencies $2f_1-f_2$ and $2f_2-f_1$, respectively. The performance effects of intermodulation have been described in [6].

D.3 Coverage effects

This section describes the coverage effect of the adjacent channel interference and the out-of-band emissions in downlink and uplink.

Downlink

The interference power in downlink is

$$P_r = P_{\mathrm{B}} - L_{\min} - L_{\mathrm{ACIR,DL}} \tag{6}$$

where P_{B} is the total transmission power of the interfering base station, $L_{\min}$ is the minimum coupling loss from the base station to the mobile station and $L_{\mathrm{ACIR,DL}}$ is the adjacent channel interference ratio described in Section entitled *Adjacent channel power* in D.2. If the required carrier-to-interference ratio is $(C/I)_{\mathrm{dB}}$ and it is assumed that the adjacent channel interference is the dominant interference mechanism, the needed carrier power is given by

$$C - P_{\mathrm{B}} + L_{\min} + L_{\mathrm{ACIR,DL}} = (C/I)_{\mathrm{dB}}. \tag{7}$$

When the maximum allowed path loss for the interfered system is $L_{\max}$ and the maximum transmission power is P_{B}, the strength of the carrier, $C = P_{\mathrm{B}}-L_{\max}$. The needed ACIR in the downlink can then be solved as

$$L_{ACIR,DL} = L_{max} - L_{min} + (C/I)_{dB}. \tag{8}$$

This gives, for example, $L_{ACIR,DL} = 135 - 80 + 20 = 75$ dB, which is a rather large value for the adjacent channel isolation and might cause dead zones around the interfering site. This was the worst case calculation, which assumes that the interfering cell is located at the cell edge of the interfered system. This is not a typical case but in an urban area, for example, the areas for highest path loss requirements locate inside the buildings, where there are no adjacent operator sites. Of course, this leads to problems in the case where one operator has an indoor system and another has only outdoor systems. In that case, a large guardband between operators is required. In the case of co-sited base stations, where $L_{max} = L_{min}$, equation (8) gives that $L_{ACIR} = (C/I)_{dB}$.

Uplink

The additional interference power from the adjacent frequency channel decreases the sensitivity of the base station receiver and therefore decreases the coverage area. The uplink interference is large when the interfering mobile station is very close to the base station receiver and is allocated into the closest frequency channel. The maximum required isolation between carriers can be calculated by using this worst case scenario. This corresponds to the case where interfered and interfering systems are not co-sited and the interfering system is using high powers in the close proximity of the interfered receiver. The effect of intermodulation has been ignored in this analysis. The received interference power at the receiver is then given by

$$P_r = P_M - L_{min} - L_T - L_{ACIR,UL} \tag{9}$$

where P_M is the mobile station maximum power, L_{min} is the minimum coupling loss and L_T is the average transmit power decrease due to bursty transmission (in time division multiple access (TDMA)). When P_r is set equal to the noise floor, P_N, so that the extra interference causes 3 dB degradation to the link budget, the needed $L_{ACIR,UL}$ can be calculated as

$$L_{ACIR} = P_M - L_{min} - L_T - P_N \tag{10}$$

which gives 30 dBm-80 dB-9 dB$-(-121$ dBm$) = 62$ dB. Here we have assumed that the system is noise limited where the noise power P_N is given by

$$P_N = 10 \cdot \log_{10}(kTB) + F \tag{11}$$

where k is Boltzmann's constant ($1.38 \times 10 - ^{23}$ W/Hz/K), T is the temperature of the receiver, B is the system bandwidth and F is the noise figure of the receiver. For the GSM system ($B = 200$ kHz), $P_N = -121$ dBm. When the interfered system is limited by its own system the required $L_{\text{ACIR,UL}}$ depends on the interference level. If the interference level is for example -100 dBm, the needed isolation is 30 dBm-80 dB-9 dB$-(-100$ dBm$) = 41$ dB.

In the co-sited case the power control of the interfering system usually decreases the interference problem. If the minimum power of the mobile is 0 dBm, the needed isolation in the thermal noise limited case is 0 dBm-80 dB-9 dB$-(-121$ dBm$) = 32$ dB. The power control is, in GSM, used in the case of traffic channels. In the case of control channels the mobile is usually using full power and therefore the receiver has to tolerate more interference.

If $L_{\text{ACIR,DL}}$ is much lower than $L_{\text{ACIR,UL}}$ the mobile will be blocked by the base station before it interferes with the base station in uplink. This phenomenon is depicted in Figure D.3.

By inserting the L_{min} from equation (8) into equation (9) the uplink interference level can be solved as

$$P_r = P_M - L_{\text{max}} - (C/I)_{\text{dB}} - L_T - L_{\text{ACIR,UL}} + L_{\text{ACIR,DL}}. \tag{12}$$

From this equation it is possible to calculate the uplink interference if $L_{\text{ACIR,UL}}$ and $L_{\text{ACIR,DL}}$ are known. Figure D.4 shows the uplink interference as a function of difference between $L_{\text{ACIR,UL}} - L_{\text{ACIR,DL}}$ with different maximum adjacent cell path losses, L_{max}. It can be seen that when $L_{\text{ACIR,UL}} = L_{\text{ACIR,DL}}$ the uplink interference is above the GSM noise level (-121 dBm) only when the adjacent cell size is rather small. Here it has been also assumed that the mobile transmits all the time with the full power. This is, however, not the case when the cell size is smaller and if the uplink power control works properly.

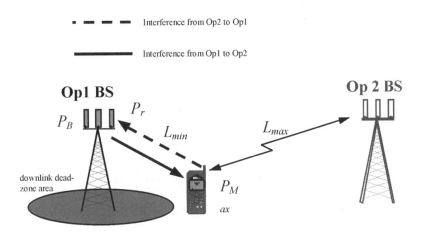

Figure D.3 Interference between operator 1 and operator 2. Operator 1 interferes with operator 2 in downlink and operator 2 interferes with operator 1 in uplink.

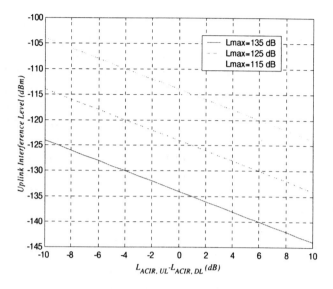

Figure D.4   Uplink interference with different adjacent cell size calculated with equation (12). $P_M = 30$ dBm, $(C/I)_{dB} = 20$ dB, $L_T = 9$ dB.

D.4 The interference from WCDMA to GSM

The capacity loss of the GSM has been simulated in the case where the GSM and the WCDMA systems were deployed at adjacent frequency bands [7]. Other standardisation references handling the GSM and UMTS blocking related items are [8,9,10,11]. The main interference mechanism that affects GSM uplink performance is the out-of-band emission of the WCDMA user equipment (UE) transmitter. The performance degradation is dependent on the transmitting power of the WCDMA UE and thus on the cell size of the interfering WCDMA system. In these simulations equal cell sizes in both systems were assumed with cell sizes up to 2400 m. The simulation results indicate that the capacity loss in GSM was between 0 to 4% with the carrier separation of 2.7 MHz between WCDMA and the closest GSM carriers. This is in the case where both systems were deployed with three sector macrocellular networks and the WCDMA network was fully loaded. This is a worst case scenario in the sense that the WCDMA base stations were located at the cell edge of the GSM system. The capacity degradation could be even lower when applying the random site locations with frequency hopping and the link adaptation schemes in the GSM system. In the co-sited base stations case, the capacity degradation of the GSM system can be considered to be negligible when deployed with WCDMA system.

D.5 Monte-Carlo simulation study (GSM/EDGE and IS-95)

This section describes the example simulation exercise that has been used to estimate the needed guardband between two systems by analysing the effect of adjacent channel interference in uplink. The selected scenario represents the deployment of IS-95 and GSM/EDGE within the same band. The interference effects of GSM to WCDMA and IS-95 to WCDMA system have not been studied here. These simulation studies can be found in [12,13]. In this example both systems are co-sited, representing the technology migration scenario. The interference levels in the uplink of IS-95 have been calculated by summing the interference contributions from randomly located GSM/EDGE mobiles. The used carrier frequency, mobile station powers and the path loss of the GSM/EDGE mobile have been generated randomly. The path loss data of realistic network scenarios have been calculated by using appropriate, environmental specific propagation models. An Okumura–Hata propagation model has been utilised in macro cells and a ray-tracing propagation model [14] in micro cells. The mobile station power has been generated according to polynomial function by mapping the coupling loss value to mobile Tx power value for each mobile in the network. The function has been obtained by fitting the measured mobile station transmission power values from the real GSM/EDGE network. The original experimental data, which have been retrieved from numerous urban and suburban measurements, and the fitted polynomial, are shown in Figure D.5. The used carrier for each mobile has been generated randomly within the total bandwidth of 2.4 MHz (12 GSM/EDGE carriers).

The used emission spectrum of the GSM/EDGE mobile, the IS-95 receive filtering as well as the resulting ACIR function calculated according to equation (4) are shown in Figure D.6.

Figure D.5 Mobile stations transmit power as a function of path loss.

Figure D.6 Filter characteristics of the IS-95 BS and out-of-band emissions of the GSM/EDGE MS (left) and the respective adjacent channel interference ratio (ACIR) (right). These are example functions respecting typical implementation values.

The target interference level in the IS-95 cell is -103 dBm in these simulations, which means approximately 5 dB noise rise when the noise figure is assumed to be 5 dB. The capacity of the interfered IS-95 network has been defined as the number of users for which the interference is above the target level at maximum 5% of the snapshots. Figure D.7 shows the interference level as a function of number of users for each simulated snapshot. Each simulation case, which means a certain number of IS-95 users, includes 2500 independent snapshots. The solid curve shows the loading not interfered network. In the figure the capacity is 10 users with the interference and 13.5 users without the interference. The capacity reduction is thus $100 \cdot (13.5 - 10)/13.5 = 25.93\%$.

Figure D.8 shows the capacity reduction as a function of the channel separation between GSM/EDGE and IS-95 carriers. The GSM/EDGE bandwidth is 0.2

Figure D.7 Interference level as a function of users in a IS-95 cell with (dots) and without (solid line) the adjacent system interference.

Figure D.8 The capacity reduction as a function of the carrier separation. Left: seven GSM/EDGE users per IS-95 cell. Right: two GSM/EDGE users per IS-95 cell.

MHz and the IS-95 bandwidth is 1.25 MHz, so that the channel separation without any guardbands is $0.2/2 + 1.25/2 = 0.725$ MHz. It can be seen from the results that the capacity reduction is larger in micro cells than in macro cells. This is because in micro cells, where the BS antenna is close to ground level, the minimum coupling loss is around 60 dB whereas in macro cells it is around 80 dB. This leads to larger interference levels and also to larger capacity reductions in micro cells. With the example ACIR function, the needed guardband is just around 100 kHz in the macro cell case and 350 kHz in the micro cell case, in order to obtain very low capacity reduction.

D.6 *Summary*

This appendix presented some radio network planning aspects associated with the coexistence of different systems in the same geographical area. The main reason for adjacent channel interference is the combined effect of non-idealities in the receiver filtering and the out-of-band emissions. The intermodulation distortion can be also relevant in some cases.

The system has to be able to tolerate a certain amount of additional interference from the adjacent carriers due to limitations in mobile terminal and base station design. The intra-operator interference can be minimised with co-siting which decreases the near–far problem. One aspect to consider is the GSM/EDGE common control channels planning, since they usually transmit at full power and therefore might cause interference problems. It is preferable to allocate the common channels far away from the adjacent system in order to minimise the interference effects.

The deployment in the same band by a single operator of GSM/EDGE with other radio technologies, such as IS-136, IS-95 and WCDMA, has proven to be feasible without the requirement of significant guardbands in the co-sited scenario. Therefore, operators can deploy multiple technologies in the same site grid without any major performance degradation.

In the non-co-sited case the near–far problem might exist which means that when the mobile is far away from its own site it can be very close to the adjacent site. In this situation the interference increases both in uplink and in downlink directions. The downlink interference might cause dead zones around the base station and the uplink interference causes desensitisation in the BS reception. The needed guardband between carriers depends on the particular systems and filter characteristics.

References

[1] National Archives and Records Administration, Code of Federal Regulations: 24.238 CFR, 47, Parts 20–39, http://www.access.gpo.gov/nara/cfr.

[2] http://www.fcc.gov/.

[3] Hamied Khalid, Labedz Gerald, 'AMPS Cell Transmitter Interference to CDMA Mobile Receiver', Proceedings of Vehicular Technology Conference, Vol. 3, 1996, pp. 1467–1471.

[4] Schilling D. L., Garodnick, J., Grieco, D., 'Impact on Capacity to AMPS Jamming CDMA/CDMA Jamming AMPS in Adjacent Cells', Proceedings of Vehicular Technology Conference, 1993, pp. 547–549.

[5] Lu Y. Edward, Lee William C. Y., 'Ambient Noise in Cellular and PCS Bands and its Impact on the CDMA System Capacity and Coverage', Communications, 1995. ICC '95 Seattle, 'Gateway to Globalization', 1995 IEEE International Conference on Volume: 2, 1995, pp. 708–712.

[6] Dong Seung Kwon, Heon Jin Hong, Sang Gee Kang, 'CDMA Mobile Station Intermodulation Interference Induced by AMPS Competitor Base Station', Proceedings of IEEE 4th International Symposium on Spread Spectrum Techniques and Applications, Vol. 1, 1996, pp. 380–384.

[7] GPP TR 25.885: 3rd Generation Partnership Project; Technical Specification Group TSG RAN; UMTS1800/ Work Items Technical Report (Release 5 (1900) 220.

[8] GPP TS 25.104 V4.0.0, UTRA (BS) FDD; Radio Transmission and Reception, 2000.

[9] GPP TR 25.942, V2.1.1, RF System Scenarios, 1999.

[10] 3rd Generation Partnership Project (3GPP), Technical Specification Group (TSG), UTRA FDD, Radio Requirements: TS 25.101 and TS 25.104.

[11] European Telecommunications Standards Institute (ETSI): GSM 05.05, http://www.etsi.org.

[12] Laiho J., Wacker A., Novosad Tomás, 'Radio Network Planning and Optimisation for UMTS', Wiley, Chichester, 2001, 512 pp.

[13] Heiska Kari, Posti Harri, Muszynski Peter, Aikio Pauli, Numminen Jussi, Hämäläinen Miikka, 'Capacity Reduction of WCDMA Downlink in the Presence of Interference from Adjacent Narrowband System', IEEE Trans. Veh. Technol., Vol. 51, No. 1, January 2002.

[14] Hefiska Kari, Kangas A., 'Microcell Propagation Model for Network Planning', Proceedings of Personal Indoor and Mobile Radio Communications, PIMRC'96, pp. 148–152.

Appendix E

Simulation Tools

Martti Moisio

E.1 Introduction

This appendix describes first the fundamentals of simulation and network level simulator (sometimes called also *system*simulators) functionalities. The simulators can be divided into two main categories: *static* simulators and *dynamic* simulators. The properties of these two simulation approaches are discussed and finally, the global system for mobile communications (GSM)/enhanced data rates for global evolution (EDGE) simulator used in this book are described together with the applied cellular models, algorithms and parameters.

E.2 Static simulations

Static simulators do not model the movement of the mobiles and there is no correlation between consecutive simulation steps. Typical static simulation methodology is the following:

1 Create mobiles in random places in the network. The number of mobile stations (MSs) depends on the wanted load.
2 Select serving base stations (BSs) for each MS according to the received signal level.

3   Adjust transmission powers of links according to, e.g. path loss criteria. After this, the interference situation in the network will be frozen.
4   Execute needed radio resource management (RRM) algorithms, e.g. select coding scheme according to link quality.
5   Collect statistics (carrier/interference (C/I), bit error rate (BER), block error rate (BLER), throughput, etc.).
6   Repeat steps 1–5 (typically, a few thousand times) to get enough statistical confidence.

The procedure above is repeated for each wanted operation point (network load). It should be noted that steps 3 and 4 might include also iterative processes (needed for, e.g. power control with certain C/I target (typical in code division multiple access (CDMA) simulations [1]). Finally, network level key performance indicators (KPIs) can be plotted against offered load. BER/BLER/throughput performance indicators are achieved through mapping tables from link simulations, using long-term average C/I vs. BER/BLER curves. Normally, fast fading is not taken into account in a static network simulator, its effect is included in the link level curves.

The main benefit of the static simulation tool is its execution speed. Also, the results are easier to understand and interpret due to simple models and especially lack of dynamics related to radio link control and RRM algorithms. However, these same issues constitute the main drawbacks of the static simulation approach: lack of dynamics, correlation and feedback loops which form an essential part of cellular system characteristics, especially when dealing with packet data. Typically, a static simulator gives optimistic upper-level results. Especially when studying the performance gain related to radio link algorithms, a dynamic simulator is needed.

E.3 Basic principles of dynamic simulation

The fundamental property of a dynamic simulator is *causality* and *time correlation* between events in the simulated network. When an event (e.g. handover (HO) or PC command) is triggered, it has its cause in the past and consequence in the future. With this kind of modelling of the system dynamics, also the effect of feedback loops (input depends on the output) that are present in many algorithms can be assessed. Maybe the most important building blocks of this kind of dynamics are *simulation time* and *user mobility* that are not relevant in static simulations. These two features bring the needed determinism in the simulated system – in each simulation step the time is increased and mobiles move along their paths. However, the movement of mobiles is not any way necessary for a simulator to be a dynamic one – the important thing is the time correlation of the signal with the evolving simulation time.

Dynamic simulators can be divided into two basic categories: they can be either *event-driven* or *time-driven*. In an event-driven simulator, the functionality is based on events (MS changes position, PC command is executed, etc.) that are processed according to their order of appearance. Events may generate *tasks*, that may be handled only after some time period or after some other event has occurred. This approach resembles the behaviour of typical protocols. It can save simulation time especially if the event inter-arrival time is relatively long, because the simulator can jump directly to the next pre-calculated event.

The simulator used in this book is a discrete time-driven simulator, where the simulation time step is fixed. This approach is convenient because the time steps can be selected according to the GSM frame structure (time division multiple access (TDMA) frame numbering). Due to the detailed modelling and high frequency and volume of the events (e.g. receiving the radio burst), the event-based approach would produce more implementation overhead leading probably to longer simulation times. Also, the time-driven approach makes the actual program code more understandable for the software developer. However, event-based methodology is used for certain parts of the program, where it was found to be suitable (PC/HO task handling in base station controller (BSC), general packet radio system (GPRS)/enhanced GPRS (EGPRS) radio link control (RLC) protocol).

An important practical property of the dynamic simulator (actually, any simulator) is that it must be possible to reproduce the results, provided that the input remains the same. Otherwise, the analysis of the system behaviour is almost impossible, not to mention the debugging of the program code. The network simulator should not just be able to produce average performance values for the whole network, often the behaviour of a single mobile is interesting. It should be possible to select any connection in the network and trace its behaviour. These requirements mean that the simulator must have pseudo-random statistical properties. In practice, the output remains exactly the same with the same input and it can be changed by changing the seed of the random number generator(s). Of course, final simulations must be long enough so that the effect of random number seeds becomes negligible.

A state-of-the-art dynamic simulator should fulfil at least the following requirements:

- Short enough time resolution to capture the effect of short-term changes in the radio link (multipath fading, frequency hopping, power control, discontinuous transmission, packet scheduling, fast link adaptation using hybrid automatic repeat request (ARQ) schemes, etc.). In GSM, a suitable time step unit is one TDMA frame, i.e. 4.615 ms – this means that every burst in the transmitting link is simulated. It can be assumed that the C/I ratio of the link is fixed for the GSM burst period (actually, if this assumption is not valid, the receiver cannot estimate the channel impulse response from the training sequence). Individual bits are not considered in a system level simulator, bit level modelling (modulation, channel coding, etc.) is handled in a dedicated link level simulator. Statistical look-up tables produced by the link simulator are then used to calculate bit and block error rates. It is very important to model the stochastical, non-deterministic processes to produce correct distributions, not just correct long-term averages. For example, when the look-up table produces certain probability for block error, a random experiment must be made against uniform distribution to produce the final result (whether the block was correct or not).

- Multiple cells with enough co-channel and adjacent-channel interfering tiers. To reduce the boundary effects of finite simulation area, a wrap-around methodology [2,3] is recommended, especially if the number of frequency clusters is small (large reuse factor and/or small number of cells).

- Modelling user-specific quality with the concept of *call* – whether it is a traditional circuit switched voice call, file downloading or Web browsing session. This is important, since finally the only relevant quality is the user-specific quality. The concept of call includes the important effect of mobility in a cellular environment – a moving MS experiences different signal strength and quality conditions depending on its location and it may be served by several different BSs during its lifetime (hence, the effect of HOs is taken into account). Also, the call may be transferred to a different layer, or it may even have to adapt to different base transceiver station (BTS)/ transceiver (TRX) configuration (e.g. some BTSs/TRXs may not support EDGE).

- The software must be efficient enough to allow enough statistically valid calls to be simulated within a reasonable time. As a rule of thumb, the number of simulated calls must be of the order of thousands. Also, to minimise the unavoidable distorting effects in the beginning and end of the simulation, the actual simulated real (busy hour) network time should be long enough. If we consider the typical case of call arrivals to the network as a Poisson process with exponential service times, the total simulation length must be long enough compared with the expectation value of the service time. It should be noted that this is not required by the Poisson process itself (which is a memory-less process), but it is to have enough normally ended calls according to the wanted call length distribution. Traffic warm-up should be used to shorten the time when the level of desired offered load is reached – however full traffic warm-up is not always possible because the load saturation point is not always known, especially in packet data services. Based on the experience with the simulator used in this book, at least 15 min of busy hour time is needed to reach full statistical confidence.

It should be noted that a large number of simulated calls do not compensate the requirement for long enough simulation times. Analogously, the simulation time must be further increased with especially low load in order to have high enough number of calls.

E.3 Description of the simulator and basic simulation models used in this book

The simulator[1] used in this book follows the principles and requirements in the previous section. This section describes the simulator functionality and modelling more accurately. For all the people involved in performance analysis, it is important to understand the modelling principles of present-day network level simulators. Also, this section (as the previous one) might give some hints for those who are interested in developing system level simulators of their own.

Software and programming issues

To ensure easier maintenance, reliability and modularity, C + + [4] object-oriented language was selected. The elements of the GSM network (BSC, BTS, MS, radio block, etc.) form a natural basis for C + + class hierarchy. The fine features of C + + language, like polymorphism and class inheritance are very useful when designing a GSM system simulator. Figure E.1 shows an example

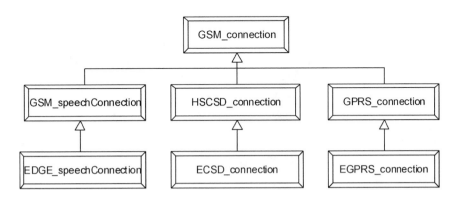

Figure E.1 An example of the hierarchical class structure of the *SMART* GSM simulator.

[1] The simulator is called SMART, originally developed at Nokia Research Center, Helsinki, Finland.

of the class structure in the simulator. The base class contains all common GSM-connection (call)-specific functionality and the derived class implements new ones and/or overloads the base class methods whenever needed. Each connection has N pieces of *links* that carries the radio blocks and bursts from transmitter to receiver. For example, a GSM speech connection class has one downlink and one uplink, while a GPRS connection can have several links in one direction.

As is clear for people familiar with object-oriented programming, polymorphism enables run-time dynamic binding (through virtual functions in C + +) of the methods. This is a very useful feature when thinking about the modelling of a hierarchical GSM system.

Basic functionality of the simulator

Figure E.2 shows a schematic view of the basic functionality of the simulator. For each simulation step, the mobiles are first moved and received powers are calculated for both up- and downlink. After this, the interference situation in the network is frozen for that time step and C/(I + N) values can be calculated for each active transmission link. The link object binds the receiving and transmitting objects together and contains all the necessary information (antenna gains, transmission powers, MS/BS positions, fading phase, etc.) in order to calculate the power (field strength) in the receiver antenna. All the

Figure E.2 Basic functionality of the simulator.

transmission links in the network are stored in one interference matrix (one for downlink, one for uplink) where each matrix element contains all the links transmitting in that physical channel. This kind of effective and simple data structure is important for the next step in the simulation loop – interference calculation – which is the most important and time-consuming function of a cellular simulator. Received power C (in watts) is calculated for each link according to following formula:

$$C = TxP^*G^*L_d^*L_S^*L_F$$

where

TxP = transmission power (in watts)
G = transmitter and/or receiver gain
L_d = distance path loss
L_S = slow fading component (each MS has its own correlated slow fading process)
L_F = fast fading component (each MS–BS pair has its own correlated frequency dependent fast fading process).

All the co-channel and adjacent channel interferers in the whole network area are taken into account. The carrier-to-interference ratio is calculated with the following formula:

$$\mathrm{CIR}_{f_N} = \frac{C_{f_N}}{\sum_k I_{f_N} + ACP1 \times \sum_l (I_{f_{N-1}} + I_{f_{N+1}}) + N_0}$$

where

CIF_{f_N} = carrier-to-interference ratio on carrier frequency f_N
C_{f_N} = signal strength at carrier frequency f_N
$\sum_k I_{f_N}$ = sum of all k interfering signals at carrier frequency f_N
$ACP1$ = adjacent channel protection on the first adjacent carrier frequency
N_0 = receiver noise floor.

After the CIR calculation, radio bursts and blocks for each link are received. As the last step in the simulation loop, all the necessary RRM algorithms are executed (most of the algorithms are executed quite seldom, typically on slow associated control channel (SACCH) frame basis) and the network time (TDMA frame number) is increased.

Link level interface

The simulator uses non-averaged burst level mapping between system and link level simulators [5,6,7,8]. The mapping consists of two phases – first, burst-wise C/I is mapped on to raw bit error rate (in the simulated cases presented here, only two mappings are needed; one for Gaussian minimum shift keying (GMSK)

Figure E.3   Example of frame error probability mapping. Left: raw data. Right: final mapping matrix.

and one for octagonal phase shift keying (8-PSK)). Next, mean and standard deviation of the bursts over the interleaving period are calculated. Based on extensive link level simulations, a frame error probability (FEP) mapping matrix has been created where each matrix element represents estimated FEP for a block having certain mean and std of errors.

Figure E.3 shows an example of the mapping procedure in the second phase. On the left the raw data are shown, i.e. frame error probabilities on the (stdBER, meanBER)-plane. On the right the final mapping matrix is shown. In the final matrix, the intensities of received blocks on the plane are taken into account. Certain (stdBER, meanBER) combinations are very rare or even impossible. If enough samples are received in certain matrix element, it is statistically valid and can be taken into account in the final mappings. Missing points are extrapolated or interpolated from the available data whenever relevant.

E.4 Description of the basic cellular models

Unless otherwise stated, the models in Tables E.1, E.2 and E3 have been applied to all the simulations in this book.

Table E.1   Common simulation models

Model	Description	Comment
Cellular layout	Hexagonal grid, three-sector sites	
Path loss model (macro cell)	L [dB] $= 128.1 + 37.6 \log_{10}(R)$	R in km, minimum coupling loss of 70 dB taken into account
Call arrival process	Poisson	
Call length distribution	Exponential	For speech
Channel allocation	Channel with least amount of interference selected	Idle slot uplink interference (non-ideal) measured at BTS
Cell selection	Path loss based	
Inter-cell handover	Path loss based with hysteresis and penalty timer	
Intra-cell handover	If RXLEV good but RXQUAL bad	
BTS antenna diversity	Yes	Simple selection diversity
Synchronisation	No	Slot synchronisation exists from interference point of view, but each site has its own frame offset

Table E.2   Common simulation parameters

Parameter name	Value	Unit	Comment
Time resolution in simulations	4.615	ms	Every traffic burst is simulated
Cell radius	1000	m	Corresponds to 3 km site-to-site distance
BS antenna height	15	m	No tilting
BS antenna gain	14	dBi	3 dB point
BS antenna beamwidth	65	deg	Horizontal antenna pattern used
BS Tx power (max.)	43	dBm	20 W
MS antenna height	1.5	m	Body loss not taken into account
MS speed	3 or 50	km/h	MS wrapped to other side when network border is reached
Slow fading correlation distance	50	m	
Slow fading standard deviation	6	dB	
Carrier frequency	900	MHz	
Adjacent channel protection	18	dB	First adjacent taken into account
System noise floor	−111	dBm	
Handover margin	3	dB	
Handover interval	10	SACCH frames	Minimum period between inter-cell HOs
DTX silence/talk periods	3.65	s	When DTX is applied it is used in both directions. When UL is silent, DL is active and vice versa

Table E.3 Some of the most important (E)GPRS simulation parameters

Parameter name	Value	Unit	Comment
RLC window size for GPRS	64	RLC block	Max. for GPRS
RLC window size for EGPRS	384	RLC block	Max. according to multislot class
Downlink TBF establishment delay	240	ms	
RLC acknowledgement delay	220	ms	
Minimum C/I for CS-2	-2 (NH) / 7 (FH)	dB	LA related
Minimum C/I for CS-3	3 (NH) / 12 (FH)	dB	LA related
Minimum C/I for CS-4	9 (NH) / 17 (FH)	dB	LA related
Expiry Limit Of T3190	5	s	
Expiry Limit Of T3195	5	s	
N3105max	8		
Cell reselection hysteresis	6	dB	
Cell reselection penalty timer	4	SACCH multiframes	
Number of TRXs/cell	1-3		
Max number of timeslots per TBF	3		
Max number of TBFs per TSL	6		
Total simulation length	200 000	TDMA frames	~15 min. real time
MS speed	3	km/h	Mobility model according to [9]

References

[1] Wacker A., Laiho-Steffens J., Sipilä K., Jäsberg M., 'Static Simulator for Studying WCDMA Radio Network Planning Issues', IEEE 49th Vehicular Technology Conference, Vol. 3, 1999, pp. 2436–2440.

[2] Hytönen T., 'Optimal Wrap-around Network Simulation', Helsinki University of Technology Institute of Mathematics: Research Reports 2001.

[3] Lugo A., Perez F., Valdez H., 'Investigating the Boundary Effect of a Multimedia TDMA Personal Mobile Communication Network Simulation', IEEE 54th Vehicular Technology Conference, Vol. 4, 2001, pp. 2740–2744.

[4] Stroustrup B., 'The C + + Programming Language', Special Edition, Addison-Wesley, Reading, MA, 2000.

[5] Malkamäki E., Ryck F. de, Mourot C., Urie A., 'A Method for Combining Radio Link Simulations and System Simulations for a Slow Frequency Hopped Cellular System', IEEE 44th Vehicular Technology Conference, Vol. 2, 1994, pp. 1145–1149.

[6] Hämäläinen S., Slanina P., Hartman M., Lappeteläinen A., Holma H., Salonaho O., 'A Novel Interface between Link and System Level Simulations', Proc. ACTS Mobile Telecommunications Summit, Aalborg, Denmark, October 1997, pp. 599–604.

[7] H. Olofsson, M. Almgren, C. Johansson, M. Höök, F. Kronestedt, 'Improved Interface between Link Level and System Level Simulations applied to GSM', Proc. ICUPC ' (1997) 97.

[8] Wigard J., Nielsen T. T., Michaelsen P. H., Morgensen P., 'BER and FER Prediction of Control and Traffic Channels for a GSM Type of Interface', Proc. VTC'98, pp. 1588–1592.

[9] Universal Mobile Telecomunications System (UMTS): Selection Procedures for the Choice of Radio Transmission Technologies of the UMTS (UMTS 30.03 Version 3.2.0), ETSI Technical Report 101 112 (1998-04).

Appendix F

Trial Partners

This appendix contains information about network operators who have run field trials in co-operation with Nokia, the employer of all the contributors of this book. The following information refers only to the trialled areas

	Trial partner	Used frequency spectrum in trial	Average number of TRXs in trial	Used features during trial
Network 1	Optus/Brisbane	5–8 MHz	2.5	FH + PC + DTX
Network 2	Major Chinese city	5–7 MHz	4	FH
Network 3	CSL/Hong Kong	6–7 MHz	3–4	FH + PC + DTX
Network 4	Major European capital	–	–	–
Network 5	Major Chinese city	5–7 MHz	4	FH + PC
Network 6	City in US	4.4 MHz	2	FH + PC + DTX
Network 7	Radiolinja/Finland	–	–	E-OTD location method
Network 8	Sonofon/Denmark	–	–	FH + PC
Network 9	Major European metropolis	8–10 MHz	3–4	FH + PC + DTX
Network 10	AWS/Several major cities	–	–	–
Network 11	Radiolinja/Estonia	–	–	CI, CI + TA, and CI + TA + RXLEV location methods

Index